Field Guidelines for Genetic Experimental Designs in High-Throughput Sequencing

Ana M. Aransay • José Luis Lavín Trueba
Editors

Field Guidelines for Genetic Experimental Designs in High-Throughput Sequencing

Springer

Editors
Ana M. Aransay
Genome Analysis Platform
CIC bioGUNE
Derio, Spain

Centro de Investigación Biomédica en Red de Enfermedades Hepáticas y Digestivas (CIBERehd)
Madrid, Spain

José Luis Lavín Trueba
Genome Analysis Platform
CIC bioGUNE
Derio, Spain

ISBN 978-3-319-31348-1 ISBN 978-3-319-31350-4 (eBook)
DOI 10.1007/978-3-319-31350-4

Library of Congress Control Number: 2016940242

Printed on acid-free paper

This Springer imprint is published by Springer Nature
The registered company is Springer International Publishing AG Switzerland

Preface

High-throughput sequencing (HTS), also named next-generation sequencing (NGS) or massive parallel sequencing (MPS), is an amazingly speedy evolving world. Since 2005, when the first HTS equipment was released to the market by 454 Life Sciences, there have been dozens of companies developing a variety of methods that offer distinct characteristics, and therefore, each protocol should be applied wisely. Being aware of the wide range and complexity of the reported HTS strategies, we observed that there is a lack of bibliographic support when scientists need to choose the most suitable methodology or combination of platforms and to define their experimental designs to achieve unambiguous aims.

Genomics core facilities can give limited advice on which technology fits one's purposes and the number of cloud-based HTS data analysis pipelines, to process output raw data in a standard mode, is rapidly increasing. Ideally, scientists that request this sort of services should have clear clue questions concerning wet-lab procedures and data analysis. Thus, the purpose of this guideline is to collect in a single volume all aspects that should be taken into account and the reasons behind when HTS technologies are being incorporated into a scientific research project, and it is directed to both, specialist, but primarily to newcomers.

Accordingly, the book encloses a brief introduction on HTS technologies challenges, followed by 14 chapters with proficient discussions and recommendations to select the best among all the available workflows for sample processing, alignment of results, algorithms at downstream data analysis, etc., and the minimum number of samples that should be characterized in each assay for accurately sequencing and interpreting genomes, sets of RNA molecules, DNA methylated regions, nucleic acids interacting with targeted proteins, metagenomes, metatranscriptomes, and/or single-cell contents. Moreover, examples of several successful strategies are analyzed to make the point of the crucial features.

Whole genome sequencing (WGS) wet-lab procedures and data analyses are portrayed in Chap. 2, followed by a description of how to face the characterization of partial genomes (i.e., genes of interest) in a number of samples in Chap. 3. In addition, a detailed variety of sequencing library preparation approaches and results examination pipelines to catalogue transcriptomes, sets of noncoding RNAs and small RNAs as well as ribosome networking RNAs under singular conditions, are depicted within Chaps. 4–8. Furthermore, ways of studying epigenetic events such as DNA methylation and interactions of DNAs or RNAs with targeted proteins are illustrated in Chaps. 9, 10, and 11, respectively. Chapters 12 and 13 discuss the appealing world of classifying environmental (e.g., microbial communities) genomes and transcriptomes by means of metagenomics and metatranscriptomics. Likewise, the hot topic of single-cell DNA and RNA content characterization is considered in Chaps. 14 and 15. The last chapter of the book, Chap. 16, is a detailed protocol on how to submit HTS data to public repositories as required when this sort of results are being published.

As a special feature, this book includes a sort of quick reference guide as appendix for each chapter, where readers can, at a glance, access a figure representing the main steps of the wet-lab and bioinformatic workflows as well as a table that gathers information about the experimental design recommendations for the techniques described and another one referred to the bioinformatic recommended analysis software together with the results yielded by each program. The intention of this section is to grant rapid access to a summary of the principles of each of the methodologies described.

Considering that HTS technologies can be applied to a vast variety of biological questions and are used by scientists working in unlike fields such as biology, medicine, or ecology, and in a wide range of taxonomical levels (mammals, plants, bacteria, viruses, etc.), we hope that this book will be a precious resource for all scientist that lack skills in HTS and pretend to incorporate such technologies into their research.

Derio, Spain Ana M. Aransay
Derio, Spain José Luis Lavín-Trueba

Contents

Contributors

Altuna Akalin, Ph.D. Bioinformatics Platform, Berlin Institute for Medical Systems Biology, Max Delbrück Centre, Berlin, Germany

Luis David Alcaraz Departamento de Ecología de la Biodiversidad, LANCIS, Instituto de Ecología, Universidad Nacional Autónoma de México, Coyoacán, Cd. Mx., México

Tyler S. Alioto, B.S., Ph.D. Centro Nacional de Análisis Genómico, Centre de Regulació Genòmica, Barcelona, Spain

Ana M. Aransay, Ph.D. Genome Analysis Platform, CIC bioGUNE, Derio, Spain

Centro de Investigación Biomédica en Red de Enfermedades Hepáticas y Digestivas (CIBERehd), Madrid, Spain

Vladimir B. Bajic, Ph.D. Computer, Electrical and Mathematical Sciences and Engineering Division, Computational Bioscience Research Center, King Abdullah University of Science and Technology (KAUST), Thuwal, Saudi Arabia

Tuncay Baubec, Ph.D. Epigenomics and Chromatin Biology Lab, Institute of Veterinary Biochemistry and Molecular Biology, University of Zurich, Zurich, Switzerland

Tara Dutka Laboratory of Muscle Stem Cells and Gene Regulation, NIAMS, Bethesda, MD, USA

Brant C. Faircloth, Ph.D. Department of Biological Sciences and Museum of Natural Science, Louisiana State University, Baton Rouge, LA, USA

Bastian Fromm, Ph.D. Department of Tumor Biology, Institute for Cancer Research, Norwegian Radium Hospital, Oslo University Hospital, Oslo, Norway

Toni Gabaldón, Ph.D. Bioinformatics and Genomics Programme, Centre for Genomic Regulation (CRG), Barcelona, Spain

Universitat Pompeu Fabra (UPF), Barcelona, Spain

Institució Catalana de Recerca i Estudis Avançats (ICREA), Barcelona, Spain

Elizabeth M. Glass Argonne National Laboratory, Argonne, IL, USA

Travis C. Glenn Department of Environmental Health Science, University of Georgia, Athens, GA, USA

Stephanie M. Greenwald Institute for Genomics and Systems Biology, Argonne, IL, USA

Sushma Nagaraja Grellscheid, Ph.D. School of Biological and Biomedical Sciences, Durham University, Durham, UK

James Hadfield, B.Sc., Ph.D. Cancer Research UK Cambridge Institute, University of Cambridge, Cambridge, UK

Markus Hafner Laboratory of Muscle Stem Cells and Gene Regulation, NIAMS, Bethesda, MD, USA

Matthias Harbers, Ph.D. Division of Genomic Technologies, RIKEN Center for Life Science Technologies, Yokohama, Kanagawa, Japan

Karolina Heyduk Department of Plant Biology, University of Georgia, Athens, GA, USA

Abdullah M. Khamis, M.Sc. Computer, Electrical and Mathematical Sciences and Engineering Division, Computational Bioscience Research Center, King Abdullah University of Science and Technology (KAUST), Thuwal, Saudi Arabia

Albert Lahat, B.Sc. School of Biological and Biomedical Sciences, Durham University, Durham, UK

José Luis Lavín Trueba, Ph.D. Genome Analysis Platform, CIC bioGUNE, Derio, Spain

Kenneth J. Livak, Ph.D. Fluidigm Corporation, South San Francisco, CA, USA

Folker Meyer Argonne National Laboratory, Argonne, IL, USA

Farshad Niazi, M.D. Department of Molecular Biology and Microbiology, Case Western Reserve University School of Medicine, Cleveland, OH, USA

Christopher O'Sullivan National Center for Biotechnology Information, U.S. National Library of Medicine, Bethesda, MD, USA

Sarah Owens Argonne National Laboratory, Argonne, IL, USA

Mariana Peimbert Departamento de Ciencias Naturales, Universidad Autónoma Metropolitana Unidad Cuajimalpa, Cuajimalpa, Cd. Mx., México

Aishe A. Sarshad Laboratory of Muscle Stem Cells and Gene Regulation, NIAMS, Bethesda, MD, USA

Rory Stark, B.A., M.Sc., M.Phil., D.Phil. Cancer Research UK Cambridge Institute, University of Cambridge, Cambridge, UK

Jessica D. Stephens Department of Plant Biology, University of Georgia, Athens, GA, USA

Keith E. Szulwach, Ph.D. Fluidigm Corporation, South San Francisco, CA, USA

William L. Trimble Institute for Genomics and Systems Biology, Argonne, IL, USA

Jonathan Trow National Center for Biotechnology Information, U.S. National Library of Medicine, Bethesda, MD, USA

Saba Valadkhan, M.D., Ph.D. Department of Molecular Biology and Microbiology, Case Western Reserve University School of Medicine, Cleveland, OH, USA

Anze Zupanic, Ph.D. Department of Environmental Toxicology, Eawag – Swiss Federal Institute for Aquatic Research and Technology, Dübendorf, Switzerland

Chapter 1
The High-Throughput Sequencing Technologies Triple-W Discussion: Why Use HTS, What Is the Optimal HTS Method to Use, and Which Data Analysis Workflow to Follow

José Luis Lavín Trueba and Ana M. Aransay

1.1 Evolution of the HTS Platforms and the Spawn of New Research Applications

High-throughput sequencing (HTS) technologies have conquered the genetic, genomic, and epigenomic worlds during the last decade. At the moment of writing this manuscript, there are more than 23,000 indexed references considering HTS techniques at the PubMed repository, focused on an incredible diversity of topics and species: from biomarkers definition for complex human diseases to ancient prokaryote taxonomic identification and evolutionary tree resolution. Promising, novel real-time nanopore sequencers output longer and longer reads in an extraordinarily speedy mode, allowing even de novo complete microbial genomes (Check Hayden 2015; Quick et al. 2014). Furthermore, the possibility of sequencing single-cell genomes and transcriptomes (reported as method of the year 2013 by *Nature Methods*; see *Nature Methods* Issue from January 2014 (Editorial 2014)) opens a novel, very exciting perspective for basic and medical research.

Since the first massive parallel sequencer was available in 2005 (Margulies et al. 2005), prices to run HTS projects have been reduced significantly, making possible to sequence a human genome by about $1000. However, no matter how cheap, easy,

J.L. Lavín Trueba, Ph.D. (✉)
Genome Analysis Platform, CIC bioGUNE,
Bizkaia Technology Park - Building 801A, 48160 Derio, Spain
e-mail: jllavin@cicbiogune.es

A.M. Aransay, Ph.D.
Genome Analysis Platform, CIC bioGUNE,
Bizkaia Technology Park - Building 801A, 48160 Derio, Spain

Centro de Investigación Biomédica en Red de Enfermedades Hepáticas y Digestivas (CIBERehd), Madrid, Spain
e-mail: amaransay@cicbiogune.es

A.M. Aransay, J.L. Lavín Trueba (eds.), *Field Guidelines for Genetic Experimental Designs in High-Throughput Sequencing*,
DOI 10.1007/978-3-319-31350-4_1

and quick HTS technologies become if there is a lack of specific questions to be answered and, accordingly, of a precise experimental design, all the resources employed in those projects will keep on being a waste. According to High-Throughput "Next-Generation" Sequencing Facilities Statistics web (http://omicsmaps.com/stats), there are 7400 sequencers from seven different companies registered around the world that have worked or are working on 69,444 sequencing projects (see https://gold.jgi.doe.gov/index and (Reddy et al. 2015)) of very different nature, from which about 36,000 are incomplete or just started.

As core facility members, we are aware of the lack of detailed information for most "materials and methods" sections within the articles that consider HTS data. Thus, in order to start the planning of any HTS project, ideally, experts on sample collection and science behind the project with clear aims and HTS wet lab and data analyses specialists should meet to share/discuss their points of view and make the most of each strategy. This communion is not always possible, and, consequently, more projects than expected are run in an inappropriate mode, resulting in big amounts of public money thrown away. To avoid these events, we have worked on the present detailed guideline, in which all the ins and outs of each currently used HTS approach are considered.

1.2 Guide to Effectively Select a High-Throughput Sequencing Technique Fitting Your Research Objectives

To make a conceptually dense book, like this, easy to read and use as reference, we have organized it into seven main thematic blocks where chapters are tightly related to each other; nevertheless, each chapter has its own distinctiveness and, therefore, can be followed independently. Here we present each of the sections and the challenges discussed in the chapters included in it.

1.2.1 Reading the Book of Life, DNA Sequencing

During the past decade, the first complete genomes of different types of organisms such as bacteria, fungi, plants, or animals were sequenced (Consortium CES 1998; International Human Genome Sequencing C 2004). First sequencing projects involved the collaboration between research groups, institutions, and sequencing facilities to afford such projects. High-throughput technologies changed this fact and made cost and time effective to perform genomic scale analysis, to study a variety of genomic characteristics. Therefore, whole-genome sequencing and resequencing become affordable to single laboratories or research projects (Anonymous 2014).

The first thematic block of this book is referred to DNA sequencing, specifically to genome sequencing and resequencing. Chapter 2 deals with whole-genome sequencing whose primary goal is to produce a high-quality genome assembly to serve as a reference for an organism or a closely related phylogenetic group.

Moreover, it is considered as a tool to grant access to the genetic information of living beings and understand their essence. Due to the current technological developments in sequencing technologies and the bioinformatics procedures developed in parallel, this technology is becoming so affordable that several single-organism (or even cells) genomes can be sequenced as part of a single research project (Jarvis et al. 2014). As a result of the rapid advances in the field, this chapter focuses on general principles that will have a more general applicability instead of merely displaying an overview of current methodologies that will likely soon become obsolete. In addition, various genome resequencing methods with a focus on target enrichment are examined in Chap. 3. A part of these methods can be applied to non-model organisms with few genetic resources available (Jarvis et al. 2014). The precise method to use for the organism of interest depends on several factors that are addressed in this chapter. Additionally, experimental design considerations, bioinformatic pipelines, and proper reporting of results for target enrichment are also carefully explained.

1.2.2 Transcribe to Survive, RNA-Sequencing Methods

The transcriptome of a cell is dynamically changing along time while adapting to variable environment conditions (Nagalakshmi et al. 2008; Wilhelm et al. 2008) (whether if it is an external environment like microbes or cells forming tissues or organs inside complex organisms). The recent developments of high-throughput sequencing (HTS) enable to achieve a relatively high base coverage of cDNA sequences, obtained from RNA samples. Comprehensive overviews on transcriptomes can be obtained today by combination of those new sequencing technologies with large-scale cDNA library preparation forming the basis to different approaches for transcriptome profiling. This fact enables to look at events like posttranscriptional modifications or alternative gene splicing. In addition to mRNA transcripts, there are several other RNA populations included in total RNA extracts, for instance, microRNA (miRNA), transfer RNA (tRNA), and long noncoding RNA (lncRNA). Sequencing methods for those different RNA species are covered in this book through the second thematic block of chapters (Chaps. 4–7) introduced in this section.

In Chap. 4, the use of full-length coding DNA (cDNA) preparations in combination with shotgun RNA-seq and RNA profiling directly from RNA (transcriptome profiling) (Hestand et al. 2010) are exhaustively explained, in addition to the use of cap analysis gene expression (CAGE) for high-throughput mRNA detection and genomic determination of transcription start sites (TSS). Real examples from studies on transcriptional regulation of gene expression are used to illustrate the transcriptome-profiling strategies covered in the chapter. To extend transcriptome-profiling strategies, RNA splicing is also included in an individual chapter, since a high proportion of human genes undergo these splicing events (Wang et al. 2008). This is a regulated biological mechanism where a single gene can give rise to multiple transcripts through alternative processing of primary RNA transcripts, RNA (Wang et al. 2008). RNA sequencing enables the analysis not only of differential gene expression but also isoform-level changes in gene expression from the same original data, although differential splicing detection requires deeper sequencing coverage. Chapter 5 widely covers the set of bioinformatic tools necessary to analyze and study splicing from RNA-seq data. Moreover, those tools are classified depending on the step of the analysis they are designed to carry out, and counseling is given on which one should be implemented depending on the focus of the study.

Among noncoding RNAs, miRNAs have become key players in different fields ranging from disease diagnosis (as biomarkers) to phylogenetic studies where they are also used to monitor evolutionary history and developmental relationships among organisms, as they present highly conserved structural features, and changes in their regulation may unleash different health conditions (Bartel 2009). Chapter 6 sheds light on this methodology focusing on the bioinformatic prediction and annotation steps, pointing out current available software and database strengths and weaknesses. There is another group of noncoding RNAs, which are bigger in size, the lncRNAs, which constitute a major fraction of the output of the genome in higher eukaryotes (Carninci et al. 2005). Analysis of lncRNAs expression from RNA-seq data is challenging because of the particularities this RNA category has, such as low expression level, abundance of repeat element-derived sequences, loci overlap between transcripts, high percentage of non-polyA molecules, and scarcity of splicing events (Zhang et al. 2014). Therefore, although wet lab protocols are mostly common to those used for RNA-seq, bioinformatic analyses are required to be aware of the peculiarities of lncRNAs. To that aim, it is required to use the tools developed exclusively for this case, and some other shared with ordinary RNA-seq, but tuning working parameters accordingly. Chapter 7 accounts for the singularity of lncRNAs and gives advice on the best alternatives to select when dealing with this novel family of RNAs.

1.2.3 Translation by Interaction, RNA-Protein Interactions

In eukaryotic cells, mRNA levels do not perfectly match with protein expression levels. This means that regulation, at translation and protein stability levels, has an important effect on the result of gene expression in those cells (Tome et al. 2014). RNA-protein interactions are essential to cellular homeostasis and management of RNA metabolism in the cell. Consequently, the inspection of RNA-protein interactions is underpinned in this third thematic block.

The sequencing of mRNA fragments protected by the translating ribosome via HTS is a method aimed at narrowing the gap between the mRNA molecule and the protein (Zupanic et al. 2014). In Chap. 8, different ways where ribosome profiling has been applied and guides readers through state-of-the-art experimental procedures, focusing on alternative protocols, are detailed. Correspondently, RNA-protein networks studies can take advantage of HTS technologies to improve methodologies looking for a better understanding of posttranscriptional gene regulation. As part of this block, photoactivatable ribonucleoside-enhanced cross-linking and immunoprecipitation (PAR-CLIP) (Spitzer et al. 2014) together with other currently available techniques is reviewed in Chap. 11. These are recently developed techniques that require a detailed discussion of experimental and data analysis procedures for readers to understand, implement in their current research, and even being able to contribute to the refinement of, for instance, current data analysis pipelines, in a close future.

1.2.4 Teaching What to Read, DNA Regulation Analysis

Gene regulation is the process of turning genes on and off, providing that the correct genes are expressed at precise times (Lister et al. 2013). Genes require some kind of interface that enables them to interact with the environment and respond to the stimuli felt in order to grant the organism survival. This task is accomplished through gene regulation, which includes a variety of mechanisms, from chemically modifying genes (e.g., methylation) to using regulatory proteins to turn genes on or off (DNA-protein interaction) (Jaenisch and Bird 2003).

DNA methylation is a relevant epigenetic modification of DNA affecting gene expression, which is typically related to a repression of transcription (Baubec et al. 2015; Maurano et al. 2015). Processes such as cancer and aging are regularly associated to variations in DNA methylation patterns which impel researchers to invest time and resources in understanding the contribution of this alteration to human health. High-throughput sequencing allows interrogating the status of DNA methylation all throughout the genome at a nucleotide-level resolution, uncovering the distribution and fluctuations of this modification between health and disease conditions. Chapter 9 considers the most accepted technologies dealing with DNA methylation, offering detailed advice from the wet lab as well as the in silico sides of the technique. Another mechanism in this thematic block, besides DNA modification, is DNA-protein interactions. This term does not only include those of transcription factor proteins to specific binding sites on DNA but also proteins related to transcriptional regulation (e.g., methylases) or transcription events (e.g., polymerases). Chapter 10 describes how to design, implement, and analyze data derived from chromatin immunoprecipitation sequencing (ChIP-seq), to elucidate different aspects involving an array of biological issues concerning DNA-protein interactions and modulation of transcriptional regulation (Mohammed et al. 2015).

1.2.5 Sequencing Communities Rather than Single Organisms, the Meta-Sequencing

As part of the fifth separated thematic block, another HTS approach, which is widely extended to perform the sequencing of "raw" environmental samples, is meta-sequencing (which includes metagenomics and metatranscriptomics, so far).

By definition, "meta-sequencing" aims to obtain information from DNA/RNA extracted from environmental samples or mixtures of microorganisms (Eisen 2007; Leininger et al. 2006). This methodology enables to obtain genomic/transcriptomic information about the full community of organisms present in the sample at once, hologenome, which is quite interesting in order to get an idea of the species composition of the sample, the genetic coding potential, and the probable metabolic functions (carried out by the organisms from the sample) and, depending on the approach, even to decipher which genes are transcribed under the conditions the sample was isolated.

The first of the meta-sequencing procedures covered in this book is metagenomics in Chap. 12. In that chapter the method is defined and topics like recommendations on the best sequencing platform to use depending on the kind of metagenomic study to carry out, whether the aim is to characterize the species composition of the sample (16s rRNA sequencing) or to also infer the coding potential of the sequenced community (whole-genome shotgun sequencing). Recommendations about important issues/features related to the experimental design such as the number of samples to sequence per comparative group, the sequencing depth per sample, or the bioinformatic analysis pipeline to choose for the different kinds of study (Wilke et al. 2013) are also discussed. The second meta-sequencing procedure addressed in this volume is metatranscriptomics, in Chap. 13. In this case the aim will be to study community-wide gene expression, under strictly determined conditions, by whole-genome shotgun RNA sequencing (Simon-Soro et al. 2014). The metatranscriptome case is elaborated, requiring strong experimental design, wet laboratory, and bioinformatic skills. This chapter provides step-by-step counseling for both wet lab and in silico analyses while highlighting some of the more common complications met in this type of experiments.

1.2.6 From Bulk to Individual Cells, Single-Cell Approaches

There is a trend toward HTS studies of single-cell genomes and transcriptomes, rather than what could be referred as "bulk cell" characterization (Trapnell et al. 2014; Wills et al. 2013). It is interesting to point out that the ability to first isolate individual cells in order to examine their nucleic acid content has led to significant advances in areas such as the examination of tumor structure, the accurate identification and characterization of specific cell types, and the screening of the transcriptome from uncommon cell types (e.g., circulating tumor cells), among others. Those facts are covered in Chaps. 14 and 15, where DNA and RNA single-cell sequencing techniques are discussed. In this manuscript, the advantages of sequencing multiple single-cell genomes in opposition to bulk cell samples are addressed along with a state-of-the-art report on the technical development of the procedures and steps required to carry out such kind of experiments.

1.2.7 Last Step but Not the Least Important, Uploading Data to Public Sequence Repositories

Chapter 16 covers a mandatory step for any HTS-related experiment that intends to be published: data upload to a public repository such as The Sequence Read Archive (SRA) at the National Center for Biotechnology Information (NCBI). There, raw sequencing data and alignment information (including metadata describing different details of whole wet lab and in silico workflows) from most of the published and some still unpublished studies related to HTS are stored and correlated to unique IDs, which grant access to those data to any member of the research community interested in obtaining them. Collecting and maintaining these large data collections is one of the most valuable actions to perform reporting the scientific progress in fields related to biosciences, by the current human civilization.

1.3 Remember the Past, Appreciate the Present, and Behold the Future of Sequencing

We are living in days where technology exponentially advances in short periods of time. There is a tendency of reducing the size of devices while increasing the number of their functionalities (e.g., nowadays a smartphone is a technological Swiss knife that can reproduce music and movies, tune TV and radio, take photographs, record video, play games, and be a pocket computer… feels like carrying the full catalog of a 1990 multimedia store in a single device). Sequencing technologies are not an exception and enhanced performance is reached in relatively short periods of time. Meanwhile, new technologies are developed to exceed their predecessors.

Although this book presents an up-to-date catalog of sequencing techniques and their current caveats, we should expect many of them to be significantly influenced by new developments, probably changing the whole experimental strategy in the years to come. Here, we will exhibit different ways technological improvements in the sequencing systems subsequently promoted advances in computing and how that changed the bioinformatics analysis landscape.

1.3.1 Days of a Present Past: **Pre-HTS Era** *Formats and Their Current Counterparts*

From now on (in this chapter), we will use the term *pre-HTS era* to refer to methods and data produced before HTS emergence. During that period, main data formats were relatively few (compared to those on the HTS), data file size (kilobytes) was smaller in general, and processing times were remarkably shorter, so multiple trials could be performed on a dataset in a short period of time, ranging from seconds to several minutes (e.g., dealing with multiple-sequence files) in a common PC or laptop.

After the HTS expansion, computational requirements changed substantially, for instance, data storage needs increased exponentially (e.g., a single *Illumina HiSeq 2500's* 150 bp paired-end run can fill up a common laptop hard disk drive), and the same happened to computational power (the number of processing cores increased up to a minimum of 8 to fluently run the analyses); required RAM memory (over 8 GB recommended to avoid long computation periods) and high-speed bandwidth became vital for data traffic (in cases like remote access to files from a server or workstation). Plus, HTS field is strongly biased toward open-source software, which is mainly developed under Unix/Linux architecture, and so, medium level skills working under environment are required. Apart from the bottlenecks mentioned before, a certain level of programming language knowledge in Linux shell, Perl, Python, Java, or R is essential in case you need to fix any unexpected issues that might come out during your analysis workflow setup.

Pre-HTS era data file formats are easier to manage; for instance, most files can be displayed in any text editor and intuitively understand what is stored on them. This is not so simple with HTS output files. Besides, there are fewer formats and not much overlapping between them (which occurs with HTS file formats). For example, *FASTA* format was used for DNA, RNA, or protein *sequences* and *gff* format was designed to store annotation data. Other formats like *embl*, *pdb*, and *genbank*, were developed by their corresponding sequence repositories (EMBL, PDB, and Genbank), being a mixture of descriptive metadata and sequences. Moreover, multiple sequences analysis (MSA) produced a set of files to depict the multiple alignment information (*aln*, *msf*, *phylip*, or *meg*), that can be readily exported into different phylogeny software (e.g., Phylip, Treeview, or MEGA) for further processing (the description of these formats can be found at http://emboss.sourceforge.net/docs/themes/SequenceFormats.html). During a certain time-lapse, being familiar with these data types was basically sufficient for a researcher to perform a wide range of bioinformatic analysis during the *pre-NGS era*.

When HTS technologies arrived, most of those formats evolved or were substituted by others that could fulfill the requirements of the new kind of data and scientific questions, although some persisted, if the new methodologies did not compete with them. If we make an analogy to the "tower of Babel" passage from the Bible (Genesis 11:5–8 at https://en.wikisource.org/wiki/Bible_%28King_James%29/Genesis#11:5 and (Harris 2002)), HTS has come, with a fistful of sequencing technological platforms that produce heterogeneous data. This heterogeneity made visible the necessity to develop highly optimized tools through a wide range of programming languages, therefore, increasing the need for creating/adapting new file formats to successfully contain the different kinds of features that each methodology needs to account for, namely, FASTQ, SAM/BAM, BED, WIG, VCF (with their respective variations), and several others described in different sites like https://genome.ucsc.edu/FAQ/FAQformat.html#ENCODE, as well as many different tool-specific output formats (e.g. Bowtie's ".bow," BWA's ".sai," Maq's ".map," or SOAP's ".gout/.gout.trim" output files) including analogous information, as they carry out the same analysis step (Hatem et al. 2013). Something that most of these files have in common is that they contain very large data volumes and are not human readable unless opened "programmatically" (command line instructions are required to inspect their content, since ordinary text editors cannot process those data volumes), and, in general, Linux/Unix system proficiency is mandatory to inspect the information in those files. As result of this, data is displayed in a significant number of file formats (that may overlap functionalities), which prevent software compatibility per se, thus requiring bioinformaticians to interconvert data formats to enable interaction between analysis tools. Since a wide range of analyses are performed, and data flux between programs is essential, an important percentage of the processing time is spent converting data from one format to another in order to guarantee the dataflow between each step of the pipeline. Hence, format conversion becomes an onerous task that consumes a large amount of the processing time, even though very efficient tools have been developed to that end (Li et al. 2009; Quinlan and Hall 2010).

Current developments suggest that this tendency may change as new technologies under development offer much longer reads, which should allow reaching enough coverage depth of the template with a significant decrease in the number of reads (as we will explain for MinION technology in the next section) and hopefully the number of file formats if analyses are back to a certain homogeneity as in the pre-HTS era. Therefore, the final volume of data yielded should be easier to handle by final users with improvements related to saving time when transferring data between computers/servers and required bandwidth and, consequently, to the expenses of the analyses, whether those are carried out at local servers or in the cloud.

1.3.2 Days of a Present Future: Example of a State-of-the-Art Technology

We will use a state-of-the-art technology as an example of the direction of current technical developments: Oxford Nanopore Technologies (ONT) MinION sequencer (https://www.nanoporetech.com/products-services/minion-mki). This device is revolutionary since it implements a new sequencing approach, namely, charged protein nanopores, consisting of DNA molecules passing through those structures on the flowcell, where different nucleotides are detected by the voltage sensors within each nanopore, and base calling begins. Although this sequencer currently yields high error rates (around 30 % of the bases), there is a thread of opinion that argues about the possibility of those "errors" being artifacts due to nucleotide modifications or analogues embedded in the DNA strand. There are studies that may support this theory (Clarke et al. 2009; Wolna et al. 2014) and also at least a patent has been registered on this subject (Stephen and Jonas 2012). If this hypothesis was true, then, current understanding of DNA will change substantially, and maybe a deeper insight into its functionality would be achieved.

Another change referred to current sequencers is its compact form, a MinION is slightly bigger than a common USB pendrive (small and portable) and, in principle, is aimed to be used by nonexperts. ONT adopted a new data format, FASTA5 (derived from the HDF standard), which is a highly compact data format capable to "represent very complex data objects and a wide variety of metadata" and "completely portable file format with no limit on the number or size of data objects in the collection" (http://www.hdfgroup.org/HDF5/whatishdf5.html).

Such features allow it to be used as a "mobile sequencer" that can be implemented for rapid diagnostic, for example, in the case of an infectious outbreak where a patient's blood sample could be directly sequenced, on the field, to determine the microbe causing the disease (Check Hayden 2015).

There are some weak points though, which may eventually prevent MinION implementation in certain cases:

1. It requires a Windows laptop that uploads the yielded data to a server for its analysis, which makes its use "on the field" impossible in places with no network connection (although Internet is not compulsory for the sequencing step, data analysis of these results does require remote web access).
2. Even though FASTA5 is a compact format, information can only be extracted programmatically using the HDF5 library; this requires further software development by ONT to enhance analysis tools that really allow nonexperts to use their technology or the obstacles related to transforming raw sequencing data into "human interpretable" results will remain, and this technology will not live up to the expectations it created.
3. Most researchers will not be very comfortable if data can only be analyzed in the cloud due to privacy law violation or bioethical concerns, so it is mandatory to enable potential users to perform sequencing data retrieval (base calling) and analysis locally, in their computers. This option would also allow the sequencer to operate "in the field" in a Wi-Fi-independent manner/basis.
4. It is mandatory to allow the use of free license software. Otherwise, the charges for private analysis software will not be affordable and this technology will not succeed, unless the whole sequencing plus analysis expenses match other technologies whose data analysis can be carried out for free.

1.4 Up-and-Coming Challenges at the HTS World

At the moment, more than 2200 tera-base pairs are open-accessed at the main HTS datarepository(http://trace.ncbi.nlm.nih.gov/Traces/sra/sra.cgi?view=announcement) and still many gigabytes of raw data are stored on hard drive disks, which, in many cases, are regrettably worthless to be further processed due to the lack of proper experimental design and/or because of an inaccurate planning of the resources required for this sort of tasks (wet lab specialized personnel, choice of appropriate technology(ies), robust computer equipment, and/or skilled analysts).

Since most of the HTS techniques are quite complex from a procedural point of view, which leads to the difficulty of many researchers in understanding their foundations or the results that should be obtained, this book intends to serve as a guide for beginners approaching the field. Researchers must be aware of the technical difficulties they will face carrying out the wet lab work until they obtain adequate material to be sequenced and the bioinformatic requirements to carry out each particular analysis. It is very important to have a detailed idea of each technique and the inclusion criteria of the samples, according to both, and the scientific aims as well as the minimum quantity and quality of DNA and/or RNA needed for each wet lab protocol, in order to evaluate the compulsory resources and skills before embarking on performing an HTS-based project. To bypass unsuccessful cases, the techniques described in this volume are thoroughly explained, and counseling on how to choose the right method to achieve the research objective is given.

Further development of suitable workflows will be required when all the rising HTS techniques will be established in the molecular labs, which obviously will require specific tuning of not only data analysis but, very specifically, of sample preparation together with secure and ethical ways of HTS data storage in cloud computers.

Acknowledgments We are grateful to Springer for giving us the opportunity of making this HTS guideline idea a reality.

This work has been supported by The Department of Industry, Tourism, and Trade of the Government of the Autonomous Community of the Basque Country (Etortek Research Programs 2013–2015) and from the Innovation Technology Department of the Bizkaia County.

References

Anonymous (2014) Illumina sequencer enables $1,000 genome. Genet Eng Biotechnol News 34:18

Bartel DP (2009) MicroRNAs: target recognition and regulatory functions. Cell 136:215–233

Baubec T, Colombo DF, Wirbelauer C et al (2015) Genomic profiling of DNA methyltransferases reveals a role for DNMT3B in genic methylation. Nature 520:243–247

Carninci P, Kasukawa T, Katayama S et al (2005) The transcriptional landscape of the mammalian genome. Science 309:1559–1563

Check Hayden E (2015) Pint-sized DNA sequencer impresses first users. Nature 521:15–16

Clarke J, Wu HC, Jayasinghe L et al (2009) Continuous base identification for single-molecule nanopore DNA sequencing. Nat Nanotechnol 4:265–270

Consortium CES (1998) Genome sequence of the nematode C. elegans: a platform for investigating biology. Science 282:2012–2018

Editorial NM (2014) Method of the Year 2013. Nat Methods 11:1

Eisen JA (2007) Environmental shotgun sequencing: its potential and challenges for studying the hidden world of microbes. PLoS Biol 5:e82

Harris S (2002) Understanding the Bible. McGraw-Hill, New York, NY, pp 50–51

Hatem A, Bozdag D, Toland AE et al (2013) Benchmarking short sequence mapping tools. BMC Bioinformatics 14:184

Hestand MS, Klingenhoff A, Scherf M et al (2010) Tissue-specific transcript annotation and expression profiling with complementary next-generation sequencing technologies. Nucleic Acids Res 38:e165

International Human Genome Sequencing C (2004) Finishing the euchromatic sequence of the human genome. Nature 431:931–945
Jaenisch R, Bird A (2003) Epigenetic regulation of gene expression: how the genome integrates intrinsic and environmental signals. Nat Genet 33(Suppl):245–254
Jarvis ED, Mirarab S, Aberer AJ et al (2014) Whole-genome analyses resolve early branches in the tree of life of modern birds. Science 346:1320–1331
Leininger S, Urich T, Schloter M et al (2006) Archaea predominate among ammonia-oxidizing prokaryotes in soils. Nature 442:806–809
Li H, Handsaker B, Wysoker A et al (2009) The sequence alignment/map format and SAMtools. Bioinformatics 25:2078–2079
Lister R, Mukamel EA, Nery JR et al (2013) Global epigenomic reconfiguration during mammalian brain development. Science 341:1237905
Margulies M, Egholm M, Altman WE et al (2005) Genome sequencing in microfabricated high-density picolitre reactors. Nature 437:376–380
Maurano MT, Wang H, John S et al (2015) Role of DNA methylation in modulating transcription factor occupancy. Cell Rep 12:1184–1195
Mohammed H, Russell IA, Stark R et al (2015) Progesterone receptor modulates ERalpha action in breast cancer. Nature 523:313–317
Nagalakshmi U, Wang Z, Waern K et al (2008) The transcriptional landscape of the yeast genome defined by RNA sequencing. Science 320:1344–1349
Quick J, Quinlan AR, Loman NJ (2014) A reference bacterial genome dataset generated on the MinION portable single-molecule nanopore sequencer. Gigascience 3:22
Quinlan AR, Hall IM (2010) BEDTools: a flexible suite of utilities for comparing genomic features. Bioinformatics 26:841–842
Reddy TB, Thomas AD, Stamatis D et al (2015) The Genomes OnLine Database (GOLD) v. 5: a metadata management system based on a four level (meta)genome project classification. Nucleic Acids Res 43:D1099–D1106
Simon-Soro A, Guillen-Navarro M, Mira A (2014) Metatranscriptomics reveals overall active bacterial composition in caries lesions. J Oral Microbiol 6:25443
Spitzer J, Hafner M, Landthaler M et al (2014) PAR-CLIP (Photoactivatable Ribonucleoside-Enhanced Crosslinking and Immunoprecipitation): a step-by-step protocol to the transcriptome-wide identification of binding sites of RNA-binding proteins. Methods Enzymol 539:113–161
Stephen T, Jonas K (2012) Modified base detection with nanopore sequencing Patent WO 2013185137 A1, Dec 12, 2013
Tome JM, Ozer A, Pagano JM et al (2014) Comprehensive analysis of RNA-protein interactions by high-throughput sequencing-RNA affinity profiling. Nat Methods 11:683–688
Trapnell C, Cacchiarelli D, Grimsby J et al (2014) The dynamics and regulators of cell fate decisions are revealed by pseudotemporal ordering of single cells. Nat Biotechnol 32:381–386
Wang ET, Sandberg R, Luo S et al (2008) Alternative isoform regulation in human tissue transcriptomes. Nature 456:470–476
Wilhelm BT, Marguerat S, Watt S et al (2008) Dynamic repertoire of a eukaryotic transcriptome surveyed at single-nucleotide resolution. Nature 453:1239–1243
Wilke A, Glass EM, Bartels D et al (2013) A metagenomics portal for a democratized sequencing world. Methods Enzymol 531:487–523
Wills QF, Livak KJ, Tipping AJ et al (2013) Single-cell gene expression analysis reveals genetic associations masked in whole-tissue experiments. Nat Biotechnol 31:748–752
Wolna AH, Fleming AM, Burrows CJ (2014) Single-molecule detection of a guanine(C8) -thymine(N3) cross-link using ion channel recording. J Phys Org Chem 27:247–251
Zhang B, Gunawardane L, Niazi F et al (2014) A novel RNA motif mediates the strict nuclear localization of a long noncoding RNA. Mol Cell Biol 34:2318–2329
Zupanic A, Meplan C, Grellscheid SN et al (2014) Detecting translational regulation by change point analysis of ribosome profiling data sets. RNA 20:1507–1518

Chapter 2
Whole-Genome Sequencing Recommendations

Toni Gabaldón and Tyler S. Alioto

2.1 Introduction to Genome Sequencing

2.1.1 Introduction

The recent revolution in sequencing technologies has democratized genome sequencing projects. What once was a daunting endeavor reserved for large international consortia backed by strong funding bodies is now a reasonable goal for a moderately sized research project and can be performed by small teams backed by public or private sequencing and bioinformatic centers. However, the decrease in sequencing costs and the increased availability to groups of sequencing and computing platforms has also brought about the necessity of keeping up with recent developments and strategies, as the sequencing technologies and bioinformatic tools for downstream analyses keep evolving at a fast pace. Sequencing approaches are thus a moving target. However, some general principles can be drawn that can guide the design of a successful genome sequencing project. Common considerations include evaluating known information about size and genome complexity of

T. Gabaldón, Ph.D. (✉)
Bioinformatics and Genomics Programme, Centre for Genomic Regulation (CRG),
Dr. Aiguader, 88, 08003 Barcelona, Spain

Universitat Pompeu Fabra (UPF), 08003 Barcelona, Spain

Institució Catalana de Recerca i Estudis Avançats (ICREA),
Pg. Lluís Companys 23, 08010 Barcelona, Spain
e-mail: tgabaldon@crg.es

T.S. Alioto, B.S., Ph.D.
Centro Nacional de Análisis Genómico, Centre de Regulació Genòmica (CRG-CNAG),
Baldiri Reixac, 4, 08028 Barcelona, Spain
e-mail: talioto@gmail.com; talioto@pcb.ub.es; tyler.alioto@cnag.crg.eu

A.M. Aransay, J.L. Lavín Trueba (eds.), *Field Guidelines for Genetic Experimental Designs in High-Throughput Sequencing*,
DOI 10.1007/978-3-319-31350-4_2

the target genome, obtaining samples with minimal sequence polymorphism, and assessing the needs in terms of contiguity, coverage, and quality of the assembly to address the desired research questions. Here we will provide some general guidelines and recommendations for planning whole-genome sequencing project while focusing on the two most extended applications of whole-genome sequencing. Genome sequencing projects can be grossly subdivided in two broad groups: (1) de novo genome sequencing, in which the objective is obtaining a high-quality genome assembly that can serve as a reference for a species or variety, and (2) resequencing, when there is an available reference genome and the objective is to map sequence variation of an individual or a set of individuals. As we will see below, these two objectives differ in the type of sequencing strategies, in the amount of initial material, as well as in the bioinformatics processing of the data. Despite these differences, all whole-genome sequencing projects have, nevertheless, a similar overall workflow. Four main steps can be defined: (1) sample collection and DNA extraction, (2) sequencing library preparation, (3) sequencing, and (4) bioinformatics data processing. After the data has been processed, this has to be interpreted and additional analyses should be performed. These additional analyses will depend on the particular question under study and will not be the focus of this book chapter.

2.1.2 *Sample Collection and DNA Extraction*

The first crucial step for whole-genome sequencing is the isolation and quality control of the extracted nucleic acids. The ability to obtain sufficient quantity of fresh samples may sometimes be compromised by the very nature of the organisms under study. For instance, whereas it is simple to obtain enough quantities of material from organisms that can be grown in the lab or that are easily accessible in nature, others may pose serious problems. Examples of problematic materials are material from museum specimens of recently extinct (or rare) species and species that cannot be grown in the laboratory or that are intimately associated with other organisms (e.g., symbionts, obligate parasites). Once samples are collected, DNA should be extracted. The extraction of sufficient quantities of pure, intact, double-stranded, highly concentrated, and uncontaminated genomic DNA is desirable for a reliable whole-genome analysis. The collection and DNA extraction protocols will depend on the organism under study. For instance, the presence of a cell wall in plant and fungal cells makes necessary the use of physical (vortexing in the presence of beads, heating) or biochemical (e.g., cellulase or zymolyase for plants and fungi, respectively) means to break this barrier. Thus a sensible approach for planning of the sample collection and DNA extraction is to survey existing methods that have been previously used for the genetic study of that particular species. In general, standard DNA extraction methods can be used, as long as the necessary quality and quantity of DNA of the target species is produced. These requirements depend on each specific application and sequencing strategy. Sections 2.4 and 2.5 provide some specific guidelines.

2.1.3 DNA Library Preparation

The preparation of sequencing libraries from DNA comprises a series of standard molecular biology reactions, such as fragmentation, amplification, or ligation. In general terms library preparation protocols include the fragmentation of the target DNA and the selection of fragments within a determined size range using gel or bead purification. The size range of the fragments depends on the specific whole-genome sequencing and/or assembly strategy. Subsequent amplification and ligation steps ensure the addition of the specific adaptors at the 5′ and 3′ ends, required for the sequencing phase (see below). Alternatively, "tagmentation" combines the fragmentation and ligation reactions into a single step, which can greatly increase the efficiency of the library preparation. Adapter-ligated fragments are then amplified by polymerase chain reaction (PCR) and purified in gel. Preparation of high-quality libraries and obtaining high yields require a good initial material (see point above) and a careful execution of the library preparation protocol. A number of kits that ease the preparation of libraries are available, and some are provided by the company that manufactures the sequencer. Potential problems in the library preparation phase include biases in the inclusion of genomic regions into the library and the creation of chimeric fragments by artifactual ligation of fragments originating from different genomic regions (Van Dijk et al. 2014).

2.1.4 Sequencing

The principle of next-generation sequencing (NGS) is similar to that of capillary electrophoresis (Sanger) sequencing: sequencing by synthesis, in which the addition of each nucleotide is monitored while DNA polymerase copies a DNA template. However, the critical difference in NGS is that instead of sequencing a single DNA fragment, millions of fragments can be processed in parallel. In the most widely used sequencing-by-synthesis NGS technology, Illumina, DNA polymerase catalyzes the incorporation of fluorescently labeled deoxyribonucleotide triphosphates (dNTPs) into a DNA template strand during a number of cycles of DNA synthesis. At each cycle, the incorporated nucleotides are identified by fluorophore excitation. Sanger sequencing is now obsolete due to its high cost, and some of the earlier generations of NGS technologies are disappearing in favor of newer ones. For instance, Roche has announced that its support for 454 sequencing will be discontinued in 2016. This turnover of sequencing technologies is likely to continue in the coming years. The interested reader is encouraged to read a recent review of current sequencing technologies (Reuter et al. 2015).

2.1.5 *Bioinformatics and Data Processing*

The sequencing process produces a significant amount of data. For instance, a single run of an Illumina HiSeq2500 will produce 1 terabyte of data in about 6 days. The raw data is primarily provided in the form of plaintext files containing the sequences with associated quality scores. The general format used is the so-called FASTQ format which bundles a FASTA sequence file to its quality data codified as ASCII characters. The information of the quality scores is generally used for an initial quality clipping of the data, in which reads with low qualities are removed or trimmed. Subsequently, in a whole-genome analysis, there are two basic operations with this data. In de novo genome sequencing, reads are assembled into larger contigs by means of detecting sequence overlap between the reads. Alternatively, in genome “resequencing,” reads are mapped (i.e., aligned) to a reference genome sequence in order to subsequently detect the desired variations (see below). Both assembly and mapping processes may require significant computational resources. Mapping can be easily parallelized but assembly needs to consider large amounts of data simultaneously which requires access to large amounts of RAM. Currently, 1 terabyte RAM, 32 core servers are often used.

2.2 Review of Achievable Objectives

2.2.1 *De Novo Genome Sequencing*

The ultimate goal of a de novo whole-genome sequencing project is to obtain a good quality reference assembly and sequence for a representative genome of a given species. What is understood as “good quality” may vary depending on the subsequent application. Generally, one major goal of high-quality genome references is to obtain high-quality gene model annotation. If there is interest in the large-scale organization of the genome and/or the dynamics of repetitive elements, high contiguity is also needed. Ideally, one would wish for a final assembly that contains a single scaffold per chromosome, encompassing all sequence information, from telomere to telomere, and containing no sequencing or assembly errors.

2.2.2 *Resequencing*

The goal of a genome resequencing project is to annotate, for a given sample (individual, cell line, tissue, etc.), the variations (polymorphisms) in the genome with respect to the reference (or to another sample). These variations may comprise all or a subset of the following types: single-nucleotide changes, including polymorphisms (SNPs), rare variants (SNVs), or simple somatic mutations (SSMs), insertions and deletions, copy number variations (CNVs), and other rearrangements broadly categorized as structural variants (SVs).

2.3 Recommended Sequencing Platforms

Sequencing platforms are evolving continuously at a fast pace (Reuter et al. 2015). The recommendations outlined here will necessarily be limited to the current available techniques which may soon be surpassed by newer technologies. In general we will phrase our recommendations in terms of read length, throughput, and read pairing strategies. The Illumina platforms give high-quality sequence at the lowest cost per Mb. The main disadvantage is that read length is limited to shorter reads (100–300 bp) because of phasing issues and size restrictions on bridging amplification. Single-molecule sequencing (Pacific Biosciences and Oxford Nanopore Technologies) can achieve longer reads at the expense of error rate, throughput, and cost. Coverage can offset problems in high error rate, at least for de novo assembly.

2.4 Experimental Design Guidelines (Best Practices)

2.4.1 De Novo Genome Sequencing

For a de novo genome sequencing, the most crucial part is to perform the assembly. This process is based on finding sequence overlaps between reads that allow their assembly into contigs and scaffolds that represent longer sequences (Simpson and Pop 2015). The presence of sequence variants within the sequenced DNA sample complicates this process, because these variants create mismatches between reads that correspond to the same genomic locus. The source of sequence variants can originate from the presence of a genetically heterogeneous set of organisms in the sample. Thus one first recommendation is to use a genetically homogeneous source of genomic DNA. In large organisms it is easy to obtain enough material from a single individual. For smaller ones, the use of several individuals from clonal populations is preferred. In diploid organisms (or organisms with higher ploidy) sequence variants of the same locus can be present in the same organism. When possible, the use of inbred lines with reduced heterozygosity levels is recommended.

Once the appropriate source for the DNA has been selected, the next important consideration is the sequencing strategy. This will be determined mainly by the size and complexity of the target genome. For the same sequencing error rate, longer reads and higher sequencing coverage facilitate the assembly process. However, different technologies or sequencing strategies differ in throughput, read length, and error rate in a way that a combination of several of them is generally the optimal solution. To inform this process, it is highly recommended to learn from previous efforts in sequencing the genomes of highly related species and to gather as much information on the complexity of the target genome in terms of size, level of heterozygosity, and abundance of highly repetitive regions. As the number of sequencing projects increases, such guidelines and learned best practices are starting to be available for more diverse sets of organisms (Richards and Murali 2015). When this information is not available in the literature for that species or closely related ones, one sensible approach is to perform a small sequencing test involving one single run. Simple analysis of k-mers (a short DNA sequence consisting of a fixed number (K) of bases) can inform us on parameters such as estimated genome size, presence of repetitive regions, and heterozygosity, among others (Simpson 2014). A common practice in the era of Sanger sequencing was to clone a few bacterial artificial chromosomes (BACs) and shotgun sequence them first and annotate them with repeats and genes.

2.4.2 *Genome Resequencing*

Genome resequencing generally involves fewer constraints on the data than de novo sequencing. When the main objective is mostly to determine single-nucleotide polymorphisms and copy number variations, the accuracy and sequence depth of coverage is instrumental, and thus sequencing strategies that provide a higher throughput are preferred. When information on genome rearrangements is required, the design needs to include sequencing strategies that provide information of the relative position of sequences over larger genomic distances. This includes technologies providing long reads or library preparation strategies that capture long genomic fragments from which the extremes are sequenced (mate-pair (MP) or clone end sequencing). Optical mapping (e.g., BioNano Genomics and OpGen) shows potential in this arena, but is not yet standard (Howe and Wood 2015; Tang et al. 2015).

2.5 Technique Overview (Wet Lab Protocol Overview: Library Construction Recommendations)

As mentioned above, sequencing involves DNA extraction and sequencing library preparation. DNA extraction should be performed with protocols that are appropriate to the particularities of the biological material available so that a sufficient quantity and quality of DNA is obtained. A first step that precedes the preparation of the library is the quality control (QC) of the DNA samples. QC involves quantification of the amount of DNA, checking the 260:280 absorbance ratio (ratios between 1.8

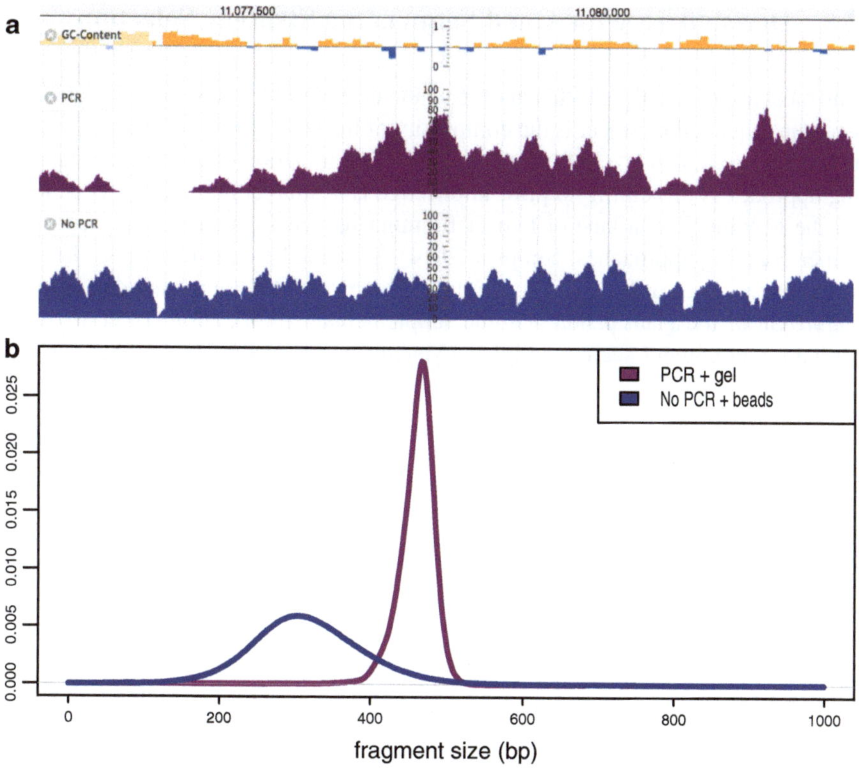

Fig. 2.1 No-PCR library preparation results in more even coverage across wide range of GC content. Panel A shows the coverage profile (both sets of reads were downsampled to 30x at the locus shown) while panel B shows the fragment-size distributions. In *magenta* is the standard PCR protocol (10 cycles of PCR) and in *blue* the no-PCR protocol. While the fragment-size distribution is not as tight, the no-PCR protocol leads to more even coverage, for the most part independent of GC content

and 2 are considered to indicate relatively pure DNA), and running an aliquot on a gel to check integrity and detect ribosomal bands. Ideally, there should be a sufficient amount of DNA to proceed with a no-PCR protocol, which reduces the GC bias effect. The difference in coverage of a particular locus affected by PCR-dependent GC bias is shown in Fig. 2.1. For Illumina SBS sequencing, sample preparation proceeds starting with DNA fragmentation (e.g., with Covaris), A-tailing, adapter ligation, and then size selection (column/beads for automation and consistency or gel for tighter size selection). An aliquot should then be run on a Bioanalyzer or similar instrument in order to choose the most promising libraries for sequencing. Longer fragments are not amplified as well by bridging PCR on the Illumina flow cell, so smaller fragments need to be removed by column purification if longer (>500 bp) fragment libraries are to be sequenced.

2.6 Decision Tree for Good Sequencing Strategy Selection

The most important aspects that anticipate the difficulty of an assembly in a de novo genome sequencing project is the complexity of the target genome, in terms of size, repeat structure, and level of heterozygosity. Determination of the correct sequencing approach is difficult if no prior knowledge is available. Fortunately, depending on the genome size, a lane or two of Illumina sequencing can be analyzed using k-mer counting approaches (Simpson 2014). This can be done using specific software (Preqc, gce) or by using the simple 17-mer counting approach described in Figure S8 of the giant panda genome supplementary information (Li et al. 2010) with a k-mer counter such as Jellyfish (Marçais and Kingsford 2011). See Fig. 2.2

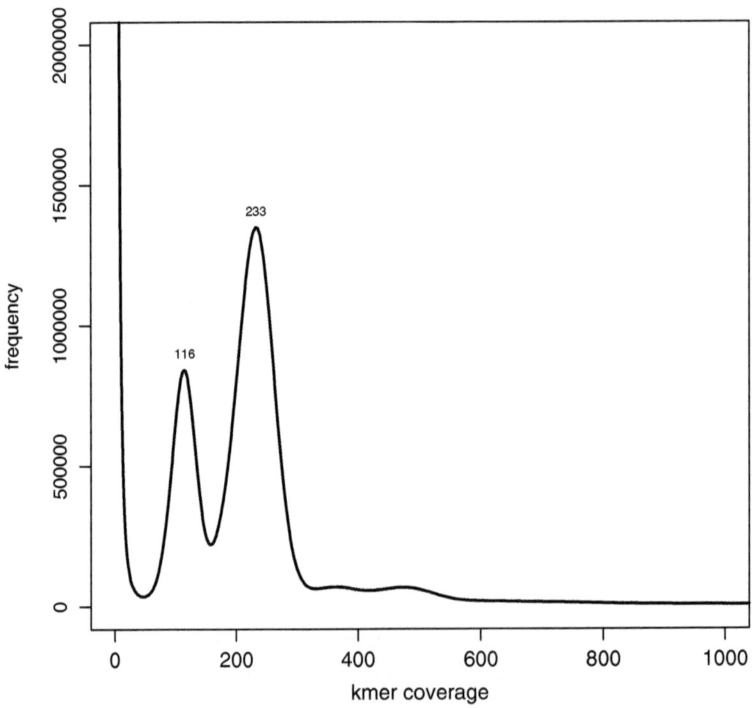

Fig. 2.2 The k-mer frequency plot for a heterozygous genome. Unique 17-mers were counted with Jellyfish. The number of unique 17-mers is plotted according to the number of times they are seen in the input set of Illumina reads (k-mer depth). The highest peak occurs at depth of one. These k-mers that appear only once in the set of reads correspond to sequencing errors. The next highest peak (at k-mer depth of 233) is the main peak, which is correlated with the depth of sequencing. In this case we see a substantial minor peak at half the depth (k-mer depth of 116), which is induced by the presence of polymorphisms. This is a diploid genome, so we only see one minor peak. In genomes of higher ploidy, it is possible to see additional peaks. To the right of the main peak, one can observe a wavelike pattern corresponding to repetitive elements. Larger peaks here are sometimes observed indicating a higher fraction of repetitive content. To estimate the genome size (without correcting for major sequencing biases like GC bias), one can simply divide the total number of k-mers by the depth of the main peak

Table 2.1 Provides several examples of sequencing and assembly strategies

Case	Sequencing strategy	Assembly strategy	Reference
Haploid fungal genome (*Penicillium digitatum*) 26 Mb	Illumina pair-end (PE) 2×50	SOAPdenovo	Marcet-Houben et al. (2012)
	Illumina mate-pairs 2×50 5 kb inserts		
Diploid fungal hybrid (highly heterozygous) (*Candida orthopsilosis*) 12.6 Mb	Illumina pair-end 2×75	SOAPdenovo	Pryszcz et al. (2014)
		REDUNDANS	
Giant panda	Illumina paired-end 2×50 and 2×75	SOAPdenovo	Li et al. (2010)
	Illumina mate-pairs 2×50 2 kb, 5 kb, 10 kb inserts		
Loblolly pine (22 Gb)	Illumina MiSeq paired-end 2×255	MaSuRCA	Neale et al. (2014)
D. melanogaster, *A. thaliana*, *S. cerevisiae*, cell line CHM1	PacBio SMRT sequencing	Celera Assembler with MHAP	Berlin et al. (2015)
E. coli	Oxford Nanopore	Nanocorrect (DALIGNER+POA), Celera Assembler, nanopolish	Loman et al. (2015)

Several different sequencing and assembly strategies are shown from examples taken from a diversity of organisms

for an example. Genome size, repeat content, and heterozygosity can all be estimated with such an approach. Table 2.1 lists some real examples that illustrate different genome complexities and the sequencing strategy that led to good quality assemblies.

One strategy that helps with highly repetitive genomes and highly heterozygous genomes (Fig. 2.3) is to divide the genome into smaller pieces by cloning fragments in BACs or fosmid vectors and sequence them either individually (antiquated Sanger-based clone-by-clone approach) or in pools (more easily managed and cost-effective on the Illumina platform). Drawbacks include cost of making the fosmid library, dividing into pools and preparing the DNA as well as the cost of sequencing, which depends on the target clone coverage. 5× clone coverage (necessary to cover 99 % of the genome) would cost at least five times as much as a standard whole-genome shotgun approach. Perhaps soon, long single-molecule reads may present a fast and cheap replacement for this approach; however, the goal remains the same—to reduce the problems caused by repeats and to deal with polymorphism. With regard to genome resequencing projects, the constraints are fewer, and the characteristics of the genome are generally known for that species, as there is a reference genome available. The genome size determines the required amount of sequencing so that variations can be called with sufficient confidence.

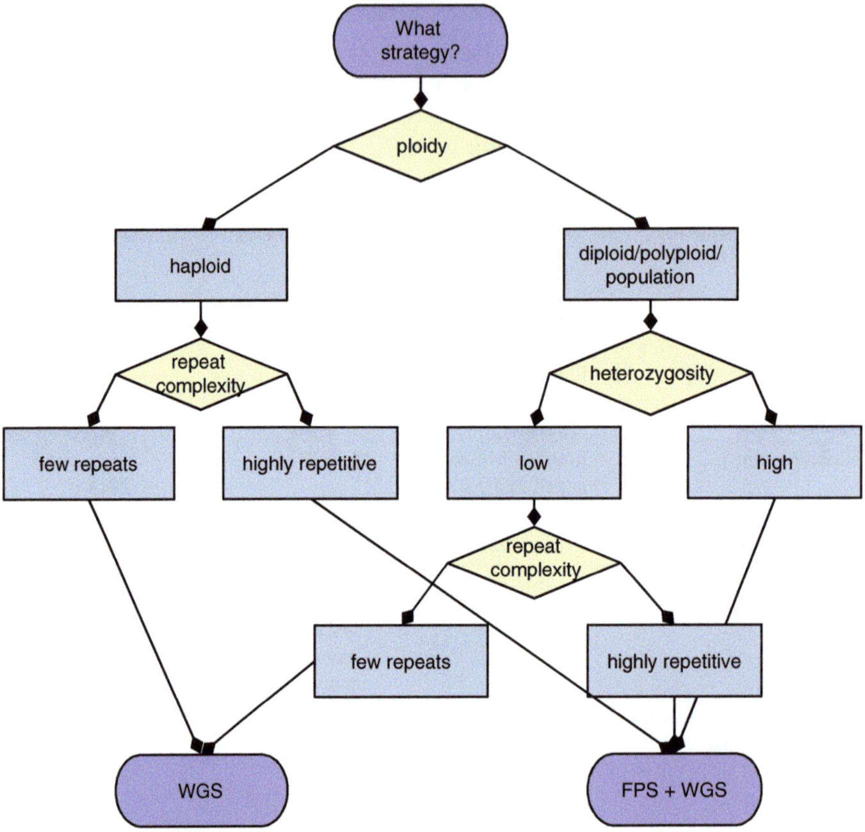

Fig. 2.3 Deciding between sequencing pools of clones vs. pure whole-genome shotgun approach. FPS = fosmid pool sequencing. WGS = whole-genome shotgun sequencing

2.7 Potential Bottlenecks of the Methodology

The sequencing itself is no longer a bottleneck for genome sequencing projects. Depending on the strategy taken, if cloning steps are involved (e.g., fosmid or BAC libraries) or if experimental sequencing library preparation is to be carried out, one can expect delays on the front end. However, the conversion of the raw sequencing data into a high-quality, finished genome assembly is generally one of the major bottlenecks in a de novo genome sequencing project. This process is complicated by the different read lengths, read counts, and error profiles that are produced by different sequencing technologies. In addition, biases in sample preparation, sequencing, and genomic alignment and assembly may result in genomic regions without coverage (i.e., gaps) and in regions with much higher or lower coverage than theoretically expected. GC-rich regions, such as CpG islands, can particularly suffer from low coverage because such regions remain annealed during the amplification step.

Highly repetitive regions, which are prominent in multicellular organisms with large genome sizes, are hard to assemble. In theory, one needs to bridge the repetitive regions by sequencing fragments that expand the whole region and its boundaries, either by using long reads or long mate-pair libraries. Due to its large size and high redundancy, some regions may remain unresolved at any given fragment size. These would need to be closed by targeted approaches that are costly and time consuming. Depending on the expected use of the assembly, this can tolerate the presence of gaps or unresolved regions, and most projects reach a compromise that would satisfy most general applications. Recently, duplicated regions, such as those deriving from tandem gene duplications, are also problematic and most assemblers would collapse these regions into a single one. The same type of regions is problematic in genome resequencing projects, for the same reasons: some regions are less covered among sequenced reads, giving rise to gaps and coverage biases. In addition, short reads may map in multiple loci leading to ambiguity in the localization of a particular variant.

2.8 Bioinformatic Analyses (Best Practices)

2.8.1 Bioinformatician Consulting for Experimental Design

It is important to consult with the team that will perform the bioinformatic analysis earlier on. Poorly designed experiments or sample collection will introduce analytical challenges in downstream analyses; to minimize these complications, bioinformatic teams can provide useful recommendations based on previous experiences. Ideally, a bioinformatic team that has previous expertise in similar analyses and that is easily accessible would be involved in the project from the beginning. Many teams doing bioinformatics research may be recruited to the project if they have a scientific interest in the project. A recommendation is to try to involve them from the start of the project and make them aware of the research interest, asking them to contribute to its solution, rather than simply using them for subsidiary help in the tedious task of "simply" processing the data. This will ensure a high level of implication and a true interest in producing the best results. An important guideline in this respect is to reward the help of bioinformatic collaborators with due recognition in terms of authorship (Chang 2015). Beyond collaborations from other groups, bioinformatic support can be obtained from core services at many large institutions or companies that specialize in bioinformatic analyses. Assessing what is the expertise of these teams in projects similar to the one at hand is crucial to ensure a successful experience. Finally, it is advisable to envision the hiring of bioinformaticians in the project. If bioinformatic expertise is lacking in the host group, these specialists could ideally be embedded (at least for some time) in teams of data analysis collaborators or cores, so that he/she benefits from expert knowledge accumulated in experienced teams.

2.8.2 *Analysis Workflow Overview (From Raw Reads QC to Functional Characterization)*

2.8.2.1 Quality Clipping, Filtering, and Error Correction

Invariably, the first step of data analysis is the quality clipping and filtering of the raw sequencing results. An efficient filtering of low-quality data will minimize problems in downstream analysis. One first filtering that must be done is to remove any partial adapter sequence that may have been sequenced. This can occur when a given sequenced fragment was shorter than the read length. In addition it is possible that concatenated adapter-only sequences have been sequenced. These sequences must be removed. Subsequently it is highly advised to perform a control of the quality of the reads which may lead to filtering or trimming reads of regions thereof that have low quality. As mentioned above raw sequencing reads are made available as FASTQ text files, in which each short read takes up four lines: the read identifier (starting with an @), the DNA sequence itself, another identifier (same as line 1, but starting with a + (or sometimes only consisting of a +)), and the Phred quality score for each base in the read. The quality score is encoded with an ASCII character code (http://www.ascii-code.com/). Illumina and other manufacturers currently (as of v1.8) use the Sanger Phred ASCII encoding offset of 33, so that the ASCII code 33 (!) is 0, and ASCII code 74 (J) is 41. Quality scores are defined as $Q_{\mathrm{phred}} = -10\log10(p)$, where p is the estimated probability of a wrong base call. So a Q_{phred} of 20 corresponds to a 99 % probability of a correctly identified base (1 % error; see Table 2.2).

One of the first evaluation routines is to assess how the distribution of quality scores and nucleotides looks like. This is generally done by summarizing and plotting the data (typically with FASTQC or a similar software). A typical plot includes the quality score per residue (see Fig. 2.4 for an example of a 250 nt HiSeq2500 read 1). Quality scores generally decrease over the length of a read (i.e., first incorporated nucleotides are determined with higher accuracy), and how fast these declines occur can vary from one sequencing run to the next. This plot will reveal whether the sequencing run maintained an overall high quality during the whole procedure or whether trimming the last residues of the reads would be advisable. Q30% (average percent of bases >Q30) is a frequently used metric to determine the overall quality of a run, while error rate (estimated by spiking in PhiX DNA as a control) is probably the most relevant metric for downstream analyses. Quality

Table 2.2 Relationships between Phred quality scores and accuracy

Phred quality score	Probability of incorrect base call	Base call accuracy
10	1 in 10	90 %
20	1 in 100	99 %
30	1 in 1000	99.9 %
40	1 in 10,000	99.99 %
50	1 in 100,000	99.999 %
60	1 in 1,000,000	99.9999 %

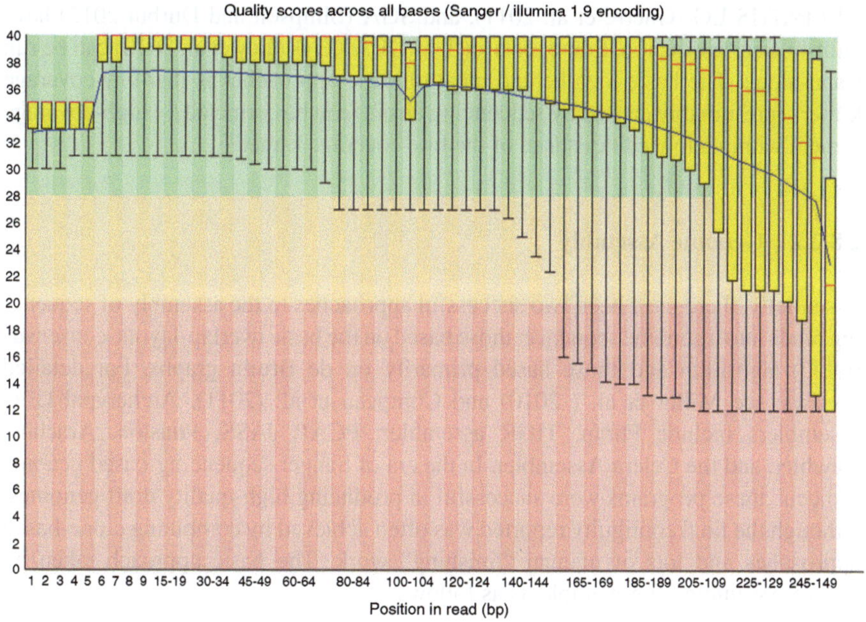

Fig. 2.4 FASTQC per-base quality report. This example is from read one of a typical 2×250 nt run of a HiSeq2500. The first few cycles typically show some sequence bias and lower quality. Sometimes a particular cycle (in this case around base 100) shows a slight dip in quality, perhaps due to a temperature fluctuation. Gradual decrease in quality is observed after 150 cycles, falling more rapidly after 200 cycles

scores and error rates are related, of course, but not perfectly, which is why some analyses recalibrate base qualities using packages such as the Genome Analysis Toolkit (GATK) from the Broad Institute (McKenna et al. 2010). Another informative plot is the base composition per residue, e.g., what fraction of A's, C's, G's, and T's has a given position in a read. A perfectly random sampling of reads along a genome should render horizontal lines for each residue, with their values in accordance to the overall base composition of the genome (e.g., with GC content). Nonuniform patterns reveal biases in the composition of the reads and may indicate strong amplification biases or the presence of sequenced adapters in the reads. In addition, it is recommended to assess the fraction of duplicate reads (identical reads present that are present in the dataset), as they may originate from primer or PCR bias, and thus a large fraction of duplicate reads may be indicative of a poor cDNA library. Several tools and packages are available for performing the quality assessment and trimming of FASTQ files. Some currently popular options include FASTX, FASTQC, Trimmomatic (Bolger et al. 2014), cutadapt, trim_galore, or PRINSEQ (Schmieder and Edwards 2011).

In addition to trimming, another way to deal with errors is to correct them. For de novo genome assembly, error correction can reduce memory consumption and lead to simpler assembly graphs. Popular assembly tools SOAPdenovo (Luo et al. 2012),

ALLPATHS-LG (Gnerre et al. 2011), and SGA (Simpson and Durbin 2012) have built-in error correction. Some tools such as QUAKE (Kelley et al. 2010) can be run stand-alone. The basic idea behind most of these approaches is that low-coverage k-mers (presumably caused by sequencing errors) can be corrected by high-coverage k-mers within a low edit distance of the low-coverage k-mer.

2.8.2.2 Genome Assembly

Essentially, there have been two successful approaches to the assembly of sequencing reads into a genome sequence: those based on the basic overlap-layout-consensus (OLC) algorithm and those based primarily on de Bruijn graphs. For detailed reviews, see Miller et al. (2010) and Compeau et al. (2011). Archetypal OLC assemblers include Phrap, TIGR assembler, PCAP, JASS, Phusion, Arachne, Newbler, and the Celera Assembler. In the era of Sanger sequencing-based genome projects, these programs were successful in producing high-quality draft genomes, although the final contiguity reported was often achieved by combining clone-based approaches and lots of manual "finishing" work. The basic approach taken by Celera Assembler, for example, is as follows:

1. Overlap
   (a) Overlaps are computed among the set of all reads ("all against all") using a BLAST-like seed and extend algorithm. *ovl* (classic) or *mer* (for 454) are used as the overlapper. Both use a seed and extend approach, but with parameters tuned to Sanger or 454 read length and error profiles, respectively. Other assemblers use similar seed approaches (like BLAST) and usually process the initial overlaps with Smith-Waterman alignment.
   (b) Such overlap computations use the majority of CPU time.
2. Layout
   (a) The genomic order or "layout" of the reads is determined by computing a Hamiltonian path in which reads are represented as vertices in a graph, the overlaps are edges, and a path is found that visits each vertex once and only once.
   (b) The CA module unitigger is used to compute initial high-confidence contigs.
   (c) Scaffolder uses additional mate-pair data to join unitigs with estimated gaps.
   (d) The layout step often uses the most memory.
3. Consensus
   (a) The optimal multiple sequence alignment is usually unattainable. Heuristics are used to guide the alignment and output a consensus. Variants can sometimes be output. Depending on the length and pairing of input data, the variants can be phased.

Practically speaking, most OLC software cannot be run efficiently on NGS data. However, the cost of hardware (CPUs and memory) has fallen and the algorithms and implementations improved so much that, for example, the Celera Assembler can now be run on Illumina data, although still not as efficiently as the k-mer graph (de Bruijn graph)-based assemblers.

With the introduction of massively parallel sequencing, which is characterized by the production of a very large number of short reads, OLC approaches became computationally infeasible, necessitating new algorithmic development. Fortunately, the mathematics had already been worked out and only required co-opting for the assembly problem. A Eulerian path, in particular the k-mer version of the de Bruijn graph, is similar to a Hamiltonian path, but where vertices are the k-mers and edges are the k—one overlaps and each edge is visited at least once. The solution to this problem is more computationally feasible and has become popular for assembling NGS data. However, the problems that complex genomes present, such as repeats and heterozygosity, become even harder to resolve. Extra attention must be paid to read trimming and error correction and to cleaning of the assembly graph (pruning tips, popping bubbles, etc.).

To generate high-quality assemblies from NGS data, one more or less follows the general workflow depicted in Fig. 2.5. It is important to preprocess the read data as described above and to do quality control checks (e.g., FASTQC) and plot k-mer frequencies to estimate genome size and complexity. Then, the overlap graph (as discussed above the more efficiently computed by de Bruijn graph) is created. To generate unitigs using a de Bruijn graph, k-mers of different lengths should be tested. K-mers that are too short will result in an assembly broken by short tandem repeats, while k-mers that are too long will result in assemblies broken at regions of low coverage. Moreover, longer k-mers often require more memory to store k-mer counts, as errors create a number of unique k-mers equal to the k-mer size each time an error occurs in a read. Then pairing information from short fragment paired-end reads and/or long fragment mate-pair reads is used to join unitigs into longer contigs and these contigs into scaffolds containing gaps of estimated size using the mean and standard deviation of the fragment lengths for each sequencing library. It is important to detect potential misassemblies along the way by trying to detect chimeras, aberrant depth (repeat) contigs, or compression/expansion errors either by determining the consistency or support of the read data aligned back to the intermediate assembly (e.g., using REAPR (Hunt et al. 2013)) or by using external information such as physical or genetic maps or alignment to phylogenetically close high-quality reference genomes. After misassembly correction, one can fill scaffold gaps using either built-in modules or stand-alone programs such as GapFiller (Boetzer and Pirovano 2012). Polishing, or fixing small errors such as single-nucleotide substitutions or indel errors like homopolymers, can be achieved using approaches nearly identical to variant calling of resequencing data. Finally, if genetic, physical, or optical maps have been generated, the assembly can be "anchored" to chromosomes/linkage groups/pseudomolecules by mapping the positioned markers onto the scaffolds and then ordering and orienting them if possible to create a final anchored assembly.

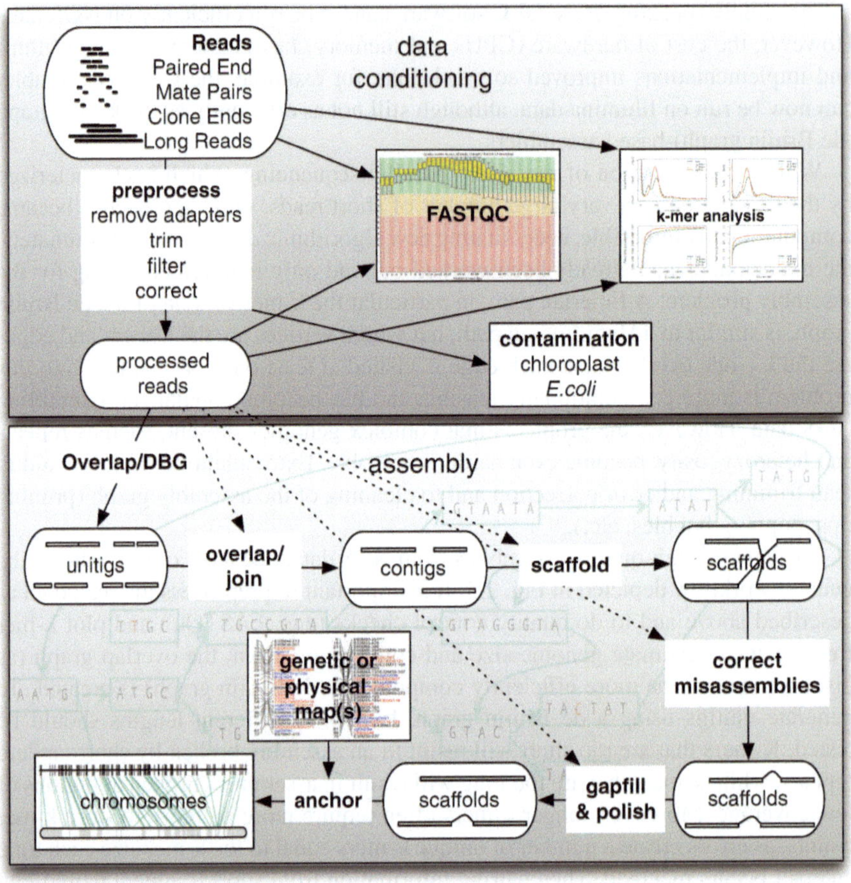

Fig. 2.5 General assembly workflow

2.8.2.3 Read Mapping and Variant Calling

Read mapping refers to the process of aligning short sequencing reads to a reference sequence, which is generally a complete genome, but can also be a transcriptome. A plethora of computer programs have been developed that map (also called align) reads to a reference sequence. These programs use different algorithms that vary in speed and accuracy (Fonseca et al. 2012). The majority of fast mapping algorithms perform indexing on the read sequences or the reference sequence, or sometimes both. Similar to Google's indexing of websites, a preprocessing of sequence data creates an index data structure that accelerates the search for a near-exact match. Depending on the nature of the index, mapping algorithms can be roughly grouped into three categories: algorithms based on hash tables, algorithms based on suffix trees, and algorithms based on merge sorting (Li and Homer 2010). Most existing algorithms belong to the first two types. All algorithms based on hash tables keep the position of each k-mer subsequence (a sequence of k residues) of the query in a table (hash table) and scans the databases for k-mer exact matches (called seeds). Algorithms based on suffix trees first identify exact matches using a data structure that stores all the suffixes of a string and then build inexact alignments around the exact matches. Different mappers diverge in their particular implementation of the strategy and in their inclusion of additional parameters that enable more efficient mapping of dissimilar types of data, for instance, the ability to perform alignments containing gaps, or split alignments, or the possibility to incorporate information from pair-end or mate-pair reads. The most immediate goal of read mapping is to create an alignment file also known as a sequence alignment/map (SAM) file. The SAM file contains one line per mapped read indicating the reference sequence and position to which it maps, as well as a Phred-scaled quality score of the mapping, among other details (Li et al. 2009). The SAM format is human readable and easier to process by conventional processing programs. The BAM format provides binary versions of most of the same data and is designed to provide higher compression.

One of the main purposes of genome resequencing is to discover genetic variation among related individuals or samples in a large scale. This inference is generally done after the mapping of the reads is completed. Again, a number of algorithms and computer programs are available that are designed to call variants from SAM/BAM files. Most are focused on the detection of single-nucleotide polymorphisms (SNPs) or small insertions and deletions. A variation from the reference sequence will result in mismatches, gaps, or a significantly different coverage, and most algorithms perform a statistical analysis of mapping results to provide a call of present variants. Most prevalent types of sequence variation, including SNPs, indels, and larger structural variants, are generally stored in a specific format denoted as variant call format (VCF). Larger variations such as copy number variants (CNVs) and genomic rearrangements are generally detected with specific programs. For instance, CNVs can be detected by methods that assess the depth of coverage, by piling up aligned reads against genomic coordinates and then calculate the depth of coverage along windows and compare it with the average coverage of the region (Consortium 2012). Genomic rearrangements can be assessed by using information of the mappings of mate-pair or pair-end reads (Xi et al. 2010).

2.8.3 Sequencing Depth (Number of Aligned Reads Required for a Reliable Analysis)

2.8.3.1 Introduction

Despite significant drops in price, sequencing costs still set limits to the total amount of sequence that can be generated. In addition, various analyses may require different minimal sequence coverage to provide reliable results. These factors are keys for the experimental design of a whole-genome sequencing project (Sims et al. 2014). Here we will provide an overview of current guidelines and precedents with respect to sequence coverage. The empirical per-base coverage (or sequencing depth) is the exact number of times that a base in the reference is covered by a high-quality aligned read in a given sequencing experiment. However, when planning a whole-genome sequencing project, we must deal with the expected coverage, which is the average number of times that each nucleotide in the genome is expected to be sequenced given a certain number and length of reads and with the assumption that reads will be randomly distributed across the genome. Lander and Waterman (1988) described this as $c = LN/G$, where L is the read length, N is the number of reads, and G is the haploid genome length. Sequencing depth is generally expressed in fold coverage units (e.g., 10× means that an average base is covered by ten reads).

Redundancy in sequencing data is necessary to overcome sequencing errors and biases. If a sequencing method would be 100 % accurate and perfectly balanced over the entire genome sequence, then a 1× depth of coverage would suffice for all downstream analyses. However, in reality, sequencing errors are not negligible. To distinguish errors from sequence variants, one needs to assess all reads mapped to a given residue. For instance, at a 1 % error rate, the combination of ten identical reads that cover the location of the variant will produce a strongly supported variant call with an associated error rate of 10^{-20}. It must be noted, however, that increased depth of coverage cannot solve other sequencing problems such as gaps or ambiguous alignments in repetitive regions. Thus, sequencing depth must be considered in combination with alternative sequencing strategies (e.g., paired-end, mate-pairs).

2.8.3.2 De Novo Sequencing

The required depth in a de novo genome sequencing project is determined by several factors including the sequencing method and strategy, read length, the assembly approach, and the complexity in terms of repetitive regions of the genome (length, similarity, and abundance of the repetitive regions). For instance, Sanger-based sequenced genomes such as dog and human provide good reference assemblies at low coverage (7–10×), whereas much higher sequencing depths (~73×) using short reads rendered poor assembly qualities in the giant panda, a genome of similar size and complexity to that of dog (Lindblad-Toh et al. 2005; Li et al. 2010). For Illumina data, the depth and library types need to be matched to the assembly algorithm, which can have very specific requirements. For example, ALLPATHS-LG requires a 2×100 PE library of fragment length 180 bp (20 bp overlap) at >50× coverage and at least one MP library of 3 kb fragment length also at 45–50× coverage. Larger mate-pair libraries are necessary for more contiguous assemblies. It can also take advantage of long PacBio reads at about 50× coverage. This software is being replaced by DISCOVARdenovo, which requires 50–80× coverage by a single 450 bp fragment PE library sequenced in 2×250 PE mode on a HiSeq2500. Of course additional scaffolding with MP libraries or other means can and should be carried out with stand-alone scaffolding software.

SOAPdenovo and ABySS are more flexible in the number of input libraries and coverage. ABySS is able to use distributed memory and thus has more flexibility in terms of the number of reads you give it. However, best results are achieved when at least 100× coverage in PE reads (all PE libraries combined) is used for the initial de Bruijn graph construction, with a minimum of 20–30× per library for scaffolding. Higher coverage can give better scaffolding results, but with diminishing returns.

For PacBio-only assemblies, one can use the MHAP algorithm (Berlin et al. 2015) that is now available as part of the Celera Assembler. Required coverage is a minimum of 50–70×. This strategy is able to reconstruct whole chromosome arms of the *D. melanogaster* genome. A similar approach for Oxford Nanopore Technologies two-directional reads has been implemented in a pair of packages called *nanocorrect* and *nanopolish* (Loman et al. 2015). At least 25× coverage is necessary, with higher depth likely to yield better results. For both technologies, the error rate is typically too high to run self-alignments with more traditional aligners, a step necessary for calculating overlaps; thus they utilize new alignment algorithms (the MinHash Alignment Process (MHAP) and DALIGNER (https://github.com/thegenemyers/DALIGNER), respectively) that are roughly based on the idea of shared k-mer content.

2.8.3.3 Resequencing

Early resequencing studies of humans using Illumina short-read approach showed that the required sequencing depth to detect most of the SNPs and short indels was 15× when they were homozygous and 33× if they were heterozygous (Bentley et al. 2008). Subsequent studies have provided similar estimates, and thus depths exceeding 30× have become the de facto standard in resequencing analyses (Ajay et al. 2011). The use of low base qualities and nonuniform coverage may challenge the detection of variants, so these numbers should be considered after filtering reads by quality and assuming a uniform coverage over the genome. For the detection of CNVs, uniformity of sequencing coverage is instrumental to avoid false positives. In addition, accurate inference of break points and absolute copy number estimation improve with increasing read depth.

2.8.4 Difficulties of the Bioinformatic Analyses

Although an increasing number of user-friendly solutions are becoming available, the difficulty of the bioinformatic analyses required remains high. Attempts to undergo a genomic analysis without the required expertise can lead to frustration and dangerous misinterpretations of the data. Thus, it is highly advisable to include in the team the necessary human resources with sufficient expertise. As mentioned above, this can be achieved through collaborations with bioinformatic teams, service cores, or companies.

2.8.5 *Expected Results*

2.8.5.1 De Novo Sequencing

The expected result for a de novo genome sequencing project is a high-quality genome assembly, which is annotated to some satisfactory level. The quality of the assembly in terms of contiguity depends on the expected use. As mentioned above, the optimal target is an end-to-end, one chromosome one contig, no-gap containing accurate sequence. However, such an objective has only been accomplished for small genomes, and larger genomes containing repetitive sequences are generally incomplete, despite extensive effort. As an example, the human genome still contains hundreds of large, unresolved gaps that correspond to repetitive or heterochromatic regions. Fortunately, not all applications of de novo genome sequencing require full completion of the assembly. For instance, protein-coding regions of the genome, which remain the main focus of de novo genome sequencing, are generally well recovered. However, a highly fragmented genome may split genes across different contigs. If the interests lie on higher-scale properties of the genome such as gene order, high contiguity in the assembly is required, although the presence of undetermined sequences may be allowed. Finally, some analyses are highly demanding on the assembly completion, for instance, when the focus is in determining the content and distribution of transposable elements.

2.8.5.2 Resequencing

The expected results for a genome resequencing analysis would be a comprehensive catalog of genetic variations in individuals, samples, or populations with respect to a given reference. This includes single-nucleotide variants, small insertions and deletions (indels), larger structural variants (such as inversions and translocations), and copy number variants (CNVs).

2.8.6 *Effective Result Reporting*

2.8.6.1 De Novo Sequencing

Genome assemblies are reported and shared as a set of files including:

1. A set of FASTA files corresponding to contigs, scaffolds, and/or chromosomes. Scaffold FASTA files are the most common and useful of these.
2. One or more AGP files describing the structure of the assembly with contigs as the building blocks. An AGP (acronym for "A Golden Path") is a commonly

used file format for describing assemblies. This format was originally conceived by the International Human Genome Sequencing Consortium and used to describe the genome assembly of human. It is now the most commonly used format for specifying assembly information (http://www.ncbi.nlm.nih.gov/projects/genome/assembly/agp/AGP_Specification.shtml). There can be multiple AGP files: one for scaffolds, another for superscaffolds, and another for pseudomolecules/chromosomes/linkage groups.

2.8.6.2 Assembly Metrics

A wide range of basic statistics are available that serve to describe the quality of a given assembly. The most basic one is the total size of the assembly (assembly size), which reports the total number of bases contained in the genome. When compared to the estimated or known size of the target genome, this metric can be transformed into the coverage of a given assembly over the genome of interest. Another set of useful metrics inform on the contiguity of the assembly, that is, whether the assembly is formed by many, small contigs or by few large ones. These statistics can refer to contigs or scaffolds, being the simplest metric the total number of contigs and scaffolds in that assembly. Rather than the mean contig length, a metric known as the N50 is often used to describe the contiguity of an assembly. It is defined as the length N for which at least 50 % of all bases in the sequences are contained in sequences of length N or longer. An easy way to compute it is to order your sequence lengths from longest to shortest and compute the cumulative sum of their lengths; when the sequence is reached, which brings the sum to greater than or equal to half of the total length of the assembly, the N50 equals the length of that sequence. The metric can also be computed for other proportions of the assembly, for example, N10 or N90 (where 90 % of the assembled bases are in scaffolds/contigs of length N90 or longer). When comparing multiple assemblies or assembly methods on a genome with an accurate size estimate, the assembly length can be substituted by the estimated genome length to give NG50 (NG10, NG80, NG90, etc.) values.

As many would point out, contiguity is good to have but not at the expense of correctness. It would be easy to make an assembly of one single contig by joining all sequences end to end, yet it would be highly inaccurate. Aggressive scaffolding requiring low support can inflate N50 values and the expense of more misassemblies. Thus other metrics should be considered. Gene content (both the completeness of the gene set and the connectivity of exons) is a very important point to consider. The CEGMA (Parra et al. 2007) or BUSCO (Simão et al. 2015) pipelines which search for a conserved set of core eukaryotic genes in draft genomes can report on both completeness of the genome and its connectivity. Several other analysis suites aim to provide a more complete picture of quality. FRCurve (Vezzi et al. 2012) can be run on assemblies to which at least one paired-end library and, optionally, one mate-pair library have been mapped and provided in BAM format. QUAST (Gurevich et al. 2013) is another useful tool for plotting a number of contiguity and gene content metrics.

2.8.6.3 Genome Resequencing

Efficient reporting of a resequencing study includes making available the raw reads, the variant calling files (VCFs, (Danecek et al. 2011)), as well as a statistical analysis that will depend on the focus of study (detection of disease variants, population structure, etc.). Quality metrics for call sets are lacking. Pipelines can be benchmarked (e.g., using the Genome in a Bottle materials (https://www-s.nist.gov/srmors/view_detail.cfm?srm=8398)), but individual call sets, unless independently validated with an orthogonal technology, cannot. As such, it is important to report base frequencies, base qualities, mapping qualities, allele frequencies, strand bias, positional bias, etc. so that the data may be reanalyzed at a future date by more up-to-date pipelines, perhaps tuned to return few false positives or few false negatives, depending on the goal of the resequencing experiment. It must be noted that variant/mutation calling procedures may vary depending on the frequency of the alternate allele.

2.8.6.4 Repositories to Upload Research Results Data for Publication

The European Nucleotide Archive (ENA (Leinonen et al. 2011)) is Europe's primary nucleotide-sequence repository. It comprises the Sequence Read Archive (SRA) where raw reads from different sequencing experiments can be submitted. The European Genome-phenome Archive (EGA) is the appropriate repository for human resequencing data. Raw data (FASTQs), alignments (BAMs), and genotypes and structural variants (VCFs) can all be submitted. Access is governed by a data access committee.

2.9 Main Remarks and Conclusions

To summarize, successful whole-genome sequencing requires the ability to think ahead and develop a strategy that accomplishes the goals of the project. Specifically, we recommend the following:

Before the Project Starts

- Survey existing genomic literature in search of required information (genome complexity, heterozygosity, size)
- Study previous projects on similar organisms.
- In the absence of related studies, consider a sequencing test to obtain preliminary data on genomic characteristics.
- Plan the sequencing strategy according to the assembly/analysis strategy that you will use afterward.
- Make a concerted effort to obtain high-quality DNA material, from samples of minimal polymorphism if possible (for genome assembly).
- Engage collaborators that will participate in the analysis from the beginning.
- Consider data storage and processing costs in addition to library preparation and sequencing costs.
- Balance cost with desirable depth of sequencing, most useful library fragment sizes, and longest reads possible (for genome assembly). Underfunding a project will achieve suboptimal results. In some cases, additional sequencing can save a project; however, depending on the strategy, it may have been a waste.

During the Project

- Revise and optimize as you go. If a strategy is not working, try to diagnose the problem and fix it as early as possible.
- Coordinate the work of the different teams involved, avoid redundant analysis, and establish clear dependencies and workflows.
- Freeze assembly and annotation at the time downstream analyses and start to avoid multiple recomputations due to constant minor updates.

After the Project

- Use efficient reporting and standard formats.
- Submit assemblies, annotations, raw data, and main analyses to public repositories.

Annex: Quick Reference Guide

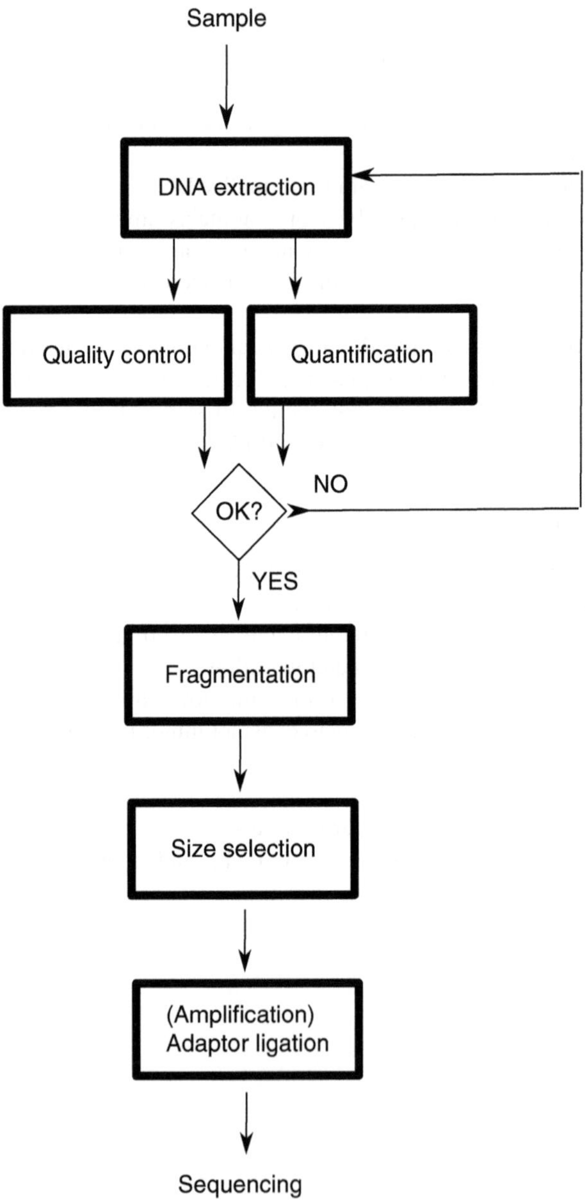

Fig. QG2.1 Representation of the wet lab procedure workflow

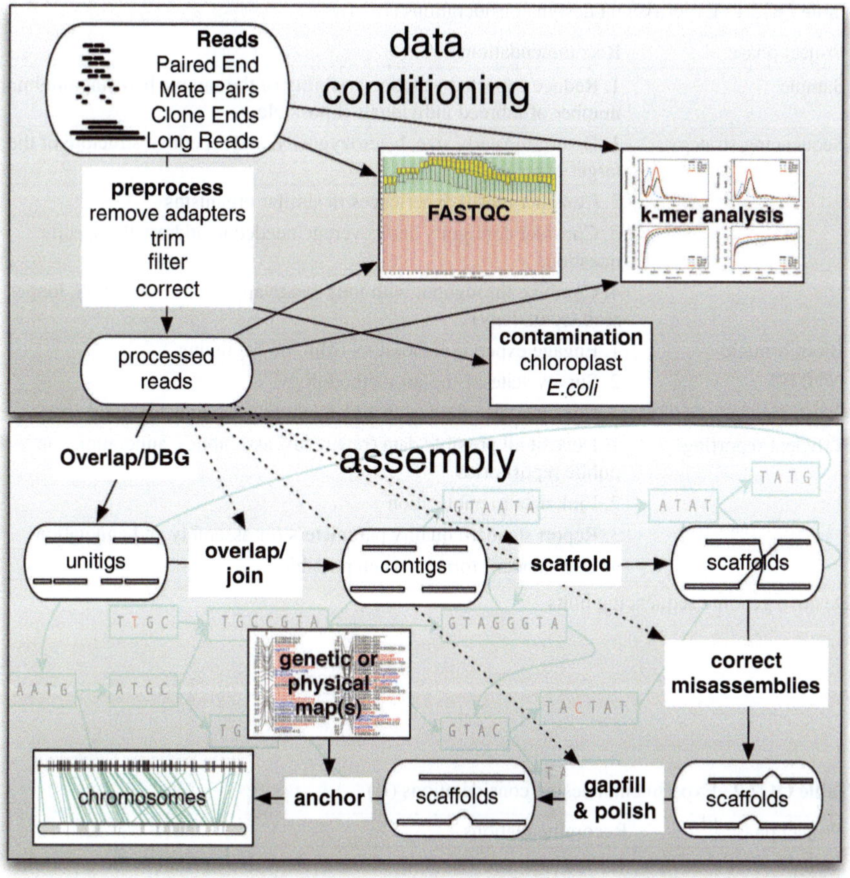

Fig. QG2.2 Main steps of the computational analysis pipeline

Table QG2.1 Experimental design considerations (I)

Project phase	Recommendations
Sample	1. Reduce expected genetic variability of the sample by using minimal number of inbreed individuals if possible
Sequencing strategy	1. Determine early size, heterozygosity, and repetitive structure of the target genome
	2. Consider recent experiences in similar organisms
	3. Consider contiguity and coverage needed to address the specific questions
	4. Combine throughput with long-range approach (FOSMIDS, longer read technology)
Bioinformatic analyses	1. Engage expert collaborators from the beginning
	2. Survey state-of-the-art methodology
	3. Consider specificities of the project (e.g., high heterozygosity)
Efficient reporting	1. Deposit all possible data (raw reads, assemblies, annotations) in public repositories
	2. Link data to publication
	3. Report standard quality parameters for assembly and annotation
	4. Use standard formats when possible

De novo genome sequencing hints

Table QG2.2 Experimental design considerations (II)

Project phase	Recommendations
Sample	1. Plan balanced sampling of a sufficient size to address the questions driven by the project
Sequencing strategy	1. Consider required sequencing depth depending on size of the target genome and required coverage for efficient variant calling
	2. Consider whether determination of structural variants is needed and use required strategy (e.g., pair-end, mate-pair libraries)
Bioinformatic analyses	1. Engage expert collaborators from the beginning
	2. Survey state-of-the-art methodology
	3. Consider specificities of the project (e.g., high heterozygosity)
Efficient reporting	1. Deposit all possible variation data in public repositories
	2. Link data to publication
	3. Use standard formats when possible

Whole-genome resequencing hints

Table QG2.3 Available software recommendations

Software	Function	Input	Reference	Result output	Result format
FASTQC	Quality control	FASTQ files	http://www.bioinformatics.babraham.ac.uk/projects/fastqc/	Quality control tables/text	text
SOAP denovo2	Genome assembly	FASTA, FASTQ files	Luo et al. (2012) SOAPdenovo2: an empirically improved memory-efficient short-read de novo assembler. GigaScience 1:18.	Genome assembly	Scaffold sequences
Velvet	Genome assembly	FASTA, FASTQ files	Zerbino, D. R. (2010) Using the Velvet de novo Assembler for Short-Read Sequencing Technologies. Current Protocols in Bioinformatics. 31:11.5:11.5.1–11.5.12.	Genome assembly	FASTA, Graph
MaSuRCA	Genome assembly	FASTQ files	Zimin et al. (2013) The MaSuRCA genome assembler *Bioinformatics* 29 (21): 2669–2677	Genome assembly	FASTA
ABySS	Genome assembly	FASTA, FASTQ, qseq, SAM files	Simpson et al. (2009) ABySS: a parallel assembler for short-read sequence data. Genome Res. 19(6):1117–23	Genome assembly	FASTA
SPADES	Genome assembly	FASTA, FASTQ, BAM	Bankevich et al. (2012). "SPAdes: a new genome assembly algorithm and its applications to single-cell sequencing" *Journal of Computational Biology* 19: 455–477	Genome assembly	FASTA, FASTG

Table displaying a selection of the recommended software available for the computational analysis of data yielded by this technique
This table has been generated by the editors for the quick reference guide corresponding to this chapter

References

Ajay SS, Parker SCJ, Abaan HO, Fajardo KVF, Margulies EH (2011) Accurate and comprehensive sequencing of personal genomes. Genome Res 21:1498–1505
Bentley DR, Balasubramanian S, Swerdlow HP, Smith GP, Milton J, Brown CG, Hall KP, Evers DJ, Barnes CL, Bignell HR et al (2008) Accurate whole human genome sequencing using reversible terminator chemistry. Nature 456:53–59
Berlin K, Koren S, Chin C-S, Drake JP, Landolin JM, Phillippy AM (2015) Assembling large genomes with single-molecule sequencing and locality-sensitive hashing. Nat Biotechnol 33:623–630
Boetzer M, Pirovano W (2012) Toward almost closed genomes with GapFiller. Genome Biol 13:R56
Bolger AM, Lohse M, Usadel B (2014) Trimmomatic: a flexible trimmer for Illumina sequence data. Bioinformatics 30:2114–2120
Chang J (2015) Core services: reward bioinformaticians. Nature 520:151–152
Compeau PEC, Pevzner PA, Tesler G (2011) How to apply de Bruijn graphs to genome assembly. Nat Biotechnol 29:987–991
Consortium T 1000 GP (2012) An integrated map of genetic variation from 1,092 human genomes. Nature 491:56–65
Danecek P, Auton A, Abecasis G, Albers CA, Banks E, DePristo MA, Handsaker RE, Lunter G, Marth GT, Sherry ST et al (2011) The variant call format and VCFtools. Bioinformatics 27:2156–2158
Fonseca NA, Rung J, Brazma A, Marioni JC (2012) Tools for mapping high-throughput sequencing data. Bioinformatics 28:3169–3177
Gnerre S, Maccallum I, Przybylski D, Ribeiro FJ, Burton JN, Walker BJ, Sharpe T, Hall G, Shea TP, Sykes S et al (2011) High-quality draft assemblies of mammalian genomes from massively parallel sequence data. Proc Natl Acad Sci U S A 108:1513–1518
Gurevich A, Saveliev V, Vyahhi N, Tesler G (2013) QUAST: quality assessment tool for genome assemblies. Bioinformatics 29:1072–1075
Howe K, Wood JM (2015) Using optical mapping data for the improvement of vertebrate genome assemblies. Gigascience 4:10
Hunt M, Kikuchi T, Sanders M, Newbold C, Berriman M, Otto TD (2013) REAPR: a universal tool for genome assembly evaluation. Genome Biol 14:R47
Kelley DR, Schatz MC, Salzberg SL (2010) Quake: quality-aware detection and correction of sequencing errors. Genome Biol 11:R116
Lander ES, Waterman MS (1988) Genomic mapping by fingerprinting random clones: a mathematical analysis. Genomics 2:231–239
Leinonen R, Akhtar R, Birney E, Bower L, Cerdeno-Tárraga A, Cheng Y, Cleland I, Faruque N, Goodgame N, Gibson R et al (2011) The european nucleotide archive. Nucleic Acids Res 39:D28–D31
Li H, Homer N (2010) A survey of sequence alignment algorithms for next-generation sequencing. Brief Bioinform 11:473–483
Li H, Handsaker B, Wysoker A, Fennell T, Ruan J, Homer N, Marth G, Abecasis G, Durbin R (2009) The sequence alignment/map format and SAMtools. Bioinformatics 25:2078–2079
Li R, Fan W, Tian G, Zhu H, He L, Cai J, Huang Q, Cai Q, Li B, Bai Y et al (2010) The sequence and de novo assembly of the giant panda genome. Nature 463:311–317
Lindblad-Toh K, Wade CM, Mikkelsen TS, Karlsson EK, Jaffe DB, Kamal M, Clamp M, Chang JL, Kulbokas EJ, Zody MC et al (2005) Genome sequence, comparative analysis and haplotype structure of the domestic dog. Nature 438:803–819
Loman NJ, Quick J, Simpson JT (2015) A complete bacterial genome assembled de novo using only nanopore sequencing data. Nat Methods 12:733
Luo R, Liu B, Xie Y, Li Z, Huang W, Yuan J, He G, Chen Y, Pan Q, Liu Y et al (2012) SOAPdenovo2: an empirically improved memory-efficient short-read de novo assembler. Gigascience 1:18

Marçais G, Kingsford C (2011) A fast, lock-free approach for efficient parallel counting of occurrences of k-mers. Bioinformatics 27:764–770

Marcet-Houben M, Ballester A-R, de la Fuente B, Harries E, Marcos JF, González-Candelas L, Gabaldón T (2012) Genome sequence of the necrotrophic fungus Penicillium digitatum, the main postharvest pathogen of citrus. BMC Genomics 13:646

McKenna A, Hanna M, Banks E, Sivachenko A, Cibulskis K, Kernytsky A, Garimella K, Altshuler D, Gabriel S, Daly M et al (2010) The genome analysis toolkit: a MapReduce framework for analyzing next-generation DNA sequencing data. Genome Res 20:1297–1303

Miller JR, Koren S, Sutton G (2010) Assembly algorithms for next-generation sequencing data. Genomics 95:315–327

Neale DB, Wegrzyn JL, Stevens KA, Zimin AV, Puiu D, Crepeau MW, Cardeno C, Koriabine M, Holtz-Morris AE, Liechty JD et al (2014) Decoding the massive genome of loblolly pine using haploid DNA and novel assembly strategies. Genome Biol 15:R59

Parra G, Bradnam K, Korf I (2007) CEGMA: a pipeline to accurately annotate core genes in eukaryotic genomes. Bioinformatics 23:1061–1067

Pryszcz LP, Németh T, Gácser A, Gabaldón T (2014) Genome comparison of Candida orthopsilosis clinical strains reveals the existence of hybrids between two distinct subspecies. Genome Biol Evol 6:1069–1078

Reuter JA, Spacek DV, Snyder MP (2015) High-throughput sequencing technologies. Mol Cell 58:586–597

Richards S, Murali SC (2015) Best practices in insect genome sequencing: what works and what doesn't. Curr Opin Insect Sci 7:1–7

Schmieder R, Edwards R (2011) Quality control and preprocessing of metagenomic datasets. Bioinformatics 27:863–864

Simão FA, Waterhouse RM, Ioannidis P, Kriventseva EV, Zdobnov EM (2015) BUSCO: assessing genome assembly and annotation completeness with single-copy orthologs. Bioinformatics 31:3210

Simpson JT (2014) Exploring genome characteristics and sequence quality without a reference. Bioinformatics 30:1228–1235

Simpson JT, Durbin R (2012) Efficient de novo assembly of large genomes using compressed data structures. Genome Res 22:549–556

Simpson JT, Pop M (2015) The theory and practice of genome sequence assembly. Annu Rev Genomics Hum Genet 16:153

Sims D, Sudbery I, Ilott NE, Heger A, Ponting CP (2014) Sequencing depth and coverage: key considerations in genomic analyses. Nat Rev Genet 15:121–132

Tang H, Lyons E, Town CD (2015) Optical mapping in plant comparative genomics. Gigascience 4:3

Van Dijk EL, Jaszczyszyn Y, Thermes C (2014) Library preparation methods for next-generation sequencing: tone down the bias. Exp Cell Res 322:12–20

Vezzi F, Narzisi G, Mishra B (2012) Feature-by-feature--evaluating de novo sequence assembly. PLoS One 7:e31002

Xi R, Kim T-M, Park PJ (2010) Detecting structural variations in the human genome using next generation sequencing. Brief Funct Genomics 9:405–415

Chapter 3
Targeted DNA Region Re-sequencing

Karolina Heyduk, Jessica D. Stephens, Brant C. Faircloth, and Travis C. Glenn

3.1 Different Types of Re-sequencing Methodologies

Multiple re-sequencing approaches have been developed and reviewed (McCormack et al. 2013a; Lemmon and Lemmon 2013). Below, we briefly summarize the major re-sequencing methods, indicating their advantages and disadvantages (Table 3.1) and the scale at which they are most appropriate (Fig. 3.1). For all methods, we assume that sequencing coverage will be reasonably deep to achieve high accuracy (Table 3.2), especially at heterozygous sites. All methods are usually paired with DNA sequence tags (also known as barcodes, indexes, or molecular identifiers, MID tags; see Faircloth and Glenn 2012) to identify individual samples from a pool of samples. We assume that lower costs will increase how widely the techniques will be adopted, and that total costs of ≤$100 US/sample, including personnel costs, are highly desirable.

Karolina Heyduk and Jessica D. Stephens are contributed equally with all other contributors.

K. Heyduk • J.D. Stephens
Department of Plant Biology, University of Georgia,
2502 Miller Plant Sciences, Athens, GA 30602, USA
e-mail: heyduk@uga.edu; jdstephe@uga.edu

B.C. Faircloth
Department of Biological Sciences and Museum of Natural Science,
Louisiana State University, 202 Life Sciences Bldg., Baton Rouge, LA 70803, USA
e-mail: brant@faircloth-lab.org

T.C. Glenn (✉)
Department of Environmental Health Science, University of Georgia,
150 East Green St, Athens, GA 30602, USA
e-mail: travisg@uga.edu

A.M. Aransay, J.L. Lavín Trueba (eds.), *Field Guidelines for Genetic Experimental Designs in High-Throughput Sequencing*,
DOI 10.1007/978-3-319-31350-4_3

Table 3.1 Advantages and disadvantages of DNA re-sequencing methods

Re-sequencing approaches	Advantages	Disadvantages
Whole genome re-sequencing	– Easy to implement in lab	– Must have a reference genome from same or closely related species
	– Most complete data	– Large genomes require a lot of sequencing
	– Many robust software options	– Large genomes require more computational effort
Transcriptome sequencing (RNA-seq)	– Already reduced complexity of genome	– Differences in expression across tissue types, developmental stage, etc.
	– Template for future marker design	– Data may violate population genomics assumptions
	– Good platform for comparison across species/individuals	– Expensive
		– Computationally intensive
PCR amplicon sequencing	– Works well with limited starting material	– Issues with sequence diversity on Illumina platforms
	– Cost efficient with few loci	– Assay development time and costs increase with number of loci
Restriction-site-associated DNA markers (RADseq)	– Discovery, development, and screening of markers are time and cost efficient	– Loci are dominant
	– Established bioinformatics pipelines	– Significant variation in reproducibility
	– Cost-efficient method	– Can result in large amounts of missing data
		– Issues with paralog determination and coverage due to untargeted loci
Target enrichment	– Less likely to have allelic loss	– Need prior genetic resources to design baits
	– Baits can target areas of interest	– Bait design can be challenging
	– Useful for large, complex genomes	– Higher up-front costs (e.g., library prep, bait design) relative to RADseq
	– Upstream methods help avoid paralogs, complexity, and repetitive regions	

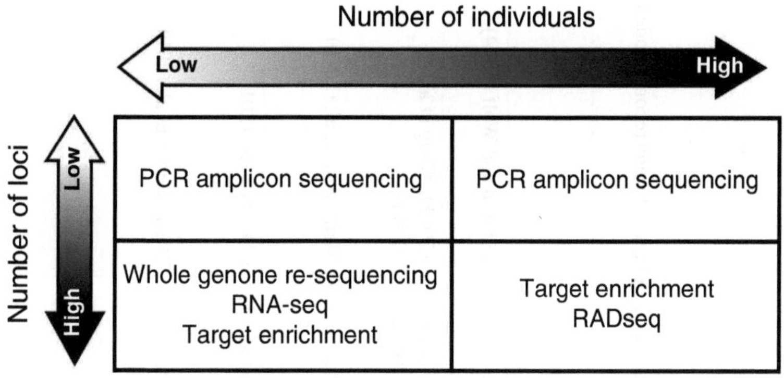

Fig. 3.1 Methods for re-sequencing based on number of individuals and loci for analyses

3.1.1 Whole Genome Re-sequencing

Whole genome re-sequencing (WGRS) is the easiest method to implement in the lab, offers the most complete data, and has excellent software support due to its widespread use in human genomics (for a review of software, see Bao et al. 2011). While WGRS studies are being published in nonhuman systems, these are mostly limited to agriculturally important crops (rice, Xu et al. 2010; soybean, Li et al. 2013) or model organisms (*Arabidopsis*, www.1001genomes.org; *Mus*, Keane et al. 2011; *Drosophila*, Zhu et al. 2012). The lack of WGRS studies are due to the inherent problems associated with WGRS; these include (1) a required reference genome from the same or a closely related species, (2) the amount of sequencing is directly proportional to genome size (i.e., big genomes require a lot of sequencing), and (3) computational efforts increase as a power function of genome size (i.e., large genomes require much more computational effort than small ones)—all of which increase costs. As of 2015, it is possible to re-sequence a human genome at 30× coverage for ~$1000 on Illumina HiSeq 4000s (www.illumina.com). Thus, it is possible to sequence *Drosophila*-sized genomes for a cost approaching $100/sample, but most other non-model and large-genome organisms remain uneconomical for WGRS efforts.

3.1.2 Transcriptome Sequencing

Transcriptome sequencing (RNA-seq) has the advantage of using the cellular transcriptional machinery to naturally reduce the complexity of genomes and enrich for functional elements. There are multiple advantages of focusing on genome reduction through transcriptomics. For example, transcript profiles for polymorphism comparisons are predicted to be similar if using the same tissue across

Table 3.2 Recommended amount of starting nucleic acids, major constraints of methods, minimum average recommended sequencing depth, and recommended read lengths (Illumina) for DNA re-sequencing methods

Technique	Recommended starting material (ng)	Constraints	Average Sequencing depth	Recommended sequencing run	Reference(s)
Whole genome re-sequencing	500	Cost of sequencing	6–30×	HiSeq or NextSeq PE75–PE150	Sims et al. (2014), Ekblom and Wolf (2014)
Transcriptome sequencing	1000	Sample material, cost of libraries, cost of sequencing	≥10 million reads per sample	HiSeq or NextSeq PE100–PE150	Ozsolak and Milos (2011), Wang et al. (2009), Wang et al. (2011)
PCR amplicon sequencing	20	Combinatorial tags, pool with diverse libraries	20×	MiSeq PE250–PE300	Feng et al. (2016), Mamanova et al. (2010)
Restriction-site-associated DNA markers (RADseq)	100	Consistency	10×	HiSeq or NextSeq SE75–PE150	Davey et al. (2011), Davey et al. (2013)
Target enrichment by sequence capture	500	Probes (information to design and cost)	30×	HiSeq or NextSeq PE100–PE150	Mamanova et al. (2010), Mertes et al. (2011)

individuals or species. There are large-scale initiatives attempting just that through a consortium of universities (plants, 1KP project, http://www.onekp.com; insects, 1KITE, http://www.1kite.org; eukaryote microbes, Marine Microbial Eukaryote Transcriptome Sequencing Project, http://www.marinemicroeukaroytes.org). Another benefit of transcriptome sequencing is that the assembled template can be used to develop markers for future studies (Ekblom and Galindo 2011).

RNA-seq has several disadvantages. First, differences in gene expression will vary depending on which tissues are collected, developmental stage of tissue, time of day, and nutritional status of individuals; this can limit comparison of orthologous loci across samples. Variation between libraries can be mitigated, however, by pooling several life stages, tissues, etc. during cDNA library preparation (Hahn et al. 2009). Second, RNA-seq requires significant sequencing depth to account for loci that are weakly expressed. Third, models relating to demographic history and population structure generally assume neutral evolutionary processes, which may be violated by transcribed genes and thus may cause problems with downstream analyses for these types of studies. Finally, RNA-seq currently costs one to a few hundred dollars per sample; thus, sampling a large number of individuals and species can be costly for reagents and sequencing and can increase computational time requirements for transcriptome assembly and subsequent analysis (Wang et al. 2009; Ozsolak and Milos 2011).

3.1.3 PCR Amplicon Sequencing

PCR can be used to produce amplicons that are sequenced using MPS. This has most frequently been done for 16S metagenomics (Wang and Qian 2009; Haas et al. 2011) and specific disease panels (Easton et al. 2015), but many other applications of this technique have been developed (Faircloth and Glenn 2012). Amplicon sequencing has the advantage of working from very limited amounts of starting material, building on well-known techniques, and can be done for well under $100 US per sample if the number of target loci is limited. The major disadvantages of amplicon sequencing are that (1) costs increase significantly as the number of target loci increases, (2) amplicons generally need to be combined with other samples to increase sequence diversity on Illumina platforms and to take advantage of capacity, and (3) assay development time and costs increase significantly as the number of target loci increases; thus, amplicon sequencing is generally limited to surveying only a very small portion of the genome.

3.1.4 Restriction-Site-Associated DNA Makers (RADseq)

RADseq uses restriction enzymes to reduce genome complexity and isolate a smaller, repeatable fraction of the genome and is combined with MPS to genotype thousands of genetic markers without having prior genetic information for the organism(s) under study. Multiple flavors of RADseq have been developed, making use of one, two, three, or more restriction enzymes (Davey et al. 2011; Puritz et al. 2014). The method used is often selected based on the genome size of the organism and the predicted amount of coverage resulting from the enzyme combination selected. RADseq was developed for and has been extensively utilized for questions pertaining to genetic mapping and population genomics (Davey et al. 2011; Puritz et al. 2014). RADseq data have also been used for phylogenetic assessments (Rubin et al. 2012; Cariou et al. 2013; Wagner et al. 2013), but these are often in small, species-level phylogenies. A major advantage of RADseq is that discovery, development, and screening of markers generally happens in only one or two rounds of MPS, making RADseq time efficient and cost-effective (Davey and Blaxter 2010). In addition, there are well-developed downstream bioinformatics pipelines to handle these data (e.g., Stacks—Catchen et al. 2013; PyRAD—Eaton 2014). Although RADseq is inefficient in its use of MPS data (i.e., most data are discarded), because MPS data are cheap, most RADseq projects still achieve costs well below $100 US/sample. Thus, RADseq represents a generally reasonable approach for acquiring genotype information dispersed across large genomes.

Unfortunately, RADseq also suffers from several disadvantages. First, RADseq loci are untargeted (i.e., any fragment of DNA with the restriction site(s) will be obtained). Thus, the loci may be less evenly spread across a genome than desired and may miss important portions simply due to chance or bias (Davey et al. 2013). Second, RADseq loci are dominant—substitutions that cause the loss of restriction sites create null alleles (Gautier et al. 2012; McCormack et al. 2013a). Thus, RADseq is not recommended for deeper-level phylogenetics because variation in restriction sites that occurs across divergent taxa yields large amounts of missing data across a given taxonomic sample (McCormack et al. 2012). Third, most RADseq users experience significant variance in reproducibility among taxa or projects, which can cause many samples to fail quality control, increasing the number of samples that must be repeated. Fourth, the variance inherent in RADseq (Davey et al. 2013) frequently results in sparse data matrixes. Finally, RADseq also presents challenges post-sequencing when trying to determine whether fragments are paralogs and have appropriate coverage, because they were not targeted (McCormack et al. 2013a).

3.1.5 Target Enrichment

Target enrichment approaches (also known as sequence capture and gene capture) use baits (also known as probes) to specifically pull out fragments of interest from a genomic library, keeping the fragments of interest while fragments that do not hybridize to the baits are washed away (Mamanova et al. 2010). In contrast to RADseq, target enrichment has higher up-front costs, both for library preparation and the cost of baits and capture, but is more efficient than RADseq because specific targeted areas make up large portions of the data (Grover et al. 2011). Target enrichment is less likely than RADseq to suffer from allelic loss (null alleles) because alleles with one to several substitutions are recovered at a higher rate across individuals and species. In addition, target enrichment baits can be designed to target a variety of genomic locations including intergenic regions assumed to evolve under neutral processes, making this method ideal for population-level questions. Target enrichment is also useful for organisms with large, complex genomes (such as plants or amphibians) because targeting specific regions can avoid repetitive elements. These strengths of target enrichment result from *a priori* upstream methods to eliminate potentially paralogous sequences, regions of low complexity, and repetitive regions while focusing on those targeted regions of interest and returning data having high coverage across these regions. Moreover, baits can be designed to target regions of varying size depending on different treatments of the data during library preparation and the MPS platform used (McCormack et al. 2013a).

Disadvantages of target enrichment include: (1) prior genetic resources are needed to design baits (e.g., genomes, genomic regions, or transcriptomes of related species); (2) bait design can sometimes be challenging when targeting genomic regions that are highly variable within and among species (e.g., introns, immune-coding loci); and (3) most target enrichment studies to date have focused on using genomic libraries of randomly sheared DNA, which are more expensive to create than RADseq libraries and result in less coverage of targeted bases per sequence. Below, we discuss study design and bioinformatic methods to ameliorate many of these disadvantages, with a focus on target enrichment for population genetic and phylogenetic studies.

3.2 Experimental Design Considerations

As with any study, understanding the biology of the organism(s) of interest is critically important to study design and downstream analyses. For instance, knowing whether the organism under study has undergone recent gene/genome duplications, whether the organism is polyploid, and/or whether the lineages being studied frequently hybridize can have a dramatic influence on data collection and subsequent inference. Paralogs, hybridization, and horizontal gene transfer can influence gene tree discordance for phylogenetic analyses. In addition, many programs have a long list of assumptions or may not properly model aspects of the study system if the proper number of samples has not been sequenced. As an example, *BEAST is an excellent program for coestimating gene trees and their underlying species tree using a Bayesian MCMC procedure; however, the authors of *BEAST recommend the use of at least two individuals per species to properly estimate population parameters (Heled and Drummond 2010). Knowing this prior to sequencing can help better inform experimental design and simplify downstream analyses.

When considering the correct number of individuals per species to sample, in a phylogenetic context, it is mostly based on preference, study system, sample availability, and downstream analyses. If the study system has frequent hybrids or taxonomic designations below the species level, then one may consider including multiple exemplar individuals for a given species to examine reciprocal monophyly within species. In this case, a phylogenetic program that assigns individuals to species and then infers the phylogeny of the species may be more appropriate than having a phylogeny where every individual represents a lineage. Moreover, some phylogenetic programs require that every gene has a representative sequence from an out-group (Table 3.3). Therefore, it may be advantageous to include multiple exemplar individuals of the out-group species to increase the likelihood of capturing a high number of targets in the out-group. This is especially important to consider if the out-group was not used in the bait design and is distantly related to the in-group species, which would result in more sequence variability in regions targeted by the hybrid enrichment baits between out-group and in-group members. Whenever possible, it is recommended that multiple individuals per species are sequenced, as it not only helps analyses but safeguards against species or population dropout due to unexpected low sequence coverage or low enrichment efficiency of any particular sample. While multiple exemplars per species or populations are beneficial to both phylogenetic and population genomic inferences, if the taxonomic sample is large, then it may not be cost-effective or computationally efficient to include multiple individuals per species.

In contrast to phylogenomic studies, the number of individuals used for population genomic studies is more contingent on capturing rare alleles within a population. Having prior knowledge of the system (i.e., population size, generation time, etc.) can better inform this decision. Ideally sampling a larger number of individuals per population is better, but sample size is dependent upon sample availability, number of populations, number of sequence tags needed for pooling samples, and overall sequencing costs, including the benchwork costs and amount of sequencing required to obtain adequate coverage. Obtaining samples for population-level work can also be more difficult. However, for both phylogenetic and population-level sequencing, DNA from preserved samples (i.e., herbaria, zoological collections, etc.) have been successfully sequenced using target enrichment methods (e.g., Carpenter et al. 2013; Enk et al. 2014; Comer et al. 2015; McCormack et al. 2015). The ability to use fragmented DNA for target enrichment greatly facilitates the sequencing of larger sets of individuals.

When deciding on the number of loci to target, it is best to plan on some modest proportion of the loci being dropped from analysis due to low coverage or poor enrichment across taxa. Thus, designing baits for a large amount of target loci will help to keep the final number of loci analyzed at the desired level, even after filtering poorly covered targets. The number of targeted loci that may actually be used for analysis varies among studies, ranging from 35 % to close to 100 % (Heyduk et al. 2016; McCormack et al. 2013b; Stephens et al. 2015a). These numbers can vary depending on biology and evolutionary history of the focal organisms, the phylogenetic scope or population divergence among the samples, and the number of samples that will be included (e.g., if a locus needs to be present in at least 50% of individuals to be analyzed, then increasing the number of samples makes this threshold harder to reach).

Table 3.3 Some examples of phylogenetic programs that handle gene tree discordance due to incomplete lineage sorting

Program[1]	Phylogenetic method	Multiple accessions	Missing data[2]	Requirements	Command line vs. GUI	Computational time[3]
BEST (Liu 2008)	Sequence alignment	Can assign accessions to species	Accepts missing data	None	Command line	Very slow (months)
*BEAST (Heled and Drummond 2010)	Sequence alignment	Multiple accessions recommended	Does not allow missing data for a given species	Priors	GUI	Very slow (months)
SVDquartets (Chifman and Kubatko 2014)	Sequence alignment	Cannot assign accessions to species	Accepts missing data	None	Command line	Very fast (hours to days)
STEM (Kubatko et al. 2009)	Summary method	Cannot assign accessions to species	Accepts missing data	Must have an estimate of individual gene evolution relative to each other	Command line	Fast (days)
MP-EST (Liu et al. 2010)	Summary method	Can assign accessions to species	Accepts missing data	All gene trees must be rooted	Both	Fast (days)
ASTRAL (Mirarab et al. 2014)	Summary method	Cannot assign accessions to species	Accepts missing data	Gene trees need to be fully resolved (can be unrooted)	Command line	Very fast (hours)

[1]There are many widely used programs not mentioned here that should be considered as well (e.g., NJst, STAR). Program parameters (e.g., multiple accessions, computational time, requirements) should be considered prior to re-sequencing depending on purpose of the study

[2]While certain programs can handle missing data, it should be noted that how missing data influences species tree estimation is not well known across programs. Thus, caution is warranted when including missing data in any phylogenetic analysis, although the use of complete matrixes can also introduce bias (Huang and Knowles 2014)

[3]Computation time is based on a phylogenetic analysis of ~70 individuals with ~200 loci on a Linux cluster with up to 75 CPUs using 8 or 12 core nodes

Determining the number of targeted loci may also be dependent on the system of interest and the study question. Questions pertaining to population genomics would benefit from sampling as many loci as the cost of sequencing allows to ensure detection of outlier loci which can improve parameter estimates such as effective population size and relatedness (Luikart et al. 2003). For studies that are examining population differentiation in phenotypic space, a larger number of loci are important to be able to accurately pinpoint genomic regions responsible for any local adaptation. On the other hand, genomic studies assessing population structure at a fine scale would benefit from highly informative loci. When selecting the number of loci to target for phylogenomic studies, the decision is equally situational. For example, if the study system has been historically difficult to resolve due to rapid or recent radiation and/or high levels of gene tree discordance, then including more genes or more informative genes in the analyses should improve resolution of species relationships. Although one would always prefer highly informative loci, it is difficult to predict which loci will be informative *a priori*. Lastly, computational time should be taken into account when adding more loci to any study, as many statistically robust methods (e.g., *BEAST, see "Post-sequencing") are unable to handle large datasets, and analysis time increases with each locus.

The types of genomic regions (e.g., exons, introns, etc.) collected using target enrichment can vary within or across studies. General approaches range from collecting single loci with single baits to using multiple baits to collect loci spread throughout the genome to collecting data from a single long region of interest with overlapping (tiled) baits (see bait design below) Exons are common targets, including collection of all the exons (i.e., the exome) of model organisms, but any region of the genome may be targeted by baits.

The use of ultraconserved elements (UCEs) for target enrichment is becoming popular given their applicability across extremely divergent taxa (Bejerano et al. 2004; Faircloth et al. 2012; McCormack et al. 2012). UCEs are highly conserved genomic regions that are ≥60 bp and found among widely divergent taxa (Bejerano et al. 2004; Dermitzakis et al. 2005). UCEs are appealing as targets because they are abundant, extremely conserved, straightforward to identify, and found within many groups of organisms (Stephen et al. 2008). In addition, UCEs tend to be orthologous (Derti et al. 2006) with few retroelement insertions. Finally, their utility for phylogenomic approaches is that while UCEs themselves show reduced variation, making them easy to capture, the flanking regions show much higher counts of informative sites (Faircloth et al. 2012). Several research groups have targeted conserved elements for target enrichment approaches, and much work remains to test and optimize the methods of identifying and using such loci. Here, we have focused on those methods that are open-access, because they are amenable to continued optimization and improvement by the research community.

3.2.1 Cost Reductions

The method used for re-sequencing can vary based on the number of individuals and number of loci required to address the questions of interest (Fig. 3.1). For questions that require sampling a limited number of individuals (<50) at very few loci (1–3), traditional PCR and Sanger sequencing may be the most cost- and time-effective methods. On the other end of the spectrum, a one-time study requiring many loci for few individuals might be best served by transcriptome sequencing. For studies requiring the collection of large numbers of loci from large numbers of individuals, then RADseq and/or target enrichment could be warranted. RADseq produces libraries at the lowest cost per sample, but more funds are spent on sequences that ultimately will not be used. Target enrichment significantly reduces both cost and time spent on sequencing, but methods to reduce costs prior to sequencing are important. Below we focus on ways to reduce costs for target enrichment.

Although a variety of home-brew methods are possible, commercial synthesis of target enrichment baits is the most convenient and cost-effective method for most researchers to conduct target enrichment (Fig. 3.2). Most companies that provide baits offer both premade kits and custom bait designs. A wide spectrum of baits can be accommodated, ranging from single biotinylated oligos from traditional oligonucleotide manufacturers (e.g., IDT, Life Technologies, Sigma, etc.) to companies that use high-density microarray technologies (e.g., Agilent, MYcroarray, NimbleGen, etc.) to construct massive numbers of unique baits. If <100 baits are needed, traditional biotinylated oligonucleotides are generally most economical. For example, if a study requires few loci for a large number of individuals, one might consider homemade baits complementary to the sequences of interest (e.g., for studies focusing on one pathway or known genes of interest). This methodology typically requires the bait sequence of interest to be PCR amplified, then subsequently size selected and biotinylated (see Peñalba et al. 2014 for methodological descriptions). If >1000 baits are needed, then high-density approaches for bait construction are most economical. Whole-exome capture kits for humans and model species can include hundreds of thousands of baits.

Although custom, commercial, high-bait number kits have list costs of hundreds of dollars per sample, many methods are available for reducing the costs of target enrichment when using such kits. First, it has long been appreciated that pooling sample libraries prior to conducting enrichment hybridization is an efficient way to reduce costs (Fig. 3.2; Cummings et al. 2010; Shearer et al. 2012). In this strategy, individual samples are tagged during library construction and pooled prior to target enrichment. This allows the costs of target enrichment to be divided among multiple samples. Pooling generally ranges from 2 to 96 samples per pool, with trade-offs between better coverage (i.e., less variance in capture efficiency and read depth with fewer samples per pool) and better cost savings (more samples per pool). In practice, we generally pool 4 to 12 samples prior to enrichment (Faircloth et al. 2012; Heyduk et al. 2016; Stephens et al. 2015a; http://ultraconserved.org). When pooling, samples should have similar: molarity (i.e., accounting for insert size and concentration), copy number (i.e., accounting for genome size and ploidy), and sequence

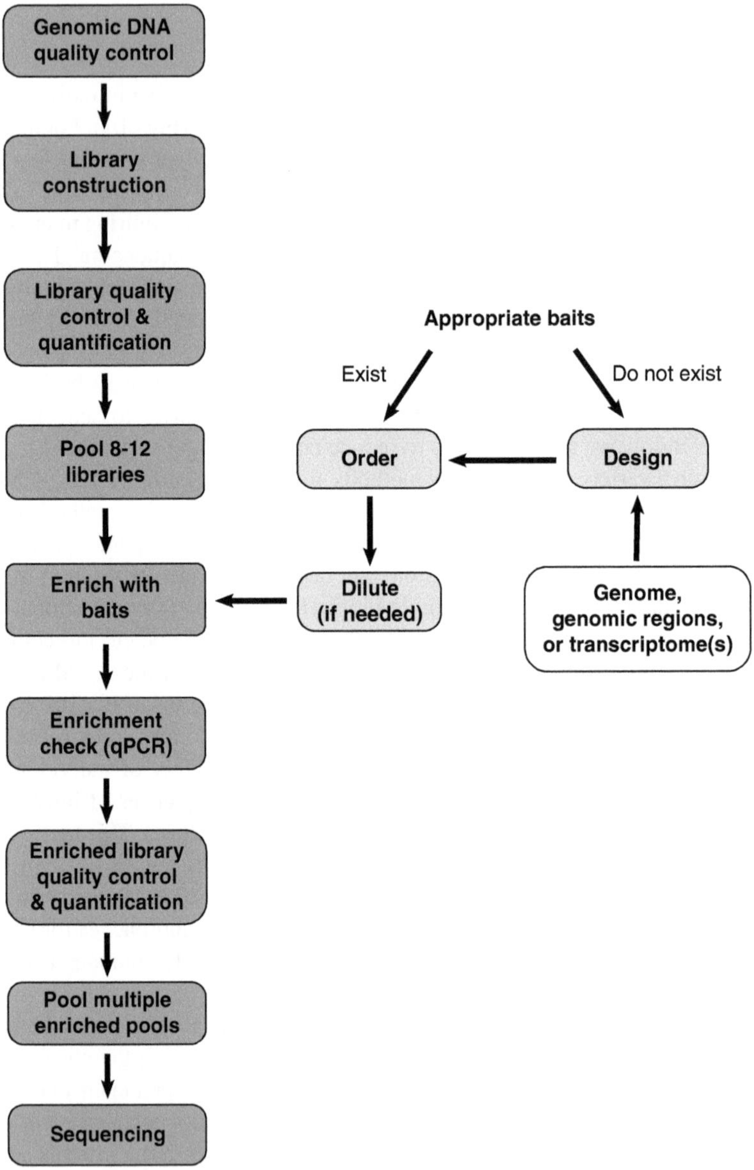

Fig. 3.2 Overview of the wet-lab workflow for target enrichment

divergence from the baits (or phylogenetic distance from the taxon used for bait design). Any of these three factors can lead to preferential capture of loci in higher number from some of the taxa in the pool (i.e., those with more targets or those with targets more similar to the baits than other individuals in the pool).

In addition to pooling prior to hybridization reactions, the quantity of baits per reaction may also be decreased if the targeted number of base pairs is significantly smaller than the protocol assumes (Faircloth et al. 2012; Heyduk et al. 2016; http://ultraconserved.org). Indeed, flooding the reaction with an overwhelming excess of baits relative to genomic targets can reduce capture efficiency rather than increase it. As a simple example, consider a project in which a researcher wishes to survey 1000 loci from 960 individuals. That research might design 2 baits per locus × 1000 loci = 2000 baits. A single custom bait kit that normally allows 12 captures, each with a 20,000 bait pool, is all that is necessary to conduct this experiment because the researcher can dilute the baits tenfold (20,000/2000 = 10; yielding enough baits for 120 captures instead of 12) and pool 8 samples per capture (120 × 8 = 960). Additional hybridization reagents will be necessary, but these can be purchased commercially or made from common reagents (Blumenstiel et al. 2010; http://ultraconserved.org).

Library preparation costs are another significant expense for target enrichment. Library costs can be reduced by decreasing reaction sizes and/or using home-brew protocols (e.g., Meyer and Kircher 2010; Fisher et al. 2011; Glenn et al. 2016; http://ultraconserved.org) rather than commercial kits. Strategically choosing a sequence tagging scheme can reduce costs as well. Illumina sequencing was once limited to a single 6 nt index. Newer methods allow two indices per fragment, employing a combinatorial approach that increases the versatility of indexing. With the dual-indexing method, n unique barcodes for each side of the fragment can be used on n^2 libraries to reduce the number, complexity, and cost of barcode oligos.

Finally, in addition to the on-target sequences captured, target enrichment methods also yield off-target bonus sequences (i.e., DNA sequence lagniappe). Off-target sequences are unavoidable because no target enrichment process is perfectly efficient. Thus, sequences that have partial similarity to the baits or were simply present in the pre-enrichment library, especially in high-copy numbers, will be present post-enrichment. As a result, high-copy DNA from chloroplasts, mitochondria, and ribosomes are commonly sequenced as off-target reads. These sequences are often informative however, and studies in both plants and animals have used these bonus sequences to assemble complete or mostly complete chloroplast and mitochondrial genomes (Weitmeier et al. 2014; Stephens et al. 2015a,b; Meiklejohn et al. 2014; Raposo do Amaral et al. 2015).

3.2.2 Workflow Bottlenecks

Sequence capture is highly effective at generating a large number of sequences for many individuals rapidly and consistently. While sequencing methods continue to improve, a number of bottlenecks exist in current workflows for sequence capture. The speed at which hundreds of libraries can be generated is limited by human labor, although protocols exist for robotic library preparation (e.g., Fisher et al. 2011; Rohland and Reich 2012). Quantification of hundreds of libraries pre-hybridization is expensive in both time and cost, depending on the method used. Most hybridization methods currently require ≥12 h for libraries to hybridize to baits. Shorter hybridization times are possible but generally require shorter baits, which require trade-offs in specificity and ability to capture library fragments with small sequence differences. Post-sequencing bioinformatic analysis is often not limited by human labor but by computational power; the same hundreds of libraries that take human hours to create may take many days and gigabytes of memory to analyze. For both pre- and post-sequencing, the number of individuals is the most influential limitation to sequence capture projects. As library protocols become more efficient and analysis programs are written to accommodate large numbers of individuals sequenced at many loci, sequence capture bottlenecks will decrease, and multi-species phylogenies and robust population genomics studies will become the norm.

3.3 Bioinformatics

3.3.1 Pre-sequencing

Initial bioinformatics work will depend on whether capture baits are being designed in-house or are available from a prior study (e.g., ultraconserved elements (UCEs), Faircloth et al. 2012). Bait design *de novo* requires genomic resources and can be conducted using genome sequences, transcriptomes, or even EST databases (Fig. 3.2). Comparative analyses of genomic data from divergent taxa can be used to design baits that will work across study systems including divergent taxa; for example, using regions that are conserved across a family will result in baits more likely to anneal to targeted regions and thus give more representative sequences per species. If the study requires examination of intra- and interspecific variation, then baits must be designed so they capture fragments with informative intraspecific sequence differences while still being able to capture targets across species (Stephens et al. 2015b), or sufficient amounts of sequence polymorphisms must accumulate in the regions immediately flanking the conserved sequences used for baits (Faircloth et al. 2012; Smith et al. 2014). This technique could also be applied to bait design for population-level questions. In particular, having genomic resources for multiple populations across the range of interest will help ensure baits are designed that maximize differences between and among populations.

Avoiding duplicated sequences is paramount to both phylogenomic and population genomic analyses, and care should be taken to exclude regions of the genome present in more than one copy (Faircloth et al. 2012, 2015). Prior to bait design, all repeat-like regions across the source data should be masked, and bait design protocols should avoid these regions. It is also recommended that potential areas for targeting should be aligned within and among species to ensure that targets are orthologous and only present in a single copy, especially in systems where polyploidy is abundant (e.g., low-copy genes across angiosperms described in Duarte et al. 2010, as done in Heyduk et al. 2016). Once targets have been determined, baits can be designed in-house (cf. http://ultraconserved.org), or target sequence information can be sent to commercial companies for bait design and synthesis. Bait sets may be designed having one bait per target or including multiple baits that are overlapped (tiled) across longer regions. Whether or how much to overlap baits depends primarily on the size of the targets, the number of baits, and research budget. Additionally, the sequence similarity of the taxa of interest will influence not only the optimal amount of overlap but also if multiple baits per locus (i.e., baits designed from multiple taxa) are necessary or desirable. Light (2×) tiling (i.e., each target nucleotide has two baits) can increase capture success even when targets are small and the target species are similar, thus decreasing sequencing costs but increasing bait costs relative to no tiling.

3.3.2 *Post-sequencing for Phylogenomics Designs*

Bioinformatics analysis post-sequencing can be quite daunting, but more pipelines and programs are being designed to handle these data. For example, those targeting UCEs can use phyluce (Faircloth 2016; https://github.com/faircloth-lab/phyluce) to go from raw reads to final alignments for phylogenetic analyses, with an added bonus of flexibility regarding how baits were designed. Throughout this process, phyluce will output relevant summary information that can be reported in a table as a supplement to the manuscript (see reporting section below). An alternative method from Heyduk et al. 2016 (https://github.com/kheyduk/reads2trees) is less streamlined than phyluce but allows for more customizable parameters throughout the bioinformatic pipeline. Together these programs and pipelines are achieving the same goal with very similar methodological steps (Fig. 3.3). First, all raw reads must be cleaned by removing Illumina adapters and trimming reads with poor quality scores. These clean reads are then used for assembly, which can either be reference based or *de novo*. Users can assemble reads through both routes and then merge similar sequences or opt to use one type of assembly program. The resulting assembled contigs can then be matched via local alignment searches (e.g., BLAST or LASTZ) against the initial targets and retained for further analyses. Contigs that match the target areas should be sorted into loci (e.g., by merging exons from the same gene), aligned, and trimmed prior to downstream analyses. A second round of duplicate removal may be necessary, depending on the target loci, because paralogous sequences may be captured or make it through as nontarget data that were not in the initial reference used for bait design.

We have seen a dramatic increase in the amount of data that can be collected using recent genomic techniques, and this trend is likely to increase as sequencing costs continue to decrease. The bottleneck with handling high-throughput data generally arises from the computational time required for their analysis and from our current understanding of phylogenomics and population dynamics. Historically, phylogeneticists would concatenate genes to estimate the species tree, but both

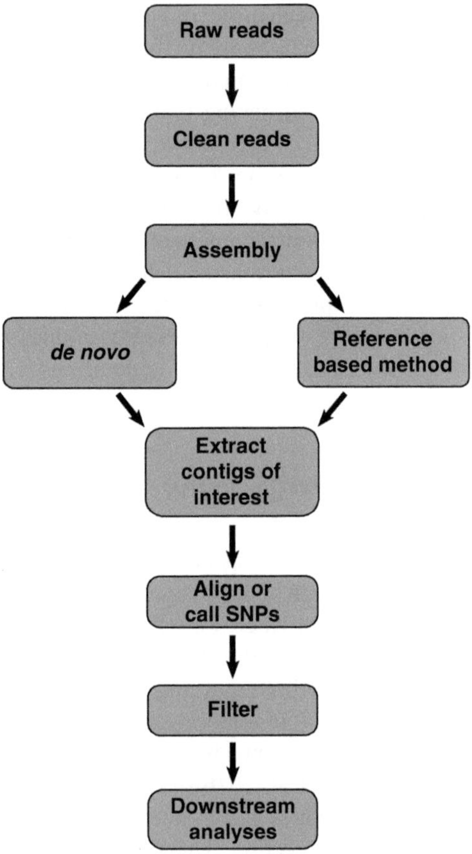

Fig. 3.3 Overview of a bioinformatic pipeline for re-sequencing data. Programs for each step should be determined based on assumptions regarding data and downstream analyses. Assembly can be conducted using multiple programs, or a single optimal assembly method can be implemented

empirical and theoretical data suggest that this is not always a robust method. Specifically, it has been known for some time that gene trees can have different histories from each other and from the species tree. Gene tree discordance can impact phylogenetic analyses, and modeling the processes that lead to discordance (i.e., incomplete lineage sorting [ILS], recombination, hybridization, etc.) has been challenging. To date the majority of phylogenetic programs can only estimate species trees when accounting for ILS. Programs are emerging to model the process of hybridization (STEM-hy—Kubatko 2009; PhyloNet—Yu et al. 2011; Yu and Nakhleh 2015), and, in general, the analysis of multilocus data is rapidly developing, making it hard for newcomers to find appropriate programs for analyzing their data. Care should also be taken to consider the biology of your taxa of interest. Therefore, we recommend that researchers consider the programs and the underlying models they are most likely going to be implementing given their system. For example, understanding the phylogenetic relationships of a recent or rapid radiation will most likely involve high levels of ILS and possibly hybridization. In this example, it may be worthwhile to sequence multiple individuals per species to increase the accuracy of parameter estimation for the coalescent models (Heled and Drummond 2010), but not all programs are capable of taking into account multiple individuals per species (Table 3.3). In addition, some programs may take an exceedingly long time (or fail) to run depending on the number of loci and number of taxa input (Table 3.3). Computational biologists are developing new ways to reduce the size and complexity of datasets for phylogenetic analyses (e.g., Bayzid and Warnow 2013), though these methods should be carefully evaluated on individual projects to assess their suitability.

3.3.3 Post-sequencing for Population Genomic Designs

Many of the difficulties described above for phylogenetic analyses hold true for population genomic analyses, as well. Pipelines for analyzing target enrichment data collected at the population level are generally lacking (but see Faircloth 2016; https://github.com/mgharvey/seqcap_pop). With a bit of legwork, one can identify genomic features of interest, including SNP and indel calls and use these data to estimate heterozygosity, FST, Tajima's D, and others, using the bcftools (https://github.com/samtools/bcftools) command line program (among others). The program requires reads to be mapped to some sort of assembly or reference genome, and it extracts and analyzes relevant information from those mappings. Note, however, that the estimates of population genomic statistics through bcftools are only as good as the reads and reference contigs that are used in mapping; duplicated loci of any kind could allow for a read to map to multiple locations and create false allele calls and erroneous estimates. Low-coverage contigs are particularly problematic because they may contain erroneous homozygous SNP calls.

3.3.4 Computational Resource Requirements

Although it is possible to run most of the individual programs on desktop computers, parallel compute clusters are highly recommended or necessary to process the data in a timely and efficient manner. For projects that have an especially large number of individuals that need to have sequence data assembled *de novo*, parallelization will greatly increase the speed at which assemblies can be completed. Similarly, for many loci, performing many calculations across all loci will be untenable without the help of parallel computing. In addition to large clusters housed at universities and research centers, researchers interested in attempting large-scale analyses can use third-party computing such as CyVerse (http://www.cyverse.org/), Amazon (www.amazon.com/hpc), and XSEDE (https://www.xsede.org/home). While parallelization greatly reduces time spent on the bioinformatics side of target enrichment, researchers should note the memory requirements for a number of programs. For example, Trinity (Grabherr et al. 2011) recommends 1 Gb of RAM per every 1 M reads; RAxML requires ~2.8 Gb for a 100 kb alignment of 50 taxa (http://www.exelixis-lab.org/software.html). Perhaps most important for consideration is the sheer size of storage space required to store raw reads, cleaned reads, assemblies, and various intermediate files that are produced during analysis. Projects with many individuals and loci can quickly use a terabyte of hard-drive space.

3.4 Results Reporting and Community Resources

3.4.1 Standards of Reporting

Sequence capture methods, no matter how baits are designed, are fundamentally similar in their attempt to reduce genomic representation in the sequenced reads. As a result, similar statistics are important for assessing the quality and efficiency of sequence capture. For example, the number of on-target contigs assembled per library, relative to how many were targeted, gives a general impression of how well hybridization worked, although this metric is slightly confounded by sequencing depth, which alone can increase the number of assembled contigs. Coverage statistics—both for assembled contigs from targeted regions and off-target regions and perhaps for exon and intron sequences separately (see Heyduk et al. 2016)—indicate whether the depth of sequencing was adequate to call polymorphisms and whether hybridization of certain baits was more efficient than others, perhaps due to sequence similarity or genomic copy variation. For studies that attempt to capture loci from taxa across broader phylogenetic distance, assessing hybridization variation in baits across taxa helps to define the phylogenetic boundary of effective capture using a particular bait set. In addition, it is often important to know how efficient capture was across the entire library—in other words, researchers might be interested in how many reads were on target or how many reads map to contigs used in the final analyses. Consistent reporting of such metrics enables comparisons of various methods and techniques across different sampling schemes and bait designs, leading to informed decision-making by researchers looking to implement sequence capture methods.

While numerical information about a given sequence capture project is useful for those looking to replicate methodology, the raw and cleaned data generated can be used by the larger scientific community as a whole. For this reason, researchers should take special care to deposit raw reads, alignments, and downstream analyses into common repositories (e.g., NCBI's Short Read Archive (http://www.ncbi.nlm.nih.gov/sra) and Dryad (http://datadryad.org/)). The bait sequences should be shared after publication as well. The time and effort put into designing effective and informative baits should be stretched beyond a single project. Indeed, some bait sets have sufficient utility that commercial companies may synthesize them in bulk, making them available to the research community at far lower cost than custom kits (http://www.mycroarray.com/mybaits/mybaits-UCEs.html).

Annex: Quick Reference Guide

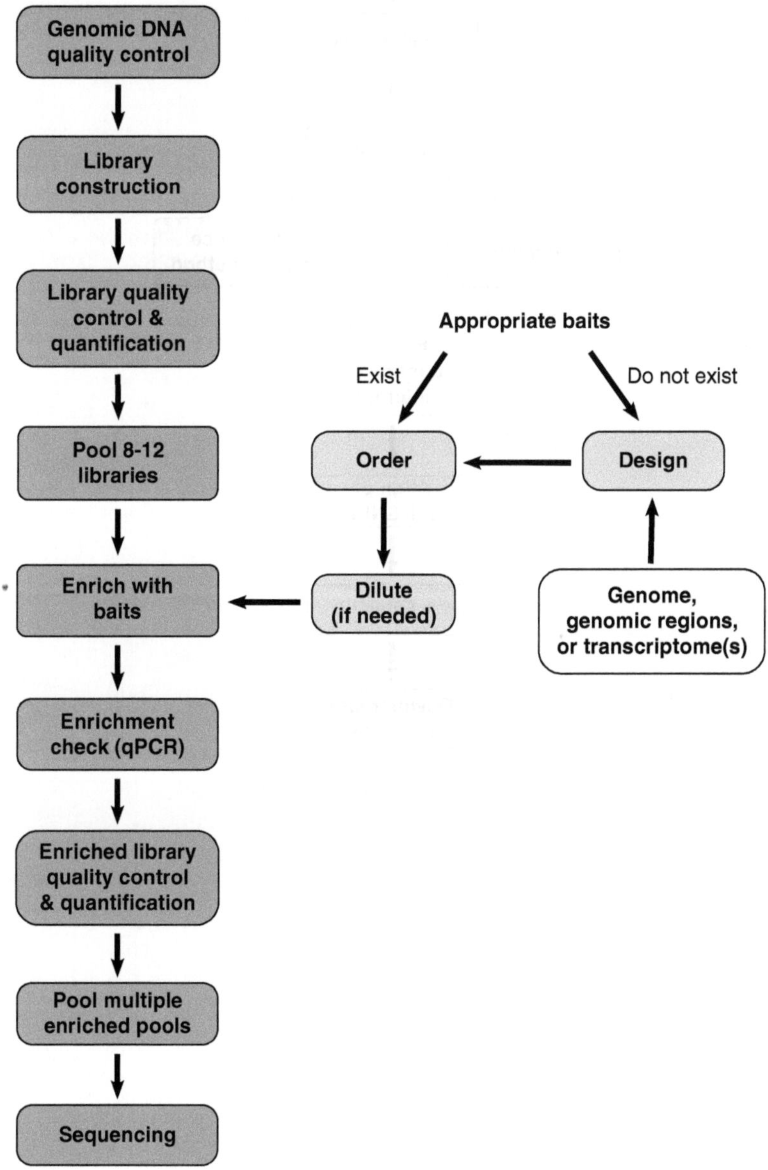

Fig. QG3.1 Representation of the wet-lab procedure workflow

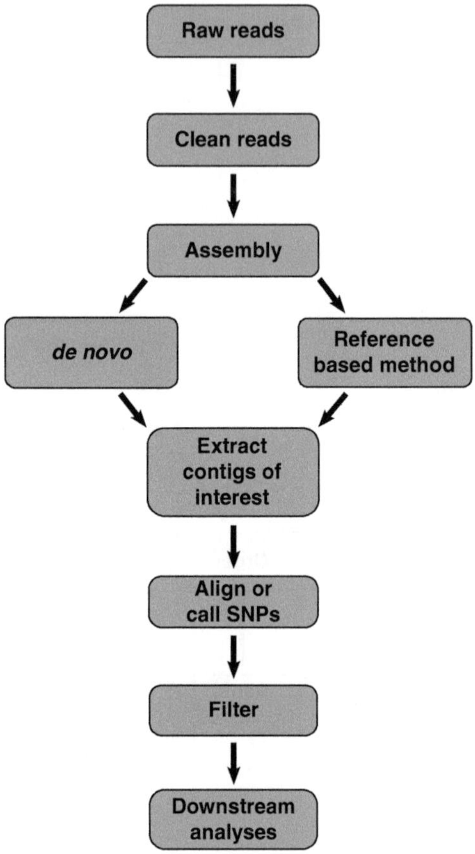

Fig. QG3.2 Main steps of the computational analysis pipeline

Table QG3.1 Experimental design considerations

Technique	Recommended starting material (ng)	Constraints	Average sequencing depth	Recommended sequencing run	Reference(s)
Whole genome re-sequencing	500	Cost of sequencing	6–30×	HiSeq or NextSeq PE75–PE150	Sims et al. (2014), Ekblom and Wolf (2014)
Transcriptome sequencing	1000	Sample material, cost of libraries, cost of sequencing	≥10 million reads per sample	HiSeq or NextSeq PE100–PE150	Ozsolak and Milos (2011), Wang et al. (2009), Wang et al. (2011)
PCR amplicon sequencing	20	Combinatorial tags, pool with diverse libraries	20×	MiSeq PE250–PE300	Feng et al. (2016), Mamanova et al. (2010)
Restriction-site-associated DNA markers (RADseq)	100	Consistency	10×	HiSeq or NextSeq SE75–PE150	Davey et al. (2011), Davey et al. (2013)
Target enrichment by sequence capture	500	Probes (information to design and cost)	30×	HiSeq or NextSeq PE100–PE150	Mamanova et al. (2010), Mertes et al. (2011)

Table that comprises relevant experimental design parameters, to carefully consider before applying this methodology

Table QG3.2 Available software recommendations

Program	Phylogenetic method	Multiple accessions	Missing data	Requirements	Command line vs. GUI	Computational time
BEST (Liu 2008)	Sequence alignment	Can assign accessions to species	Accepts missing data	None	Command line	Very slow (months)
*BEAST (Heled and Drummond 2010)	Sequence alignment	Multiple accessions recommended	Does not allow missing data for a given species	Priors	GUI	Very slow (months)
SVDquartets (Chifman and Kubatko 2014)	Sequence alignment	Cannot assign accessions to species	Accepts missing data	None	Command line	Very fast (hours to days)
STEM (Kubatko et al. 2009)	Summary method	Cannot assign accessions to species	Accepts missing data	Must have an estimate of individual gene evolution relative to each other	Command line	Fast (days)
MP-EST (Liu et al. 2010)	Summary method	Can assign accessions to species	Accepts missing data	All gene trees must be rooted	Both	Fast (days)
ASTRAL (Mirarab et al. 2014)	Summary method	Cannot assign accessions to species	Accepts missing data	Gene trees need to be fully resolved (can be unrooted)	Command line	Very fast (hours)

Table displaying a selection of the recommended software available for the computational analysis of data yielded by this technique

References

Bao S, Jiang R, Kwan WK, Wang BB, Ma X, Song YQ (2011) Evaluation of next-generation sequencing software in mapping and assembly. J Hum Genet 56:406–414

Bayzid MD, Warnow T (2013) Naïve binning improves phylogenomic analyses. Bioinformatics 29:2277–2284

Bejerano G, Pheasant M, Makunin I, Stephen S, Kent W, Mattick J, Haussler D (2004) Ultraconserved elements in the human genome. Science 304:1321

Blumenstiel B, Cibulskis K, Fisher S, DeFelice M, Barry A et al. (2010) Targeted exon sequencing by in-solution hybrid selection. Curr Protoc Hum Genet Chapter 18: Unit 18.4.

Cariou M, Duret L, Charlat S (2013) Is RAD-seq suitable for phylogenetic inference? An in silico assessment and optimization. Ecol Evol 3:846–852

Carpenter ML, Buenrostro JD, Valdiosera C et al (2013) Pulling out the 1%: whole-genome capture for the targeted enrichment of ancient DNA sequencing libraries. Am J Hum Genet 93:852–864

Catchen J, Hohenlohe P, Bassham S, Amores A, Cresko W (2013) Stacks: an analysis tool set for population genomics. Mol Ecol 22:3124–3140

Chifman J, Kubatko L (2014) Quartet inference from SNP data under the coalescent model. Bioinformatics 30:3317. doi:10.1093/bioinformatics/btu530

Comer JR, Zomlefer WB, Barrett CF, Davis JL, Stevenson DW, Heyduk K, Leebens-Mack J (2015) Resolving relationships within the palm subfamily Arecoideae (Arecaceae) using plastid sequences derived from next-generation sequencing. Am J Bot 102:888–899

Cummings N, King R, Rickers A, Kaspi A, Lunke S, Haviv I, Jowett JBM (2010) Combining target enrichment with barcode multiplexing for high throughput SNP discovery. BMC Genomics 11:641

Davey JW, Blaxter ML (2010) RADSeq: next-generation population genetics. Brief Funct Genomics 9:416–423

Davey JW, Hohenlohe PA, Etter PD, Boone JQ, Catchen JM, Blaxter ML (2011) Genome-wide genetic marker discovery and genotyping using next-generation sequencing. Nat Rev Genet 12:499–510

Davey JW, Cezard T, Fuentes-Utrilla P, Eland C, Gharbi K, Blaxter ML (2013) Special features of RAD Sequencing data: implications for genotyping. Mol Ecol 22:3151–3164

Dermitzakis ET, Reymond A, Antonarakis SE (2005) Conserved non-genic sequences—an unexpected feature of mammalian genomes. Nat Rev Genet 6:151–157

Derti A, Roth FP, Church GM, Wu C-T (2006) Mammalian ultraconserved elements are strongly depleted among segmental duplications and copy number variants. Nat Genet 38:1216–1220

Duarte JM, Wall PK, Edger PP, Landherr LL, Ma H, Pires JC, Leebens-Mack J, dePamphilis CW (2010) Identification of shared single copy nuclear genes in *Arabidopsis*, *Populus*, *Vitis*, and *Oryza* and their phylogenetic utility across various taxonomic levels. BMC Evol Biol 10:61

Easton DF, Rharoah PDP, Antoniou AC et al (2015) Gene-panel sequencing and the prediction of breast-cancer risk. N Engl J Med 372:2243–2257

Eaton DAR (2014) PyRAD: assembly of *de novo* RADseq loci for phylogenetic analyses. Bioinformatics 30:1844. doi:10.1093/bioinformatics/btu121

Ekblom R, Galindo J (2011) Applications of next generation sequencing in molecular ecology of non-model organisms. Heredity 107:1–15

Ekblom R, Wolf JBW (2014) A field guide to whole-genome sequencing, assembly, and annotation. Evol Appl 7(9):1026–1042

Enk JM, Devault AM, Kuch M, Murgha YE, Rouillard JM, Poinar HN (2014) Ancient whole genome enrichment using baits built from modern DNA. Mol Biol Evol 31:1292–1294

Faircloth BC (2016) PHYLUCE is a software package for the analysis of conserved genomic loci. Bioinformatics 32:786-788. doi:10.1093/bioinformatics/btv646

Faircloth BC, Glenn TC (2012) Not all sequence tags are created equal: designing and validating sequence identification tags robust to indels. PLoS One 7:e42543. doi:10.1371/journal.pone.0042543

Faircloth BC, McCormack JE, Crawford NG, Harvey MG, Brumfield RT, Glenn TC (2012) Ultraconserved elements anchor thousands of genetic markers spanning multiple evolutionary timescales. Syst Biol 61:717–726

Faircloth BC, Branstetter MG, White ND, Brady SG (2015) Target enrichment of ultraconserved elements from arthropods provides a genomic perspective on relationships among Hymenoptera. Mol Ecol Resour 15:489

Feng YJ, Liu QF, Chen MY, Liang D, Zhang P (2016) Parallel tagged amplicon sequencing of relatively long PCR products using the Illumina HiSeq platform and transcriptome assembly. Mol Ecol Resour 16:91. doi:10.1111/1755-0998.12429

Fisher S, Barry A, Abreu J, Minie B, Nolan J et al (2011) A scalable, fully automated process for construction of sequence-ready human exome targeted capture libraries. Genome Biol 12:R1

Gautier M, Gharbi K, Cezard T, Foucaud J, Kerdelhue C, Pudlo P, Cornuet JM, Estoup A (2012) The effect of RAD allele dropout on the estimation of genetic variation within and between populations. Mol Ecol 22:3165–3178

Glenn TC, Nilsen R, Kieran TJ, Finger JW Jr, Pierson TW, García-De-Leon FJ, del Rio Portilla MA, Reed K, Anderson JL, Meece JK, Alabady M, Belanger M, Faircloth BC (2016) Adapterama I: universal stubs and primers for thousands of dual-indexed Illumina Nextera and TruSeqHT compatible libraries (iNext & iTru). bioRxiv

Grabherr MG, Haas BJ, Yassour M, Levin JZ, Thompson DA, Amit I, Adiconis X, Fan L, Raychowdhury R, Zeng Q, Chen Z, Mauceli E, Hacohen N, Gnirke A, Rhind N, di Palma F, Birren BW, Nusbaum C, Lindblad-Toh K, Friedman N, Regev A (2011) Full-length transcriptome assembly from RNA-Seq data without a reference genome. Nat Biotechnol 29:644–652

Grover CE, Salmon A, Wendel JF (2011) Targeted sequence capture as a powerful tool for evolutionary analysis. Am J Bot 99(2):312–319

Haas BJ, Gevers D, Earl AM et al (2011) Chimeric 16S rRNA sequence formation and detection in Sanger and 454-pyrosequenced PCR amplicons. Genome Res 21:494–504

Hahn DA, Ragland GJ, Shoemaker DD, Denlinger DL (2009) Gene discovery using massively parallel pyrosequencing to develop ESTs for the fleshy fly *Sarcophaga crassipalpis*. BMC Genomics 10:234. doi:10.1186/1471-2164-10-234

Heled J, Drummond AJ (2010) Bayesian inference of species trees from multilocus data. Mol Biol Evol 27:570–580

Heyduk K, Trapnell DW, Barnett CF, Leebens-Mack J (2016) Estimating relationships within *Sabal* (Arecaceae) through multilocus analyses of sequence capture data. Biol J Linnean Soc 17(1):106–120

Huang H, Knowles LL (2014) Unforeseen consequences of excluding missing data from next-generation sequences: simulation study of RAD sequences. *Syst Biol* doi: 10.1093/sysbio/syu046

Keane TM, Goodstadt L, Danecek P, White MA, Wong K et al (2011) Mouse genome variation and its effect on phenotypes and gene regulation. Nature 477:289–294

Kubatko LS (2009) Identifying hybridization events in the presence of coalescence via model selection. Syst Biol 58:478–488

Kubatko LS, Carstens BC, Knowles LL (2009) STEM: species tree estimation using maximum likelihood for gene trees under coalescence. Bioinformatics 25:971–973

Lemmon EM, Lemmon AR (2013) High-throughput genomic data in systematics and phylogenetics. Annu Rev Ecol Evol Syst 44:99–121

Li Y, Zhao S, Ma J, Li D, Yan L, Li J, Qi X, Guo X et al (2013) Molecular footprints of domestication and improvement in soybean revealed by whole genome re-sequencing. BMC Genomics 14:579. doi:10.1186/1471-2164-14-579

Liu L (2008) BEST: Bayesian estimation of species trees under the coalescent model. Bioinformatics 24:2542–2543

Liu L, Yu L, Edwards SV (2010) A maximum pseudo-likelihood approach for estimating species trees under the coalescent model. BMC Evol Biol 10:302. doi:10.1186/1471-2148-10-302

Luikart G, England PR, Tallmon D, Jordan S, Taberlet P (2003) The power and promise of population genomics: from genotype to genome typing. Nat Rev Genet 4:981–994. doi:10.1038/nrg1226

Mamanova L, Coffey AJ, Scott CE, Kozarewa I, Turner EH et al (2010) Target-enrichment strategies for next-generation sequencing. Nat Methods 7:111–118

McCormack JE, Maley JM, Hird SM, Derryberry EP, Graves GR, Brumfield RT (2012) Next-generation sequencing reveals population genetic structure and a species tree for recent bird divergences. Mol Phylogenet Evol 62:397–406

McCormack JE, Hird SM, Zellmer AJ, Carstens BC, Brumfield RT (2013a) Applications of next-generation sequencing to phylogeography and phylogenetics. Mol Phylogenet Evol 66:526–538

McCormack JE, Harvey MG, Faircloth BC, Crawford NG, Glenn TC, Brumfield RT (2013b) A phylogeny of birds based on over 1,500 loci collected by target enrichment and high-throughput sequencing. PLoS One 8:e54848. doi:10.1371/journal.pone.0054848

McCormack JE, Tsai WLE, Faircloth BC (2015) Sequence capture of ultraconserved elements from bird museum specimens. Molecular Ecology Resources doi: 10.1111/1755-0998.12466

Meiklejohn KA, Danielson MJ, Faircloth BC, Glenn TC, Braun EL, Kimball RT (2014) Incongruence among different mitochondrial regions: a case study using complete mitogenomes. Mol Phylogenet Evol 78:314–323

Mertes F, ElSharawy A, Sauer S, van Helvoort JMLM, van der Zaag PJ, Franke A, Nilsson M, Lehrach H, Brookes AJ (2011) Targeted enrichment of genomic DNA regions for next-generation sequencing. Brief Funct Genomics 10(6):374–386

Meyer M, Kircher M (2010) Illumina sequencing library preparation for highly multiplexed target capture and sequencing. Cold Spring Harb Protoc 2010: pdb prot5448

Mirarab S, Reaz R, Bayzid MS, Zimmerman T, Swenson MS, Warnow T (2014) ASTRAL: genome-scale coalescent-based species tree estimation. Bioinformatics 30:i541–i548

Ozsolak F, Milos PM (2011) RNA sequencing: advantages, challenges, and opportunities. Nat Rev Genet 12:87–98

Peñalba JV, Smith LL, Tonione MA, Sass C, Hykin SM, Skipwith PL, McGuire JA, Bowie RCK, Moritz C (2014) Sequence capture using PCR-generated probes: a cost-effective method of targeted high-throughput sequencing for nonmodel organisms. Mol Ecol 14(5):1000–1010

Puritz JB, Matz MV, Toonen RJ, Weber JN, Bolnick DI, Bird CE (2014) Demystifying the RAD fad. Mol Ecol 23(24):5937–5942

Raposo do Ameral F, Neves LG, Resende MF Jr, Mobili F, Miyaki CY, Pellegrino KC, Biondo C (2015) Ultraconserved elements sequencing as a lowcost source of complete mitochondrial genomes and microsatellite markers in non-model amniotes. PLoS One 10:e0138446

Rohland N, Reich D (2012) Cost-effective, high-throughput DNA sequencing libraries for multiplexed target capture. Genome Res 22:939–946

Rubin BER, Ree RH, Moreau CS (2012) Inferring phylogenies from RAD sequence data. PLoS One 7:1–12

Shearer EA, Hildebrand MS, Ravi H, Joshi S, Guiffre AC, Novak B, Happe S, LeProust EM, Smith RJH (2012) Pre-capture multiplexing improves efficiency and cost-effectiveness of targeted genomic enrichment. BMC Genomics 13:618

Sims D, Sudbery I, Ilot NE, Heger A, Ponting CP (2014) Sequencing depth and coverage: key considerations in genomic analyses. Nat Rev Genet 15:121–132

Smith BT, Harvey MG, Faircloth BC, Glenn TC, Brumfield RT (2014) Target capture and massively parallel sequencing of ultraconserved elements (UCEs) for comparative studies at shallow evolutionary time scales. Syst Biol 63(1):83–95

Stephen S, Pheasant M, Makunin IV, Mattick JS (2008) Large-scale appearance of ultraconserved elements in tetrapod genomes and slowdown of the molecular clock. Mol Biol Evol 25:402–408

Stephens JD, Rogers WL, Heyduk K, Cruse-Sanders JM, Determann RO, Glenn TC, Malmberg RL (2015a) Resolving phylogenetic relationships for the recently radiated carnivorous plant genus *Sarracenia* using target enrichment. Mol Phylogenet Evol 85:76–87

Stephens JD, Rogers WL, Mason CM, Donovan LA, Malmberg RL (2015b) Species tree estimation of diploid *Helianthus* (Asteraceae) using target enrichment. Am J Bot 102:921–941

Wagner CE, Keller I, Wittwer S, Selz OM, Mwaiko S, Greuter L, Sivasundar A, Seehausen O (2013) Genome-wide RAD sequence data provide unprecedented resolution of species boundaries and relationships in the Lake Victoria cichlid adaptive radiation. Mol Ecol 22:787–798

Wang Y, Qian PY (2009) Conservative fragments in bacterial 16S rRNA genes and primer design for 16S ribosomal DNA amplicons in metagenomic studies. PLoS One 4:e7401. doi:10.1371/journal.pone.0007401

Wang Z, Gerstein M, Snyder M (2009) RNA-Seq: a revolutionary tool for transcriptomics. Nat Rev Genet 10:57–63

Wang Y, Ghaffari N, Johnson CD, Braga-Neto UM, Wang H, Chen R, Zhou H (2011) Evaluation of the coverage and depth of transcriptome by RNA-Seq in chickens. BMC Bioinformatics 12:S5. doi:10.1186/1471-2105-12-S10-S5

Weitmeier K, Straub SCK, Cronn RC, Fishbein M, Schmickl R, McDonnell A, Liston A (2014) Hyb-Seq: combining target enrichment and genome skimming for plant phylogenomics. Appl Plant Sci 2:1400042. doi:10.3732/apps.1400042

Xu J, Zhao Q, Du P, Xu C, Wang B, Feng Q, Liu Q, Tang S, Gu M, Han B, Liang G (2010) Developing high throughput genotyped chromosome segment substitution lines based on population whole-genome re-sequencing in rice (*Oryza sativa* L.). BMC Genomics 11:656. doi:10.1186/1471-2164-11-656

Yu Y, Nakhleh L (2015) A distance-based method for inferring phylogenetic networks in the presence of incomplete lineage sorting. Bioinform Res Appl 9096:378–389

Yu Y, Cuong T, Degnan JH, Nakhleh L (2011) Coalescent histories on phylogenetic networks and detection of hybridization despite incomplete lineage sorting. Syst Biol 60:138–149

Zhu Y, Bergland AO, González J, Petrov DA (2012) Empirical validation of pooled whole genome population re-sequencing in *Drosophila melanogaster*. PLoS One 7:e41901. doi:10.1371/journal pone.0041901

Chapter 4
Transcriptome Profiling Strategies

Abdullah M. Khamis, Vladimir B. Bajic, and Matthias Harbers

4.1 Introduction to Technologies

Our understanding of transcriptomes, the RNA content of biological samples, correlates by large with the progress of DNA sequencing technologies. With the development of capillary sequencers using the Sanger sequencing method, it became feasible to sequence deeply into cDNA libraries prepared from RNA pools. In the initial sequencing studies, a few thousand or even tens of thousands of randomly isolated cDNA clones were sequenced most commonly from their 3′ end to obtain short, so-called EST (expressed sequence tag) reads. Unsupervised EST sequencing for the first time gave an overview on the complexity of RNA transcripts and their presence at different biological stages. This approach was later extended by the development of full-length cDNA cloning technologies for obtaining sequence information on the entire RNA transcripts (Harbers 2008). The knowledge gained from large-scale cDNA cloning and sequencing projects led to transcriptome and genome annotations that are today the basis to all approaches to transcriptome profiling. New high-speed sequencing methods can by now provide comprehensive overviews on transcriptomes at reasonable cost and by far exceed the achievements of the early EST projects (de Klerk et al. 2014). Many of these methods allow for

A.M. Khamis, Ph.D. • V.B. Bajic, Ph.D.
Computer, Electrical and Mathematical Sciences and Engineering Division, Computational Bioscience Research Center, King Abdullah University of Science and Technology (KAUST), Thuwal 23955-6900, Saudi Arabia
e-mail: abdullah.khamis@kaust.edu.sa; vladimir.bajic@kaust.edu.sa

M. Harbers, Ph.D. (✉)
Division of Genomic Technologies, RIKEN Center for Life Science Technologies, 1-7-22 Suehiro-cho, Tsurumi-ku, Yokohama, Kanagawa 230-0045, Japan
e-mail: matthias.harbers@riken.jp

A.M. Aransay, J.L. Lavín Trueba (eds.), *Field Guidelines for Genetic Experimental Designs in High-Throughput Sequencing*,
DOI 10.1007/978-3-319-31350-4_4

quantitative measurements on individual RNA transcripts to monitor changes in the RNA content between different samples.

While working for the annotation of the honey bee (*Apis mellifera*) genome and later studying transcriptional regulation of gene expression during behavioral maturation of worker bees, we have applied full-length cDNA preparation and different high-speed sequencing methods to transcriptome analysis. Based on our experience, we describe here the use of full-length cDNA preparations in combination with shotgun sequencing in mRNA profiling (so-called RNA-Seq methods for "RNA sequencing") and RNA-Seq profiling starting directly from RNA. Moreover, we describe the use of CAGE for high-throughput mRNA detection and determination of transcription start sites (TSSs) on the genome level. For other RNA profiling methods refer to the recent review from (de Klerk and 't Hoen 2015) on RNA sequencing.

Various approaches to RNA characterization have been developed addressing different aspects of transcriptome analysis. It can be meaningful to apply multiple methods on the same samples during your studies for a better understanding of RNA structures and regulatory processes. Although beyond the scope of this chapter, there are additional methods for characterization and monitoring of short RNAs as well as not sequencing-based approaches to transcriptome profiling such as DNA microarrays or qPCR methods. Some of those are of interest to the users of high-speed sequencing methods to confirm their results by independent experimental means.

4.2 Objectives of RNA-Seq and CAGE Experiments

Ideally, we would like to obtain the full-length sequence of every mRNA transcript in a sample. Only high-quality sequence information on the entire mRNA transcripts would allow us to distinguish between splice variants derived from the same gene and to understand the extent of alternative splicing related to the complex regulation of biological processes. Our technical limitations, however, restrict today our ability to study RNA splicing, where most high-speed sequencing methods can only provide short sequencing reads of some 100–200 bp. Those reads are much shorter than an average mRNA molecule, and therefore RNA profiling methods are using tag-based approaches or RNA-Seq methods for transcriptome analysis. Tag-based approaches like CAGE obtain a single sequencing read per RNA molecule for transcript identification. Since just only one sequencing read per RNA molecule is obtained, the number of reads directly correlates with the transcript levels in the sample ("digital sequencing"). On the contrary, RNA-Seq methods obtain multiple, random reads from each transcript for a better coverage of the entire RNA sequence (Kawaji et al. 2014). RNA-Seq data sets are more complex and require additional considerations during data analysis to obtain quantitative measures on mRNA levels. To obtain quantitative data on the expression of different splice variants is still a great challenge for RNA-Seq experiments and requires very high sequencing depths.

We advise to use CAGE for basic mRNA profiling and quantification and will therefore focus on CAGE experiments in this chapter while providing additional information on more commonly used RNA-Seq to show how the processes and data compare. The 5′ end of eukaryotic mRNA molecules is protected by the 7-methylguanylate cap structure, which can be used to obtain sequences from the 5′ end of mRNAs. Different CAGE protocols are in use that utilize the so-called Cap-Trapper method or template switching in the 5′ end selection step. Short cDNA fragments complementary to the 5′ end of mRNAs are then sequenced at high throughput, where the CAGE method has been adapted to different sequencing methods. CAGE commonly achieves a much better coverage of the mRNA content for the same amount of sequencing than possible by any RNA-Seq method. The short sequencing reads allow for reliable transcript identification and quantification. Moreover, CAGE sequencing reads can be mapped to a reference genome for TSS identification, thus providing accurate information on transcriptional activity at defined genome locations. Therefore CAGE was the method of choice for genome-wide TSS mapping during the ENCODE (Consortium 2012) and FANTOM (Consortium F et al. 2014; Lizio et al. 2015) projects, which provided essential information for further analysis of regulatory regions in the human genome. CAGE experiments have been further performed on a number of other model organisms, and the method has been recognized as one of the basic approaches to study transcriptional networks.

While CAGE is effective in RNA transcript and TSS identification, the method falls short on providing further information on full-length RNA structures. Therefore, CAGE cannot provide a complete picture on the extent of RNA splicing in the sample. This limitation of CAGE can be overcome by different RNA-Seq methods for the preparation of random cDNA fragments from RNA pools. These random cDNA fragments are then sequenced at a very high throughput to obtain sufficient sequence information for covering the entire length of each transcript by multiple reads. RNA-Seq experiments can identify different splice variants mostly by using reads comprising splice junctions. However, in most cases it is very difficult to reconstruct the full-length sequences of different splice variants from RNA-Seq data. Moreover, transcript quantification is more complex than working with tag-based methods, because the multiple reads obtained from transcripts of different length require additional normalization steps during data analysis. Regardless of the complex data analysis and sequencing requirements, RNA-Seq methods are today the most commonly used approach to transcriptome profiling, where different providers offer reagent kits to conduct such experiments. Table 4.1 summarizes the main differences between CAGE and RNA-Seq methods.

4.3 Sequencing Platforms

CAGE as well as RNA-Seq methods have been adapted to different sequencing platforms, but as of today most laboratories including the work done for our projects use Illumina sequencing on a routine basis. During protocol development and feasibility studies, the smaller sequencing yields of an Illumina MiSeq instrument are suitable for library sequencing. However, for deep sequencing of RNA-Seq libraries and multiplex sequencing of many RNA-Seq or CAGE samples, we prefer the use

Table 4.1 Main differences between CAGE and RNA-Seq methods

CAGE	RNA-Seq
Sequencing from capped 5′ end of RNA molecules	Sequencing of fragments distributed randomly along RNA molecules
One read corresponds to a single transcript	Multiple reads may correspond to a single transcript
5′ Cap selection of RNA (no preference for poly(A))	Poly(A) selection of RNA, or removal of rRNA by capturing method or digestion
Effective for transcript identification and quantification	Provides better coverage of the entire RNA sequence, quantification possible
Reliable method for TSS identification	Effective for RNA splicing and detection of genetic variations

of a HiSeq instrument. Providers like Illumina offer sets of adapters for multiplex sequencing experiments, which should have been optimized for library preparation. We recommend using multiplex sequencing as a proven approach for standard experiments to cut sequencing costs. While planning the sequencing experiment, consider first the desired sequencing depth for each sample versus to the total number of reads per run to determine how many samples may be pooled per run. In our experiments, we used multiplex sequencing, and each of the pooled libraries was sequenced on one lane on an Illumina HiSeq2000 instrument for 100 cycles from each end of the fragments using a TruSeq SBS sequencing kit version 3 followed by data processing with Casava 1.8 (pipeline 1.9).

We will not discuss further on the use of other sequencing methods, although the new PacBio RS II, which is a third-generation single-molecule, real-time DNA sequencing system, provides interesting means for full-length cDNA sequencing that could lead to much better information on RNA transcripts than possible by any RNA-Seq method. Unfortunately, the present throughput of the method and the associated costs make the method unfeasible for regular transcriptome analysis. Preliminary data, however, showed that full-length cDNA sequences can be obtained on a PacBio sequencer with a reasonable success rates.

4.4 Experimental Design

Transcriptome profiling experiments should be well planned to assure that they provide meaningful results for describing a biological context. Originally, descriptive high-speed sequencing experiments have been made that just targeted at a catalog of the different transcripts present in a given sample. Such data sets are not suitable to compare different samples and to quantify mRNA levels by statistical means. Moreover, in biological studies commonly relative mRNA levels are compared between different samples. Therefore, the experimental design of the study has to give considerations to which is the most meaningful reference sample or samples to drive data analysis.

We advise to first perform some test experiments to establish a new method at the laboratory. Some three or more technical replicas should be used to confirm the reproducibility of the method, and results should be compared to published data to assure the procedures provide reliable data. There are high-speed sequencing data sets in the public domain (see below) that can be downloaded for testing bioinformatics routines before starting the analysis of your own experimental data. Once the bench protocols and analysis platform work well in the laboratory, the method should no longer be changed during the course of the study. Changing parts of a protocol and analysis platform can make it difficult or even impossible to compare the data from different experiments. For the statistical analysis of the data from biological samples, we advise to use at least three biological replicas per data point all analyzed by the same standard procedure. Biological replica should provide consistent data, although there is always some fluctuation between biological samples, because environmental conditions cannot be perfectly normalized. Therefore, you should consider even more biological replicas when working, for example, on individual wildlife animals as we had done for our work on honey bees. For large-scale projects using many different samples over an extended time, it may be useful to consider also a "technical standard", which uses a reference RNA that is repeatedly used during different rounds of library preparation. Analyzing the data obtained from such a "technical standard" can confirm the constancy of the different experiments. A reference RNA may be prepared from a cell line for easy availability; also the brain RNA has been used as a reference because of its high complexity (e.g., the human brain reference RNA is commercially available).

Different transcriptome profiling methods have also different sample requirements (compare Fig. 4.1 and Table 4.2). Therefore, the experiments may be restricted by the available RNA amounts, and choices have to be made on which aspects are the most important to be answered in the study. Sometimes, it may be necessary to pool RNA samples to reach the necessary RNA amount for doing the experiments. It should be noted, however, that pooling many RNA samples leads to a kind of "normalization" for the different RNA species. While RNAs expressed in all samples can keep their concentration constant in the pool, RNA species only present in some of the RNA samples within the pool will be diluted. Consequently, rare transcripts may be harder to find in pooled RNA samples.

Control experiments should not only consider a reference sample as outlined above, but sufficient RNA should be available from each sample to confirm the results from the sequencing experiments by other experimental means, e.g., by performing some qPCR experiments on selected targets, after the sequencing experiments have been completed.

4.5 Full-Length cDNA Preparation

Methods for the preparation of full-length cDNAs have been instrumental for building large collections of cDNA clones from different species, but are less frequently used in high-speed sequencing experiments. The most commonly used RNA-Seq

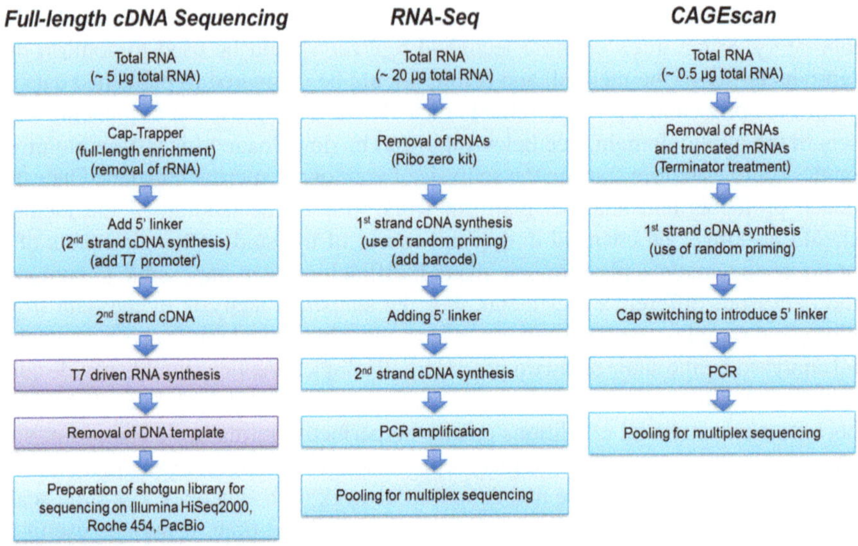

Fig. 4.1 Comparison of the different RNA requirements and processing steps for preparing CAGE and RNA-Seq libraries as used in our work on honey bee samples. We prepared RNA-Seq libraries directly from rRNA-depleted RNA and full-length cDNA. RNA prepared from the full-length cDNA was used in RNA-Seq library preparation as outlined under RNA-Seq

protocols avoid additional steps for 5′ end enrichment to streamline library preparation. However, this comes at the expense of a much lower coverage of the 5′- and 3′ end sequences of the transcripts in the sequence data. Therefore, we think that a 5′ end enrichment step can help for certain experiments to improve the coverage of RNA-Seq experiments; for experiments like CAGE that target sequence information from the 5′ end of mRNAs, the enrichment step is essential to obtain reliable data. Although sequencing costs are going down, a higher sequencing depth is not a tool to overcome shortcomings of the library preparation methods. For the evaluation of any experiment, it is important to understand the possible bias of the method(s) selected for the study and to consider the specifications of the methods when interpreting the data. Important findings should be confirmed by independent experimental means to assure the conclusions of the study and to avoid mistakes caused by library or analysis artifacts.

Different approaches for full-length cDNA selection have been described in the literature, where we focus on the so-called "Cap-Trapper" method and "template switching." Cap trapping proved to be the most effective approach to full-length cDNA cloning during different large-scale projects (Harbers 2008). After the reverse transcription reaction, the Cap structure of mRNA is selectively biotinylated to capture the cDNA/mRNA hybrids on streptavidin-coated beads. Remaining single-stranded RNA is then digested by RNase I treatment leaving only those cDNA/mRNA hybrids intact where the complementary cDNA reached the 5′ end of

Table 4.2 Experimental design for three different techniques used for transcriptome profiling experiments

Technique	Protocol	Control library	Recommended starting material	Number of replicates	Sequencing depth	Recommended sequencing platform and run
CAGEscan	Cap-Trapper method or template-switching method	Use RNA from a cell line or brain RNA	0.5 μg total RNA when using template-switching methods	3 (minimum per condition)	Requires less sequencing depth than RNA-Seq (Sims et al. 2014)	– Illumina MiSeq or HiSeq 100–200 paired end
						– HeliScope
Full-length cDNA sequencing	Cap-Trapper method		5 μg total RNA		Depends on transcriptome size, e.g., human requires ~10 million reads (Liu et al. 2014)	– Illumina HiSeq 100–200 paired end
						– PacBio RS II
Direct RNA-Seq	Removal of rRNAs		20 μg total RNA			– Illumina HiSeq 100–200 paired end

mRNAs. The single-stranded cDNA from those hybrids is then isolated after mRNA hydrolysis and can be used for cDNA cloning or direct sequence analysis.

As an alternative to the Cap-Trapping method, we are using the so-called "template-switching" reaction of the reverse transcriptase. Oligonucleotides with three rG nucleotides at their 3′ end (so-called template-switching oligonucleotides) can interact with the Cap structure at the 5′ end of mRNAs. The bound oligonucleotide is then becoming a template when the reverse transcriptase "switches" from the mRNA template to the oligonucleotide. At the end of the transcription reaction, the resulting cDNA comprises sequences complementary to the template-switching oligonucleotide. Those oligonucleotide-derived sequences can then be used for the selective enrichment of full-length cDNAs. The template-switching method allows for cDNA preparations from very small amounts of RNA giving the method a very high sensitivity even though the full-length enrichment is not as good as in the Cap-Trapper experiments.

In the examples to this chapter, we will show some data on how to use full-length enriched cDNA in RNA-Seq experiments. We advised to use the Cap-Tapper method for preparing full-length cDNA templates because the cDNA may not be used only for preparing an RNA-Seq library. Having a high-quality cDNA pool provides further means to isolated cDNAs for selected targets identified while analyzing the RNA-Seq data. Those cDNAs can be cloned and used to accurately determine their full-length sequences, which may not have been obtained correctly by assembling short RNA-Seq reads. In addition, cloned cDNA fragments can be used in functional annotation experiments.

4.6 CAGE Library Preparation and Sequencing

The original CAGE protocol was developed based on the experience working with the Cap-Trapper method in cDNA cloning projects. While using the Cap-Trapper method for selecting regions from the 5′ end of mRNAs for sequencing, the new protocol divided from the cDNA library protocol by adding a digestion step to cut off a short cDNA fragment at the end of full-length cDNAs. These short fragments or "tags" could be amplified and then sequenced at high throughput. Moreover, we shifted from using oligo(dT) priming used in full-length cDNA cloning to the use of random priming in the reverse transcription reaction. Random priming not only increases the changes to reach the 5′ end of very long transcripts but also allows for capturing tags from non-polyadenylated mRNAs that are not covered in oligo(dT)-primed cDNA preparations (there is a large number of non-polyadenylated mRNAs).

The basic CAGE protocol has been adapted over the years for use on different sequencing platforms (Murata et al. 2014; Takahashi et al. 2012a, b). The latest version of the CAGE protocol for using long sequencing reads on Illumina platforms no longer requires the tag-digestion step, but sequences cDNA fragments of different length are obtained by random priming. From such longer cDNA fragments, end sequences can be obtained from the 5′ and 3′ ends by using paired-end reads on the Illumina sequencers ("CAGEscan" method) (Plessy et al. 2010).

Although the CAGE protocol has been improved for working with small amounts of RNA, the Cap-Trapper method does not reach the same sensitivity as the template-switching method. Therefore, a new version of CAGE, denoted as "nanoCAGE," was developed that uses template switching instead of Cap trapping to obtain CAGE data even from very small amounts of RNA (50–500 ng total RNA) (Salimullah et al. 2011). We found it useful to pretreat total RNA with an exonuclease that digests 5′ end phosphorylated RNA prior to preparing a nanoCAGE library. This pretreatment reduces background signals derived from rRNAs in the library and seems to improve also the 5′ selection.

For our analysis of changes in the gene expression in honey bee brains, we wanted to compare expression levels and to have an outlook on the alternative usage of TSS between individual animals that belonged to two distinct groups based on their behavioral maturation, nurses, and foragers (Khamis et al. 2015). Because of the small RNA amounts obtained from individual honey bee brains, we decided to apply nanoCAGE for our studies. The nanoCAGE protocol allowed us to prepare cDNA fragments from individual samples that were individually tagged by specific sequencing tags. Therefore, nanoCAGE libraries from eight individual animals of each group could be sequenced in parallel in a single-Illumina HiSeq2000 sequencing reaction, and the reads for each sample were then sorted by using the sample-specific sequencing tags. Such multiplex sequencing methods are common by now to make better use of the very throughput of high-speed sequencers.

4.7 Bioinformatics Data Analyses of nanoCAGE Data

The advancement in sequencing technologies in the past decade has increased their capacity making them useful for transcriptome profiling with high sequencing depth and transcriptome coverage. A typical Illumina HiSeq sequencing platform can sequence hundreds of millions of single- or paired-end reads, each having few hundred base pairs. Consequently, suitable bioinformatics analysis methodology is required to leverage this volume of data and the information it contains. Here, we discuss common bioinformatics analysis steps to process nanoCAGE data. An overview of a nanoCAGE data analysis process is depicted in Fig. 4.2. The basic process is the same regardless whether only 5′ end sequences are provided ("nanoCAGE") or the DNA fragments have been sequenced from both ends using paired-end reads on the Illumina platform ("CAGEscan"). In our example working on the honey bee project, paired-end reads had been available for the analysis.

As shown in Fig. 4.2, the primary analysis starts by generating the sequences ("reads") and the read quality data. This is followed by generating read counts that are usually captured in a matrix whose rows represent genes and columns represent samples (e.g., in our study different ages within the lifespan of worker honey bees). Finally, biological insight and data interpretation are performed on the gene expression matrix using different analysis techniques. Figure 4.3 shows the detailed workflow for a bioinformatics analysis pipeline for typical nanoCAGE data. We provide detailed descriptions on each of these analysis steps in this chapter.

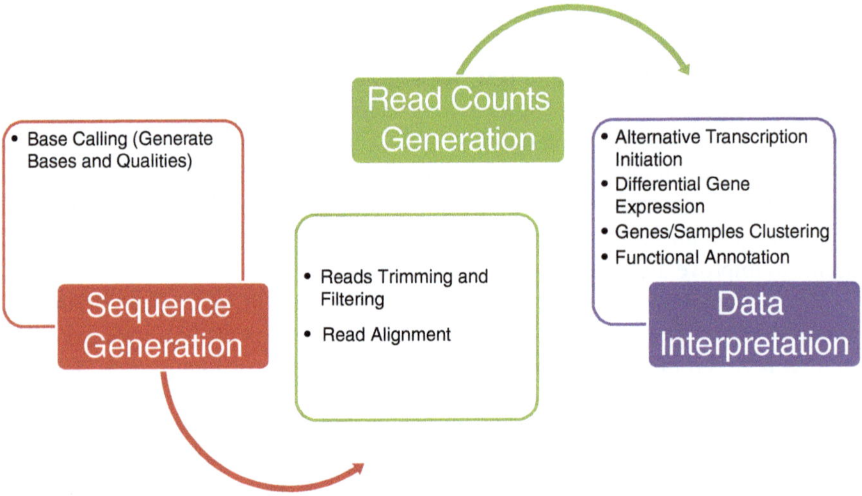

Fig. 4.2 Overview of nanoCAGE data analysis process

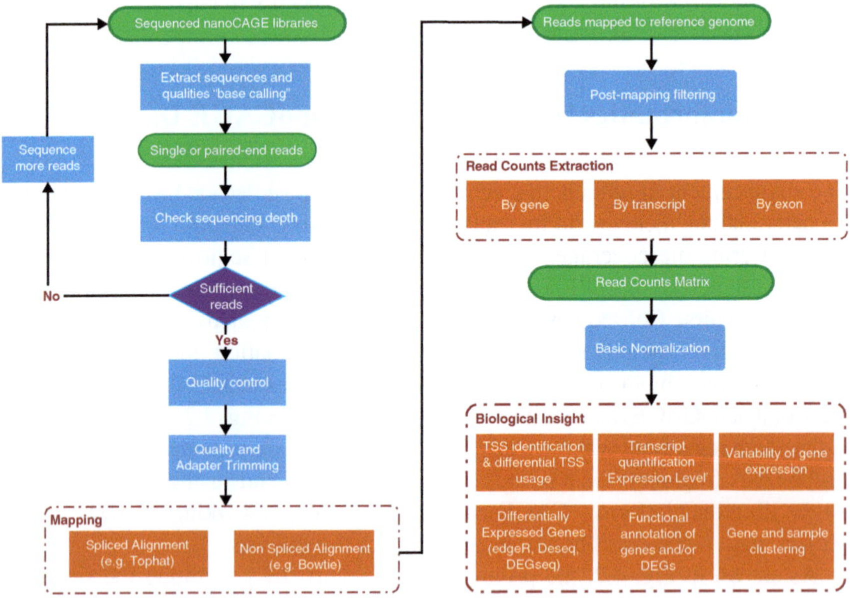

Fig. 4.3 Typical bioinformatics analysis pipeline for nanoCAGE data

4.7.1 Sequencing Depth

A crucial requirement in any transcriptome profiling experiment is to determine the minimal number of reads required to reasonably capture the profile of a particular sample. This number differs from species to species based on the genome size. For example, a minimum of ten million reads per sample is recommended to profile the transcriptome of the human genome which consists of ~3 billion bp (Liu et al. 2014). However, few hundred thousands of reads are sufficient to sequence for a typical prokaryote species. To ensure that the mapped reads provide sufficient transcriptome coverage for further analysis, their distribution can be plotted using RSeQC (Wang et al. 2012).

4.7.2 Base Calling

When the sequencing of nanoCAGE libraries has been completed, the process known as "base calling" of producing nucleotide sequences from the chromatogram peaks starts. As a result of base calling, the raw data files that contain the sequenced reads are generated, usually as SRA files. Then, in order to proceed further in the data analysis, the SRA files are validated and then extracted into FASTQ files. This is usually performed using the SRA Toolkit.

4.7.3 Quality Control

As the sequenced reads are generated, a quality control to assess the sequenced libraries is highly recommended in order to detect potential problems (e.g., artifacts, contaminations) that may affect subsequent data analysis. An example of such quality control tools is FASTQC (http://www.bioinformatics.bbsrc.ac.uk/projects/fastqc) that we have used in our honey bee data.

4.7.4 Quality and Adapter Trimming

Trimming the sequenced reads based on their quality and removing their 5′ and 3′ adapters is of particular importance in order to avoid potential problems during the mapping process of the sequenced reads to the reference genome. For example, in the study on honey bees, we removed artificial sequences having no or incorrect adapters. Then, we removed the adapter sequence (21 bp) from the first mate of the paired-end sequences and correspondingly trimmed part the second mate, such that both mates had equal lengths (79 bps). Also, sequences can be trimmed by removing low-quality portions, keeping the main part of the reads intact. Trimmomatic (Bolger et al. 2014) and Cutadapt (Martin 2011) are examples of common quality and adapter trimming tools.

4.7.5 Sequence Alignment (Mapping)

After trimming, sequences are mapped to a reference genome using one of the sequence alignment tools. There are numerous alignment tools that can align CAGE reads such as BWA (Li and Durbin 2009), Bowite (Langmead et al. 2009), Bowtie2 (Langmead and Salzberg 2012), and Tophat2 (Kim et al. 2013). In our honey bee nanoCAGE data, the 79 bp paired-end reads obtained after trimming were mapped to the honey bee reference genome (version 4.5) using Bowtie2 v2.1.0 (Langmead and Salzberg 2012) in order to calculate the estimated mean (588 bp) and standard deviation (767 bp) of the inner distance between mapped paired-end reads. Then, we used these estimated values of the mean and the standard deviation along with the CAGE reads as input to Tophat v2.0.8 64 (Kim et al. 2013) sequence alignment tool, allowing for up to two mismatches and two gaps per read.

4.7.6 Post-mapping Filtering

The sequence alignment tools attempt to map all CAGE tags to the reference genome using the user-provided parameters. As this process completes, a post-mapping filtering is required in order to remove mapped reads that were mapped incorrectly. For the CAGE tag filtering process in our data, we filtered out mapped reads that had a low-mapping quality score (MAPQ < 20) corresponding to a probability of (greater than 1 %) of being mapped incorrectly. To ensure the selection of proper alignment output, we excluded paired-end reads when: (1) mates mapped to alternate strands, (2) one mate was unmapped, (3) the mates mapped to different scaffolds, or (4) there was an inner distance greater than (mean + standard deviation) of the estimated inner distance between paired reads.

4.7.7 Read Count Matrix (Gene Expression)

As mentioned earlier, the CAGE method produces one read per RNA molecule, and therefore the transcript identification becomes much easier than with RNA-Seq methods. Consequently, the number of reads directly correlates to the transcript levels in the sample. To generate the read count matrix, the CAGE tags are mapped to the gene set of the species under study. In the case of honey bee, the CAGE tags were mapped to the honey bee gene set, OGSv3.2 (Elsik et al. 2014). A CAGE tag was considered to be associated with a gene if it intersects with the region that covers [−2000 bp upstream of 5′ end of a gene, 3′ end of a gene], but may be restricted by the end of the upstream gene at the same strand. In such cases the tag was considered to be associated if it maps to the region [3′ end of the upstream gene + 1, 3′ end of a gene]. Consequently, it is possible for multiple CAGE tags to be associated with one gene, or one CAGE tag to span two adjacent genes. Note that we include up to 2000 bp upstream of an annotated 5′ end of a coding region in the genome, because many TSSs are still unknown, and those could be located upstream of the annotated 5′ end.

As a result, a gene expression data matrix is generated using the association of tags and genes, where each row represents the expression levels for a gene and each column represents a sample (in our case we have a total of 16 samples, eight are forager bees and other eight samples are nurse bees). For the purpose of using genes that have significant expression for subsequent analysis, we retained those genes that had nonzero expression level in at least two samples of any of the nurse/forager groups and excluding other genes that did not meet this condition. There are some tools that can be used to generate the read count matrix, such as, HTSeq (Anders et al. 2015) and featureCounts (Liao et al. 2014).

4.7.8 Normalization

There are multiple technical effects that may occur during the sample preparation and sequencing process. Such effects need to be corrected or at least reduced before proceeding in the analysis of the gene expression matrix. For this purpose, numerous normalization methods have been proposed in the literature. Most of these methods aim at correcting one or two sources of technical effects. The first is the sequencing bias that leads to different library sizes (different number of reads per library leads to different coverage of the transcripts). Removing the sequencing bias enables between-sample comparison. The second is the within-sample gene-specific effects such as the gene length or GC-content effects. For a survey of common normalization methods, see (Dillies et al. 2013). Because CAGE sequencing produces one read per transcript, a simple normalization method that eliminates the sequencing bias might be sufficient. An example of these methods is the tags per million of reads (TPM), which is the number of CAGE tags divided by the total number of mapped tags, multiplied by 10^6. In our honey bee work, we normalized the gene expression by rescaling the number of tags from each sample to the minimum number of tags across all samples to reduce sequencing bias.

4.7.9 Differentially Expressed Genes

Identifying genes that are differentially expressed between two groups of samples or time points has become popular in a wide variety of applications. Such analysis is important to detect changes between different conditions leading to discovering of biologically relevant genes or even important biomarkers for clinical use. Typical differential analysis of nanoCAGE data represented in the form of an expression matrix starts by excluding genes (or transcripts) that are not (or not significantly) expressed between different conditions. Then, the data is fitted to a model, and a method is derived to identify significant differentially expressed genes across experimental conditions. EdgeR (Robinson et al. 2010), DESeq (Anders and Huber 2010), and DESeq2 (Love et al. 2014) are common tools for differential analysis of expression levels. It is highly recommended to use the raw read count matrix as an input to these tools because such tools apply their own normalization method on the raw count matrix. In our honey bee work, we used EdgeR to identify differentially expressed genes between nurse and forager bees. The raw read count matrix was used as the input to EdgeR, which applies its own normalization using trimmed mean of M-value (TMM) method.

4.7.10 Identification of Transcription Start Sites

Tag-based methods (including CAGE and nanoCAGE) provide a unique and accurate tool to identify TSSs. In our analysis of honey bee nanoCAGE data, we grouped CAGE tags for each sample independently and clustered the 5′ end positions of these tags in small clusters with a maximum width of 50 bp using Paraclu (Frith et al. 2008). Clusters with more than 50 bp in length or having fewer than five tags after rescaling were removed. We also excluded clusters having a maximum density/baseline density ratio of less than 2 (because of low signal strength which is insufficient to represent a real TSS). These CAGE clusters represent potential TSSs of genes. The common TSSs clusters among samples of a particular group may represent a common TSS for that group. The difference in common TSSs of a particular gene between different groups may provide insights on alternative TSS usage between different sample groups or experimental conditions. It is common that multiple TSSs are associated with one gene. The assignment to a given gene can be further confirmed using the 3′ end reads in paired-end reads (CAGEscan). The 3′ end reads should map onto different exons within the same gene as their positions are derived from random priming.

4.7.11 Gene and Sample Clustering Based on Expression

Sometimes it is useful to cluster the gene expression matrix based on genes, samples, or both genes and samples (i.e., biclustering) to detect gene clusters for groups of samples. For our analysis to compare and determine the differences in the brain gene expression levels between honey bee nurses and foragers, we performed two-way unsupervised hierarchical clustering using MATLAB to cluster genes and samples with Ward's method using inner squared distance (minimum variance

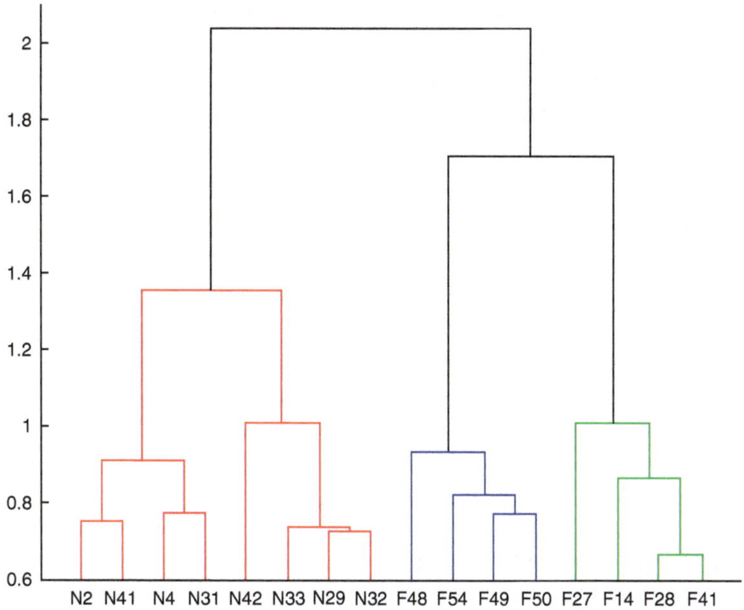

Fig. 4.4 Dendrogram plot of the hierarchical binary clustering tree for the honey bee brain gene expression data. A unique color was assigned to each group of nodes (samples) in the dendrogram whose in-group distances is less than 70 % of the maximum distance in the tree. The height of each joint point represents the distance between the two nodes being connected. The labels on the *x*-axis represent the sample ID for eight foragers (represented by "F") and eight nurses (represented by "N")

algorithm). The Euclidean distance metric and Pearson's correlation coefficient were used to measure the distances between gene profiles (rows) and between sample profiles (columns), respectively. Figure 4.4 shows that foragers and nurses were clustered into two separate groups.

In order to statistically measure how the clustering maintains the actual differences between the clustered samples, an unsupervised evaluation of hierarchical clustering, e.g., using cophenetic correlation coefficient (CPCC), can be performed. The CPCC is defined as:

$$\text{CPCC} = \frac{\sum_{i<j}(x_{ij} - x)(d_{ij} - d)}{\sqrt{\sum_{i<j}(x_{ij} - x)^2 \sum_{i<j}(d_{ij} - d)^2}}$$

where x_{ij} is the Euclidean distance between *i*th and *j*th observation, and d_{ij} is the cophenetic distance, which is the height of the link that joins the two observations in the obtained clustering dendrogram; x and d are the averages of x_{ij} and d_{ij}, respectively. CPCC is the linear correlation coefficient between the observed distances in samples and the obtained cophenetic distances from the clustering. In the hierarchical clustering of honey bee expression data, the CPCC value was 0.78, suggesting that the hierarchical clustering represents the actual differences between the two groups (nurses and foragers).

4.7.12 Variability of Gene Expression

Using the expression matrix, we can measure if samples belonging to a particular group are more variable than samples of the other group(s). This can be achieved by calculating either the per-gene or the per-sample variance between genes/samples in each of the groups. For the honey bee expression data, we evaluated differences in the brain gene expression between individual bees within the nurse and forager groups by calculating the per-gene variance in expression levels between individuals within each group. The variance was calculated on scaled expression data using the z-score, so that the expression values of each gene had a zero mean and standard deviation of 1. To examine if the variation in gene expression between forager samples was significantly different from the variation between nurse samples, we used the Wilcoxon rank-sum test between the two vectors of variances. Finally, we compared the samples using the per-sample biological coefficient of variation (the square root of the dispersion parameter for the 500 most variable genes) and the per-gene squared coefficient of variation (CV2) (the squared ratio of the standard deviation of gene expression across all group samples to the group average gene expression) (see Fig. 4.6).

4.7.13 Functional Annotation of DEGs

The functional characteristics of differentially expressed genes/transcripts between two groups of samples may help to identify key differences in function and/or behavior between these groups. The functional annotation of differentially expressed genes (DEGs) starts by extracting Gene Ontology (GO) terms for the DEGs. If no GO terms are defined for DEGs in your species of interest, the GO terms of orthologous genes from a close species are used instead. Then, there are different methods to study the enriched functions in these GO terms. One method is to study the GO enrichment using Fisher's exact test followed by multiple testing corrections (e.g., false discovery rate or Bonferroni correction). Different tools can be used to study functional annotation such as, DAVID (da Huang et al. 2009), GOrilla (Eden et al. 2009), REVIGO (Supek et al. 2011), and many other tools. In our honey bee work, we identified orthologous genes from *Drosophila melanogaster* for honey bee genes. Then, we assigned the GO terms of the orthologous genes to our DEGs. In the next step, we studied the functional annotation of the DEGs using Fisher's exact test with false discovery rate and using DAVID and GOrilla tools.

4.7.14 Repositories to Upload Data

We recommend uploading both of the raw data (e.g., SRA files or FASTQ data) and processed data (e.g., raw read count matrix or normalized matrix) to NCBI GEO (Barrett et al. 2013). This is important to enable others using the data in their analysis.

4.7.15 Recommended Tools for CAGE Data Analysis

Table 4.3 summarizes some common bioinformatics tools that can be used in the analysis of CAGE data.

4.7.16 Examples for Output of CAGE Data Analysis

The original data from our CAGE analysis studying honey bee workers will be published elsewhere by (Khamis et al. 2015), but we provide here some examples on the information that can be obtained from such experiments. As explained in the analysis pipeline, we mapped reads using Tophat and then associated the mapped reads to the nearest genes. The results given in Table 4.4 show the statistics on the mapped reads, the mapping percentage, and the percentage of mapped reads that could be associated to genes. The data show a high rate for linking mapped reads to genes in the honey bee genome, although the genome is less well annotated as compared to the human or mouse genome.

Clustering gene expression matrix provides insight on the groups of genes that discriminate two groups of samples. We performed hierarchical clustering using Ward's method of the brain gene expression profiles of nurse and forager honey bees. As shown in Fig. 4.4, the two groups of honey bee samples are clearly separated using their gene expression profiles. This indicates different regulation mechanisms that underline the two groups.

The identification of the DEGs is useful to highlight genes that are regulated differently between two groups. In our analysis of the honey bee data, we used EdgeR to identify 1058 DEGs of which 534 were overexpressed in forager group, and 524 were overexpressed in nurse group. The gene expression profiles of these DEGs (Fig. 4.5) show distinct expression patterns of the DEGs between the two groups (nurses and foragers).

Measuring the within-group variability is useful to identify if samples belong to a particular group have more complex regulation mechanisms than other group(s). Such analysis may indicate different regulatory mechanisms within the same group. In our analysis of the per-gene variance between honey bee samples, we found that there was a substantially higher degree of within-group variation in gene expression among foragers than nurses (Fig. 4.6). This finding may reflect the fact that foragers have to respond to a far more diverse set of stimuli and adapt to more variable conditions outside the hive than necessary for the nurses with a limited number of tasks on caring for the offspring.

Table 4.3 Common bioinformatics tools for high-throughput sequencing data analysis

Tool description					Effective results reporting
Software category	Software	Method	Language/ platform	Reference	Main results output/format
Sequence extraction "base calling"	SRA toolkit	Consists of independent modules to convert SRA data to different formats	Executable	http://www.ncbi.nlm.nih.gov/books/NBK158900/	Reads extracted from SRA file into (ABI SOLiD, FASTA, FASTQ, SFF, SAM, and Illumina native) formats
Quality control	RSeQC	Consists of independent modules to assess data quality and sequencing saturation	Python	Wang et al. (2012)	Variant, depending on module. (e.g., saturation curves to evaluate sequencing depth)
	FASTQC	Quality control checks on raw sequence data	Executable	http://www.bioinformatics.bbsrc.ac.uk/projects/fastqc	Summary graphs and tables in HTML report
Quality and adapter trimming	Trimmomatic	Trimming paired-end and single-end sequenced reads	Java	Bolger et al. (2014)	Trimmed reads (FASTQ)
	Cutadapt	Removes adapters and trim sequenced reads	Executable	Martin (2011)	Trimmed reads (FASTQ/FASTA)
Sequence alignment	BWA	Sequence alignment using three algorithms (one for short reads and other two for long reads)	Executable	Li and Durbin (2009)	Aligned reads (SAM)
	Bowtie/Bowtie2	Long-sequence alignment	Executable	Langmead and Salzberg (2012), Langmead et al. (2009)	Aligned reads (SAM)
	Tophat2	Sequence alignment and identify splice junctions between exons	Executable	Kim et al. (2013)	Aligned reads (SAM/BAM)

Software category	Software	Method	Language/ platform	Reference	Main results output/format
Read counts matrix generation	HTSeq	Count mapped reads to each feature (e.g., gene)	Python	Anders et al. (2015)	Read count matrix
	featureCounts		R package	Liao et al. (2014)	
Differentially expressed genes	EdgeR	Negative binomial model (balance estimated dispersion toward the mean)	R package	Robinson et al. (2010)	List of differentially expressed genes
	DESeq/DESeq2	Negative binomial model (takes the maximum estimated dispersion)	R package	Anders and Huber (2010), Love et al. (2014)	
Clustering mapped reads	Paraclu	Cluster closely located TSSs into clusters	Executable	Frith et al. (2008)	List of TSS clusters
Functional annotation	DAVID	Database and tools for gene annotation	Web-based application	da Huang et al. (2009)	Enriched GO terms and functional annotation clustering
	GOrilla	Identify enriched GO terms in ranked gene lists	Web-based application	Eden et al. (2009)	Enriched GO terms and visualized hierarchical structure of the enriched GO terms
	REVIGO	Summarize and visualize GO terms	Web-based application	Supek et al. (2011)	GO terms visualized in semantic similarity-based scatterplots

Table 4.4 The statistics of mapping nanoCAGE library reads to honey bee genome and the statistics of reads which could be associated to genes

Sample	Total number of reads	Mapped reads	Mapping percentage	Percentage of mapped reads associated to genes to the total mapped reads
F14	5,449,888	3,143,064	57.67 %	87.35 %
F27	3,586,235	2,157,798	60.16 %	89.14 %
F28	7,725,895	4,244,582	54.93 %	85.32 %
F41	9,237,207	3,130,408	33.88 %	82.64 %
F48	2,936,239	1,798,240	61.24 %	89.07 %
F49	4,355,366	3,040,200	69.80 %	89.48 %
F50	4,307,121	2,955,469	68.61 %	88.52 %
F54	2,348,738	1,623,527	69.12 %	88.54 %
N2	14,172,924	8,084,168	57.03 %	88.22 %
N4	12,209,666	7,068,767	57.89 %	88.84 %
N29	10,652,219	6,949,056	65.23 %	89.25 %
N31	10,424,626	6,820,778	65.42 %	88.84 %
N32	11,031,651	7,173,714	65.02 %	89.61 %
N33	10,998,419	7,503,984	68.22 %	89.50 %
N41	12,987,132	8,050,612	61.98 %	88.52 %
N42	10,126,459	6,529,766	64.48 %	88.12 %

The first column contains the sample ID for eight foragers (represented by "F") and eight nurses (represented by "N")

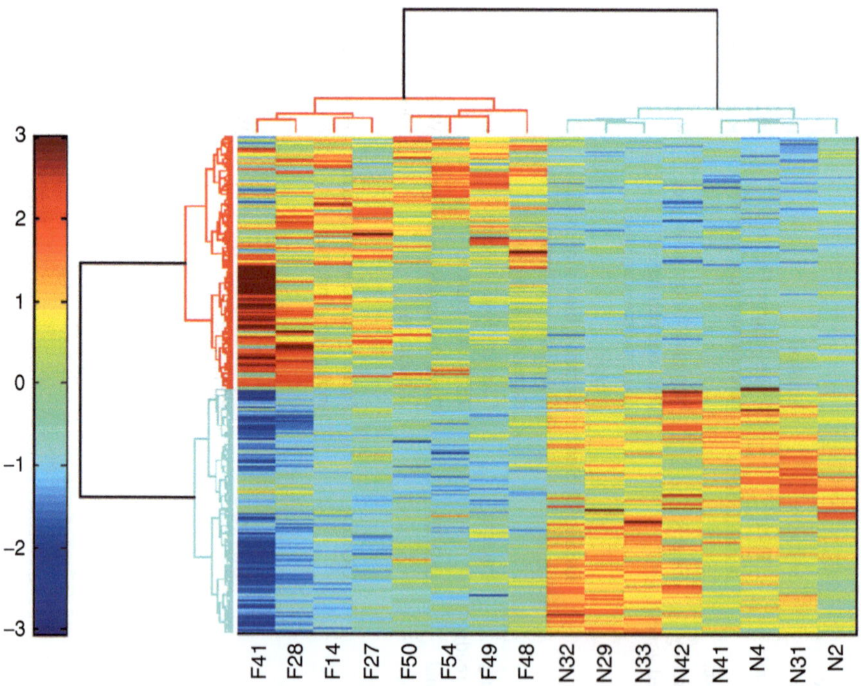

Fig. 4.5 Hierarchical clustering of 1058 DEGs using k-mean algorithm. Rows correspond to 1058 DEGs and columns represent samples. The scale bar indicates the z-scores of gene expression values. Highly expressed genes are shown in *dark red* while genes with low levels of expression are in *dark blue*

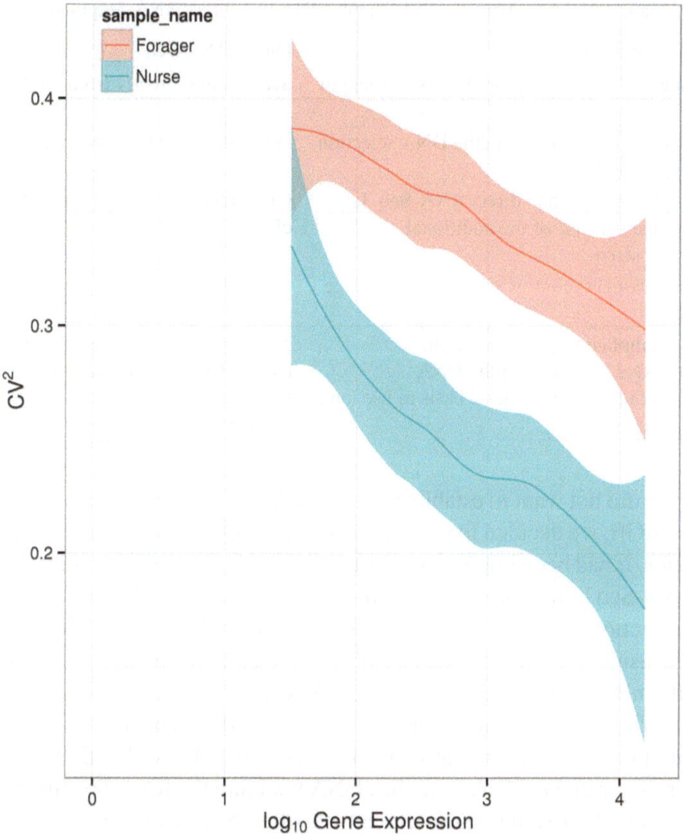

Fig. 4.6 The squared coefficient of variation (CV2) in per-gene expression for foragers and nurses. We used all genes that have at least one tag per million reads in at least two samples, which correspond to the threshold 1.5 in the log10 of per-gene expression. The *x*-axis is the log10 normalized per-gene expression level, and the *y*-axis is the squared coefficient of variance (CV2)

4.8 RNA-Seq Library Preparation and Sequencing

There are many protocols in the literature for the preparation of RNA-Seq libraries. In addition, various providers offer commercial kits for RNA-Seq library preparation. Refer to the recent review from de Klerk and 't Hoen (2015) on RNA sequencing for more information on important criteria for selecting a good method for the preparation of RNA-Seq libraries. In our honey bee work after concluding the CAGE experiment, we pooled RNA samples from nurse and forager bees into two pools, one pool for the preparation of full-length cDNA and the other pool for RNA-Seq directly prepared from RNA. Then, we prepared three libraries from each of these two pools.

Table 4.5 Main differences between full-length cDNA and direct RNA-Seq methods

From full-length cDNA	Direct RNA-Seq
Sequencing from RNAs prepared from full-length cDNA	Sequencing directly from mRNA
rRNA removed during full-length cDNA selection (Cap-Trapper method)	rRNA depleted by rRNA removal kit
Gives lower coverage than direct RNA-Seq. It loses coverage because of the additional experimental steps	Gives higher coverage than full-length cDNA
Gives better (longer) contigs with better coverage of the ends	Gives shorter contigs than full-length cDNA
Provides a template to clone interesting transcripts. Also, the full-length cDNA enrichment could be a very good basis to use long reads, e.g., on PacBio	No template for preparation of cDNA clones. Contig sequences may not be suitable for preparing cDNAs by gene synthesis

Since we did not want to establish a new RNA-Seq protocol in-house to extend our work on CAGE, we decided to use a commercial RNA-Seq kit from Epicenter for our experiments. It had been important to us to select a kit that keeps the strand orientation during RNA-Seq library preparation. We prepared RNA-Seq libraries according to the maker's directions starting from rRNA-depleted total RNA. The rRNA removal was performed using a ScriptSeq mRNA library preparation kit from Epicenter. When using full-length cDNA as a temple for RNA-Seq library preparation, we transcribed RNA transcripts from the full-length cDNA that could then be directly used for RNA-Seq library preparation using the same kit (refer to Fig. 4.1; (Khamis et al. 2015)). To avoid carrying over any DNA from the cDNA pool into the RNA-Seq library, all DNA templates were destroyed by DNase treatment after the RNA synthesis had been completed. In our experiments, the full-length cDNA had always been prepared by the Cap-Trapper method following the basic protocol for cDNA library preparation, but omitting the last steps for cloning the cDNA into a vector. Each of the pooled libraries was sequenced on an Illumina HiSeq2000 sequencer. Table 4.5 summarizes the main differences between the two types of RNA-Seq library preparation methods (with and without full-length cDNA preparation step).

4.9 Bioinformatics Data Analysis of RNA-Seq Data

As mentioned earlier in this chapter, CAGE sequencing yields one read per transcript, whereas RNA-Seq produces multiple random reads per transcript. However, the first few steps of the data analysis related to sequence generation and read count generation are identical for CAGE and RNA-Seq data (Fig. 4.7).

Most of the analysis steps for nanoCAGE (Fig. 4.3) are common with RNA-Seq (Figs. 4.7 and 4.8). This includes extraction of sequences and qualities, sequencing depth assessment, quality control, quality and adapter trimming, and read count

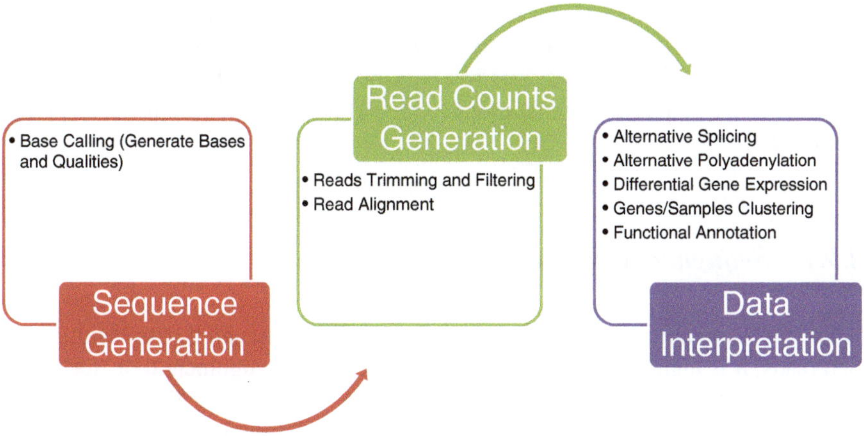

Fig. 4.7 Overview of RNA-Seq data analysis process

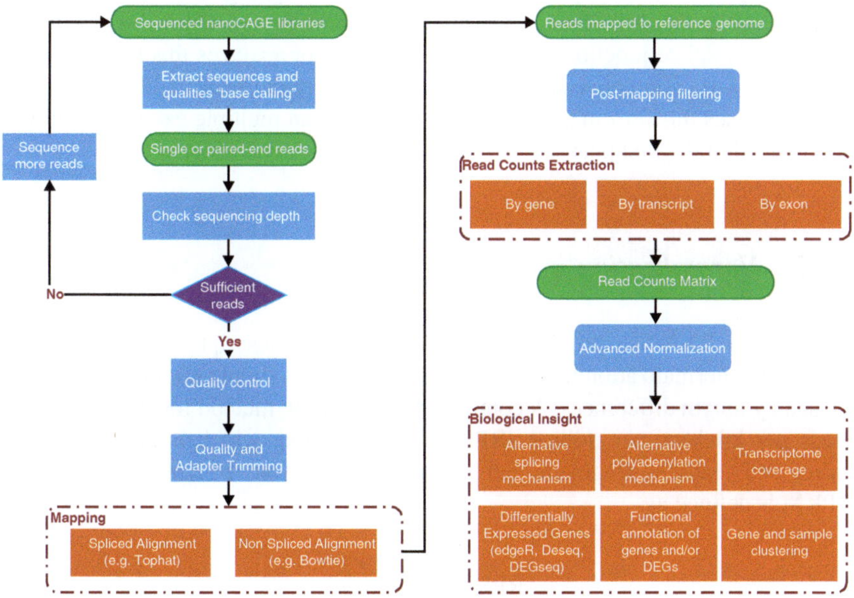

Fig. 4.8 Typical bioinformatics analysis pipeline for RNA-Seq data

extraction. However, there are few main differences. One difference is in the alignment of reads to a reference genome as it is highly recommended to use splice-aware alignment tools. Another difference is the normalization technique to be used with the read count matrix and finally the target analysis to be performed which leads to biological insights on the sequenced samples.

There are good overviews on RNA-Seq data analysis (Yendrek et al. 2012; Auer and Doerge 2010; Anders et al. 2013), and therefore we focus here on those steps we used in the analysis of our data to compare the results obtained by using two different RNA-Seq library preparations (with and without full-length cDNA preparation step).

4.9.1 Sequence Alignment (Mapping)

While the same alignment tools used to align CAGE reads can be used with RNA-Seq reads, it is highly recommended to use splice-aware alignment tools for RNA-Seq. Such tools exploit the advantage of RNA-Seq to discover splice junction sites and exon/intron boundaries and eventually identify transcript isoforms based on this information. Examples of such splice-aware mapping tools include among others Tophat (Kim et al. 2013), SOAPsplice (Huang et al. 2011), and STAR (Dobin et al. 2013). A typical RNA-Seq alignment tool starts by aligning reads to the reference genome and at the same time recording information about exon/intron boundaries and possible splice junction sites. The alignment tool analyzes this information in order to identify alternative splicing events and transcript isoforms. Then, the alignment tool uses this information to map reads that span multiple exons, which were not mapped in the first phase.

4.9.2 Normalization

Because the RNA-Seq provides multiple reads per transcript that have different lengths, the normalization technique for RNA-Seq data should eliminate the effects caused by length differences. A common normalization method for RNA-Seq is the reads per kilobase per million (RPKM) read method, which is also known for paired-end reads as fragments per kilobase per million (FPKM) reads method. RPKM is defined as:

$$\text{RPKM}_{ij} = \frac{10^9 * N_i}{T_j * L_i}$$

where N_i is the number of mapped reads to gene (transcript) i, and T_j refers to the total number of sequences reads for sample j. L_i is the total number of bp (length) of all exons within gene (transcript) i.

4.9.3 RNA Splicing Analysis

To study RNA splicing in a particular sample and to identify the alternative splicing mechanisms, we need to use RNA-Seq to sequence multiple reads that cover the entire length of each transcript. The identification of different splice variants is achieved by using reads comprising splice junctions. Multiple software tools can be used for this purpose, which include, for example, Tophat (Kim et al. 2013) and Cufflinks (Trapnell et al. 2012).

4.9.4 Further Analysis of RNA-Seq Data

Similar to CAGE data, also RNA-Seq data can be used to identify DEGs and to further annotate DEGs, for example, using GO terms. We will not describe those steps in more detail here, because the process is the same for CAGE and RNA-Seq data. Therefore refer to the description of the CAGE data analysis on "Gene and Sample Clustering Based on Expression," "Variability of Gene Expression," and "Functional Annotation of DEGs."

Similar to the data obtained by CAGE, we advise to upload RNA-Seq data in public databases like the NCBI GEO.

4.9.5 Examples for Output of RNA-Seq Data Analysis

As described above we have prepared two pools of RNA for RNA-Seq, one pool for RNA-Seq through full-length cDNA and another one for direct RNA-Seq from mRNA. We prepared the libraries from each pool and sequenced them using Illumina HiSeq2000. The number of sequenced reads is much higher in direct RNA-Seq experiments as compared to full-length cDNA (Table 4.6, column 3), suggesting that RNA-Seq from cDNA loses coverage because of the additional experimental steps; the lower number of reads could go along with lower DNA yields obtained from the libraries. We further compared the read coverage of both types of libraries over the gene body within the honey bee transcriptome in order to check if the coverage of reads is uniform and to examine if there is any 5′ or 3′ bias. Figure 4.9 shows a much better coverage at the 5′ end when using the full-length cDNA, while the coverage toward the 3′ end is more similar for both approaches. This observation suggests that full-length cDNA sequences provide better data spanning of the entire transcript length and generate a more uniform coverage over the gene body. Direct RNA-Seq data may do better at the 3′ end because of the

Table 4.6 Mapping statics using full-length CDNA and direct RNA-Seq

Library type (method)	Sample ID (3 replicas)	Total number of sequenced reads	Total number of mapped reads	Mapping percentage	Total number of clusters	Total clusters + strand	Total clusters − strand
Full-length cDNA	RNASeq1	25,447,398	16,970,573	66.6 %	80,523	41,050	39,473
	RNASeq2	21,838,868	14,685,920	67.2 %	81,431	41,362	40,069
	RNASeq3	26,658,262	18,471,350	69.2 %	72,229	36,472	35,757
Direct RNA-Seq	RNASeq4	61,406,622	37,843,220	61.6 %	125,944	62,820	63,124
	RNASeq5	37,059,810	22,659,028	61.1 %	145,041	72,352	72,689
	RNASeq6	54,376,538	32,719,221	60.1 %	129,258	64,479	64,779

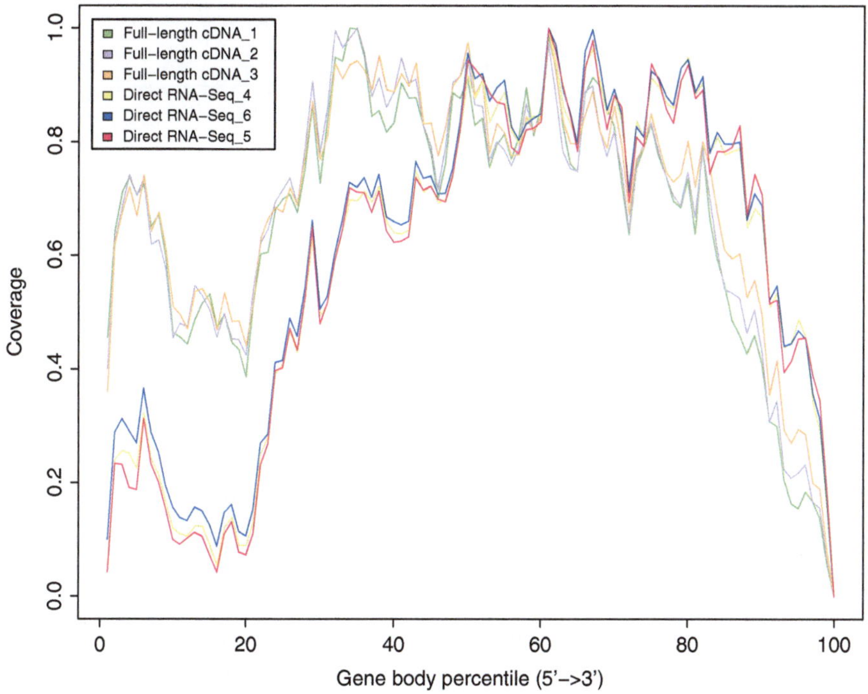

Fig. 4.9 Comparison of gene body coverage between full-length cDNA and direct RNA-Seq

Table 4.7 Association of honey bee OGS3.2 gene set (15,314 genes) with mapped reads generated by different sequencing methods

Type	Number of genes covered by sequenced reads	Percentage (out of the total 15,314 honey bee genes)
nanoCAGE	13,111	85.61 %
Full-length cDNA	13,162	85.94 %
Direct RNA-Seq	13,978	91.27 %

higher sequencing depth obtained in our experiments. Furthermore, in order to see whether the difference in the number of sequenced reads obtained by both types of libraries could explain the difference in the number of genes covered by those reads, we mapped the reads from all libraries to the honey bee genome and clustered the mapped reads into clusters of 50 bp width. While the results in (Table 4.6, columns 6–8) show that the number of clusters obtained from direct RNA-Seq is much more than those obtained from full-length cDNA, we notice that only an about ~5 % difference in the number of genes that could be associated to the sequencing reads in both methods (Table 4.7). This suggests that the high number of clusters were

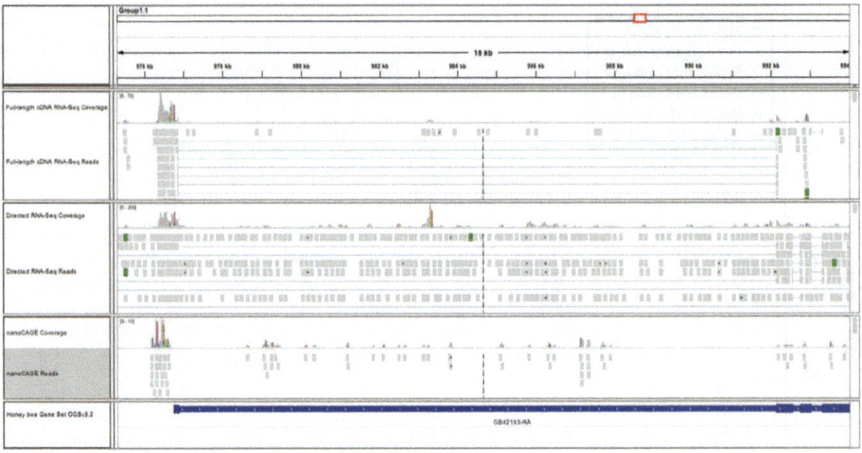

Fig. 4.10 Genome browser snapshot shows comparison in the honey bee genome coverage between nanoCAGE, full-length cDNA, and direct RNA-Seq for gene (GB42183) on scaffold (Group 1.1)

obtained because of the larger number of reads generated by the direct RNA-Seq method. We observed, however, a somewhat higher mapping rate for the reads from the libraries that had included the full-length cDNA preparation step, which could argue for a better library quality, and may compensate in part for the lower overall number of reads.

We associated the mapped reads to the honey bee OGS3.2 gene set (15,314 genes). Table 4.7 supports our previous finding of high coverage of direct RNA-Seq as compared to other sequencing methods.

We have further compared the three types of sequencing methods (nanoCAGE, full-length cDNA, and direct RNA-Seq) by monitoring the distribution of the mapped reads on the genome using IGV genome browser (Thorvaldsdottir et al. 2013). An example in Fig. 4.10 shows for gene GB42183 that the sequenced libraries of direct RNA-Seq provide higher number of small reads distributed over the gene body. However, full-length cDNA provides longer contigs (Fig. 4.10). Also, we notice that nanoCAGE provides information about the TSS positions of transcripts across the transcriptome (so-called exon painting). This had been observed before, where CAGE tags had been found at the beginning of exons. It is unclear whether all those positions represent real TSS.

Overall, while we see a high number of the annotated genes in the honey bee genome that could be covered by reads obtained from the direct use of RNA-Seq libraries (91.27 % for direct RNA-Seq as compared to 85.94 % and 85.61 % in RNA-

Seq from full-length cDNA and nanoCAGE, respectively), we think that the nano-CAGE and full-length cDNA protocols offer additional information on the expressed genes. In particular, full-length cDNA provided uniform and better distributed reads over the entire transcriptome while direct RNA-Seq had a bias toward better coverage of the 3′ end. CAGE is the only option to specifically identify TSS regions in the genome, which are underrepresented in direct RNA-Seq libraries.

4.10 Conclusions

In this chapter, we have summarized our experience on using a tag-based method like CAGE and RNA-Seq shotgun sequencing for transcriptome profiling of honey bee brain samples. The different methods were used to address specific aspects of gene regulation. While CAGE gave us for the first time an overview on TSS and transcriptional regulation, we used the RNA-Seq data to better annotate honey bee transcripts and genome sequences. We used two different library protocols for obtaining RNA-Seq data from RNA pools, where one set of data was obtained from full-length selected cDNAs. While more time-consuming to prepare than standard RNA-Seq libraries, our data show also some benefits for including a full-length cDNA selection step for preparing RNA-Seq libraries. Although much less reads were obtained from those libraries, the reads were more equally distributed over the entire transcripts. Therefore we think that this approach is suitable, where new splice variants should be identified and later characterized by full-length sequencing of individual cDNA fragments isolated from the full-length cDNA pool. Admitting the general value of RNA-Seq methods for transcriptome profiling, we would still suggest the use of CAGE methods for studies on gene regulation with a focus on promoter usage and regulatory networks. Until new sequencing methods reach the market that can obtain full-length RNA sequences at high throughput and low cost, choices will have to be made on the focus of a transcriptome profiling study and the selection of the most suitable approach. We hope this chapter provides some useful information to the readers to plan their own experiments and to consider whether to focus on the importance of splicing or a better understanding of the regulatory principles behind differential gene expression.

Acknowledgment We want to express our great thanks to Adam R. Hamilton, Yulia A. Medvedeva, Tanvir Alam, Intikhab Alam, Magbubah Essack, Boris Umylny, Boris R. Jankovic, Nicholas L. Naeger, Makoto Suzuki, and Gene E. Robinson for their great support for our honey bee project, which would have not been possible without working together with them. We further want to thank Charles Plessy and Piero Carninci for their support and encouragement for using CAGE.

Annex: Quick Reference Guide

Fig. QG4.1 Representation of the wet-lab procedure workflow

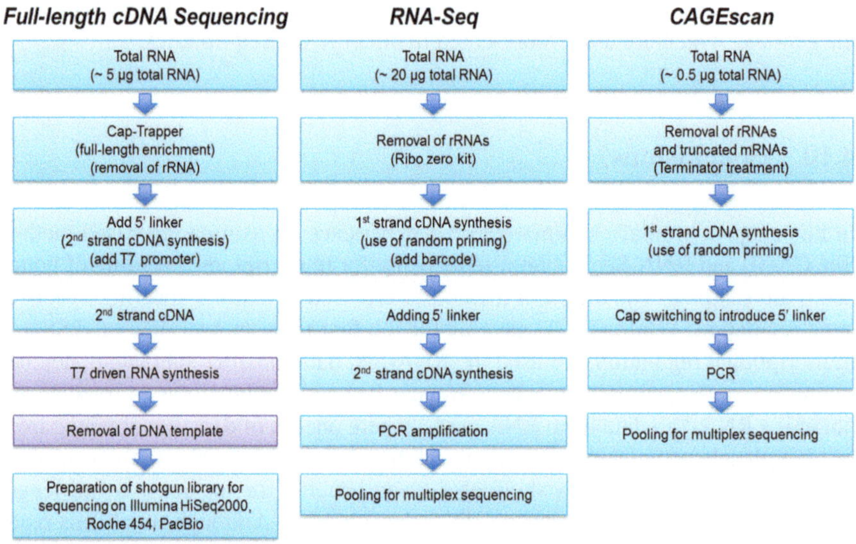

Fig. QG4.2 Main steps of the computational analysis pipeline

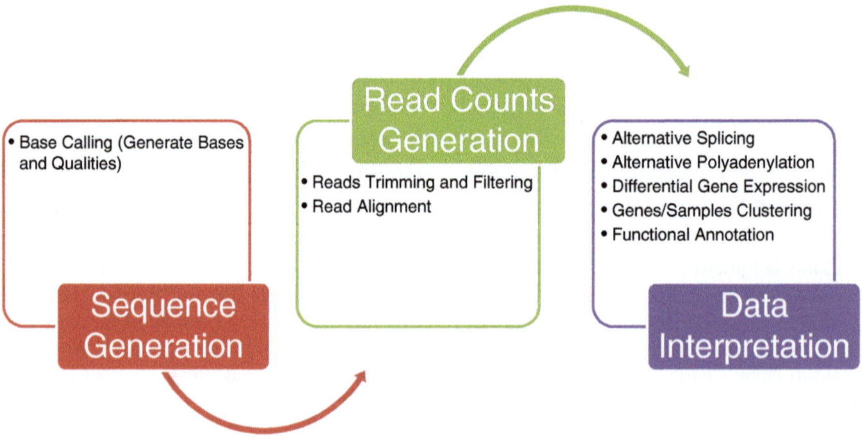

Table QG4.1 Experimental design considerations

Technique	Protocol	Control library	Recommended starting material	Number of replicates	Sequencing depth	Recommended sequencing platform and run
CAGEscan	Cap-Trapper method or template-switching method	Use RNA from a cell line or brain RNA	0.5 μg total RNA when using template-switching methods	3 (minimum per condition)	Requires less sequencing depth than RNA-Seq (Sims et al. 2014)	–Illumina MiSeq or HiSeq 100–200 paired end
						–HeliScope
Full-length cDNA sequencing	Cap-Trapper method		5 μg total RNA		Depends on transcriptome size, e.g., human requires ~10 million reads (Liu et al. 2014)	–Illumina HiSeq 100–200 paired end
						–PacBio RS II
Direct RNA-Seq	Removal of rRNAs		20 μg total RNA			–Illumina HiSeq 100–200 paired end

Table that comprises relevant experimental design parameters to carefully consider before applying this methodology

Table QG4.2 Available software recommendations

Tool description					Effective results reporting
Software category	Software	Method	Language/ platform	Reference	Main results output/format
Sequence extraction "base calling"	SRA Toolkit	Consists of independent modules to convert SRA data to different formats	Executable	http://www.ncbi.nlm.nih.gov/books/NBK158900	Reads extracted from SRA file into (ABI SOLiD, FASTA, FASTQ, SFF, SAM, and Illumina native) formats
Quality control	RSeQC	Consists of independent modules to assess data quality and sequencing saturation	Python	Wang et al. (2012)	Variant, depending on module (e.g., saturation curves to evaluate sequencing depth)
	FASTQC	Quality control checks on raw sequence data	Executable	http://www.bioinformatics.bbsrc.ac.uk/projects/fastqc	Summary graphs and tables in HTML report
Quality and adapter trimming	Trimmomatic	Trimming paired-end and single-ended sequenced reads	Java	Bolger et al. (2014)	Trimmed reads (FASTQ)
	Cutadapt	Removes adapters and trim sequenced reads	Executable	Martin (2011)	Trimmed reads (FASTQ/FASTA)
Sequence alignment	BWA	Sequence alignment using three algorithms (one for short reads and other two for long reads)	Executable	Li and Durbin (2009)	Aligned reads (SAM)
	Bowtie/ Bowtie2	Long-sequence alignment	Executable	Langmead and Salzberg (2012), Langmead et al. (2009)	Aligned reads (SAM)
	Tophat2	Sequence alignment and identify splice junctions between exons	Executable	Kim et al. (2013)	Aligned reads (SAM/BAM)
Read counts matrix generation	HTSeq	Count mapped reads to each feature (e.g., gene)	Python	Anders et al. (2015)	Read count matrix
	featureCounts		R package	Liao et al. (2014)	

Differentially expressed genes	EdgeR	Negative binomial model (balance estimated dispersion toward the mean)	R package	Robinson et al. (2010)	List of differentially expressed genes
	DESeq/ DESeq2	Negative binomial model (takes the maximum estimated dispersion)	R package	Anders and Huber (2010), Love et al. (2014)	
Clustering mapped reads	Paraclu	Cluster closely located TSSs into clusters	Executable	Frith et al. (2008)	List of TSS clusters
Functional annotation	DAVID	Database and tools for gene annotation	Web-based application	da Huang et al. (2009)	Enriched GO terms and functional annotation clustering
	GOrilla	Identify enriched GO terms in ranked gene lists	Web-based application	Eden et al. (2009)	Enriched GO terms and visualized hierarchical structure of the enriched GO terms
	REVIGO	Summarize and visualize GO terms	Web-based application	Supek et al. (2011)	GO terms visualized in semantic similarity-based scatterplots

Table displaying a selection of the recommended software available for the computational analysis of data yielded by this technique

References

Anders S, Huber W (2010) Differential expression analysis for sequence count data. Genome Biol 11(10):R106. doi:10.1186/gb-2010-11-10-r106

Anders S, McCarthy DJ, Chen Y, Okoniewski M, Smyth GK, Huber W, Robinson MD (2013) Count-based differential expression analysis of RNA sequencing data using R and Bioconductor. Nat Protoc 8(9):1765–1786. doi:10.1038/nprot.2013.099

Anders S, Pyl PT, Huber W (2015) HTSeq--a Python framework to work with high-throughput sequencing data. Bioinformatics 31(2):166–169. doi:10.1093/bioinformatics/btu638

Auer PL, Doerge RW (2010) Statistical design and analysis of RNA sequencing data. Genetics 185(2):405–416. doi:10.1534/genetics.110.114983

Barrett T, Wilhite SE, Ledoux P, Evangelista C, Kim IF, Tomashevsky M, Marshall KA, Phillippy KH, Sherman PM, Holko M, Yefanov A, Lee H, Zhang N, Robertson CL, Serova N, Davis S, Soboleva A (2013) NCBI GEO: archive for functional genomics data sets--update. Nucleic Acids Res 41(Database issue):D991–D995. doi:10.1093/nar/gks1193

Bolger AM, Lohse M, Usadel B (2014) Trimmomatic: a flexible trimmer for Illumina sequence data. Bioinformatics 30(15):2114–2120. doi:10.1093/bioinformatics/btu170

Consortium EP (2012) An integrated encyclopedia of DNA elements in the human genome. Nature 489(7414):57–74. doi:10.1038/nature11247

Consortium F, the RP, Clst, Forrest AR, Kawaji H, Rehli M, Baillie JK, de Hoon MJ, Haberle V, Lassmann T, Kulakovskiy IV, Lizio M, Itoh M, Andersson R, Mungall CJ, Meehan TF, Schmeier S, Bertin N, Jorgensen M, Dimont E, Arner E, Schmidl C, Schaefer U, Medvedeva YA, Plessy C, Vitezic M, Severin J, Semple C, Ishizu Y, Young RS, Francescatto M, Alam I, Albanese D, Altschuler GM, Arakawa T, Archer JA, Arner P, Babina M, Rennie S, Balwierz PJ, Beckhouse AG, Pradhan-Bhatt S, Blake JA, Blumenthal A, Bodega B, Bonetti A, Briggs J, Brombacher F, Burroughs AM, Califano A, Cannistraci CV, Carbajo D, Chen Y, Chierici M, Ciani Y, Clevers HC, Dalla E, Davis CA, Detmar M, Diehl AD, Dohi T, Drablos F, Edge AS, Edinger M, Ekwall K, Endoh M, Enomoto H, Fagiolini M, Fairbairn L, Fang H, Farach-Carson MC, Faulkner GJ, Favorov AV, Fisher ME, Frith MC, Fujita R, Fukuda S, Furlanello C, Furino M, Furusawa J, Geijtenbeek TB, Gibson AP, Gingeras T, Goldowitz D, Gough J, Guhl S, Guler R, Gustincich S, Ha TJ, Hamaguchi M, Hara M, Harbers M, Harshbarger J, Hasegawa A, Hasegawa Y, Hashimoto T, Herlyn M, Hitchens KJ, Ho Sui SJ, Hofmann OM, Hoof I, Hori F, Huminiecki L, Iida K, Ikawa T, Jankovic BR, Jia H, Joshi A, Jurman G, Kaczkowski B, Kai C, Kaida K, Kaiho A, Kajiyama K, Kanamori-Katayama M, Kasianov AS, Kasukawa T, Katayama S, Kato S, Kawaguchi S, Kawamoto H, Kawamura YI, Kawashima T, Kempfle JS, Kenna TJ, Kere J, Khachigian LM, Kitamura T, Klinken SP, Knox AJ, Kojima M, Kojima S, Kondo N, Koseki H, Koyasu S, Krampitz S, Kubosaki A, Kwon AT, Laros JF, Lee W, Lennartsson A, Li K, Lilje B, Lipovich L, Mackay-Sim A, Manabe R, Mar JC, Marchand B, Mathelier A, Mejhert N, Meynert A, Mizuno Y, de Lima Morais DA, Morikawa H, Morimoto M, Moro K, Motakis E, Motohashi H, Mummery CL, Murata M, Nagao-Sato S, Nakachi Y, Nakahara F, Nakamura T, Nakamura Y, Nakazato K, van Nimwegen E, Ninomiya N, Nishiyori H, Noma S, Noma S, Noazaki T, Ogishima S, Ohkura N, Ohimiya H, Ohno H, Ohshima M, Okada-Hatakeyama M, Okazaki Y, Orlando V, Ovchinnikov DA, Pain A, Passier R, Patrikakis M, Persson H, Piazza S, Prendergast JG, Rackham OJ, Ramilowski JA, Rashid M, Ravasi T, Rizzu P, Roncador M, Roy S, Rye MB, Saijyo E, Sajantila A, Saka A, Sakaguchi S, Sakai M, Sato H, Savvi S, Saxena A, Schneider C, Schultes EA, Schulze-Tanzil GG, Schwegmann A, Sengstag T, Sheng G, Shimoji H, Shimoni Y, Shin JW, Simon C, Sugiyama D, Sugiyama T, Suzuki M, Suzuki N, Swoboda RK, 't Hoen PA, Tagami M, Takahashi N, Takai J, Tanaka H, Tatsukawa H, Tatum Z, Thompson M, Toyodo H, Toyoda T, Valen E, van de Wetering M, van den Berg LM, Verado R, Vijayan D, Vorontsov IE, Wasserman WW, Watanabe S, Wells CA, Winteringham LN, Wolvetang E, Wood EJ, Yamaguchi Y, Yamamoto M, Yoneda M, Yonekura Y, Yoshida S, Zabierowski SE, Zhang PG, Zhao X, Zucchelli S, Summers KM, Suzuki H, Daub CO, Kawai J, Heutink P, Hide

W, Freeman TC, Lenhard B, Bajic VB, Taylor MS, Makeev VJ, Sandelin A, Hume DA, Carninci P, Hayashizaki Y (2014) A promoter-level mammalian expression atlas. Nature 507(7493):462–470. doi:10.1038/nature13182

da Huang W, Sherman BT, Lempicki RA (2009) Systematic and integrative analysis of large gene lists using DAVID bioinformatics resources. Nat Protoc 4(1):44–57. doi:10.1038/nprot.2008.211

de Klerk E, 't Hoen PA (2015) Alternative mRNA transcription, processing, and translation: insights from RNA sequencing. Trends Genet 31(3):128–139. doi:10.1016/j.tig.2015.01.001

de Klerk E, den Dunnen JT, 't Hoen PA (2014) RNA sequencing: from tag-based profiling to resolving complete transcript structure. Cell Mol Life Sci 71(18):3537–3551. doi:10.1007/s00018-014-1637-9

Dillies MA, Rau A, Aubert J, Hennequet-Antier C, Jeanmougin M, Servant N, Keime C, Marot G, Castel D, Estelle J, Guernec G, Jagla B, Jouneau L, Laloe D, Le Gall C, Schaeffer B, Le Crom S, Guedj M, Jaffrezic F, French StatOmique C (2013) A comprehensive evaluation of normalization methods for Illumina high-throughput RNA sequencing data analysis. Brief Bioinform 14(6):671–683. doi:10.1093/bib/bbs046

Dobin A, Davis CA, Schlesinger F, Drenkow J, Zaleski C, Jha S, Batut P, Chaisson M, Gingeras TR (2013) STAR: ultrafast universal RNA-seq aligner. Bioinformatics 29(1):15–21. doi:10.1093/bioinformatics/bts635

Eden E, Navon R, Steinfeld I, Lipson D, Yakhini Z (2009) GOrilla: a tool for discovery and visualization of enriched GO terms in ranked gene lists. BMC Bioinformatics 10:48. doi:10.1186/1471-2105-10-48

Elsik CG, Worley KC, Bennett AK, Beye M, Camara F, Childers CP, de Graaf DC, Debyser G, Deng J, Devreese B, Elhaik E, Evans JD, Foster LJ, Graur D, Guigo R, HGSC production teams, Hoff KJ, Holder ME, Hudson ME, Hunt GJ, Jiang H, Joshi V, Khetani RS, Kosarev P, Kovar CL, Ma J, Maleszka R, Moritz RF, Munoz-Torres MC, Murphy TD, Muzny DM, Newsham IF, Reese JT, Robertson HM, Robinson GE, Rueppell O, Solovyev V, Stanke M, Stolle E, Tsuruda JM, Vaerenbergh MV, Waterhouse RM, Weaver DB, Whitfield CW, Wu Y, Zdobnov EM, Zhang L, Zhu D, Gibbs RA, Honey Bee Genome Sequencing C (2014) Finding the missing honey bee genes: lessons learned from a genome upgrade. BMC Genomics 15:86. doi:10.1186/1471-2164-15-86

Frith MC, Valen E, Krogh A, Hayashizaki Y, Carninci P, Sandelin A (2008) A code for transcription initiation in mammalian genomes. Genome Res 18(1):1–12. doi:10.1101/gr.6831208

Harbers M (2008) The current status of cDNA cloning. Genomics 91(3):232–242. doi:10.1016/j.ygeno.2007.11.004

Huang S, Zhang J, Li R, Zhang W, He Z, Lam TW, Peng Z, Yiu SM (2011) SOAPsplice: genome-wide ab initio detection of splice junctions from RNA-Seq data. Front Genet 2:46. doi:10.3389/fgene.2011.00046

Kawaji H, Lizio M, Itoh M, Kanamori-Katayama M, Kaiho A, Nishiyori-Sueki H, Shin JW, Kojima-Ishiyama M, Kawano M, Murata M, Ninomiya-Fukuda N, Ishikawa-Kato S, Nagao-Sato S, Noma S, Hayashizaki Y, Forrest AR, Carninci P, Consortium F (2014) Comparison of CAGE and RNA-seq transcriptome profiling using clonally amplified and single-molecule next-generation sequencing. Genome Res 24(4):708–717. doi:10.1101/gr.156232.113

Khamis AM, Hamilton AR, Medvedeva YA, Alam T, Alam I, Essack M, Umylny B, Jankovic BR, Naeger NL, Suzuki M, Harbers M, Robinson GE, Bajic VB (2015) Insights into the transcriptional architecture of behavioral plasticity in the honey bee Apis mellifera. Sci Rep 5:11136

Kim D, Pertea G, Trapnell C, Pimentel H, Kelley R, Salzberg SL (2013) TopHat2: accurate alignment of transcriptomes in the presence of insertions, deletions and gene fusions. Genome Biol 14(4):R36. doi:10.1186/gb-2013-14-4-r36

Langmead B, Salzberg SL (2012) Fast gapped-read alignment with Bowtie 2. Nat Methods 9(4):357–359. doi:10.1038/nmeth.1923

Langmead B, Trapnell C, Pop M, Salzberg SL (2009) Ultrafast and memory-efficient alignment of short DNA sequences to the human genome. Genome Biol 10(3):R25. doi:10.1186/gb-2009-10-3-r25

Li H, Durbin R (2009) Fast and accurate short read alignment with Burrows-Wheeler transform. Bioinformatics 25(14):1754–1760. doi:10.1093/bioinformatics/btp324

Liao Y, Smyth GK, Shi W (2014) featureCounts: an efficient general purpose program for assigning sequence reads to genomic features. Bioinformatics 30(7):923–930. doi:10.1093/bioinformatics/btt656

Liu Y, Zhou J, White KP (2014) RNA-seq differential expression studies: more sequence or more replication? Bioinformatics 30(3):301–304. doi:10.1093/bioinformatics/btt688

Lizio M, Harshbarger J, Shimoji H, Severin J, Kasukawa T, Sahin S, Abugessaisa I, Fukuda S, Hori F, Ishikawa-Kato S, Mungall CJ, Arner E, Baillie JK, Bertin N, Bono H, de Hoon M, Diehl AD, Dimont E, Freeman TC, Fujieda K, Hide W, Kaliyaperumal R, Katayama T, Lassmann T, Meehan TF, Nishikata K, Ono H, Rehli M, Sandelin A, Schultes EA, 't Hoen PA, Tatum Z, Thompson M, Toyoda T, Wright DW, Daub CO, Itoh M, Carninci P, Hayashizaki Y, Forrest AR, Kawaji H, Consortium F (2015) Gateways to the FANTOM5 promoter level mammalian expression atlas. Genome Biol 16:22. doi:10.1186/s13059-014-0560-6

Love MI, Huber W, Anders S (2014) Moderated estimation of fold change and dispersion for RNA-seq data with DESeq2. Genome Biol 15(12):550. doi:10.1186/s13059-014-0550-8

Martin M (2011) Cutadapt removes adapter sequences from high-throughput sequencing reads. EMBnet J 17(1):10–12, doi: 10.14806/ej.17.1.200

Murata M, Nishiyori-Sueki H, Kojima-Ishiyama M, Carninci P, Hayashizaki Y, Itoh M (2014) Detecting expressed genes using CAGE. Methods Mol Biol 1164:67–85. doi:10.1007/978-1-4939-0805-9_7

Plessy C, Bertin N, Takahashi H, Simone R, Salimullah M, Lassmann T, Vitezic M, Severin J, Olivarius S, Lazarevic D, Hornig N, Orlando V, Bell I, Gao H, Dumais J, Kapranov P, Wang H, Davis CA, Gingeras TR, Kawai J, Daub CO, Hayashizaki Y, Gustincich S, Carninci P (2010) Linking promoters to functional transcripts in small samples with nanoCAGE and CAGEscan. Nat Methods 7(7):528–534. doi:10.1038/nmeth.1470

Robinson MD, McCarthy DJ, Smyth GK (2010) edgeR: a Bioconductor package for differential expression analysis of digital gene expression data. Bioinformatics 26(1):139–140. doi:10.1093/bioinformatics/btp616

Salimullah M, Sakai M, Plessy C, Carninci P (2011) NanoCAGE: a high-resolution technique to discover and interrogate cell transcriptomes. Cold Spring Harbor protocols 2011(1):pdb prot5559. doi: 10.1101/pdb.prot5559

Sims D, Sudbery I, Ilott NE, Heger A, Ponting CP (2014) Sequencing depth and coverage: key considerations in genomic analyses. Nat Rev Genet 15(2):121–132. doi:10.1038/nrg3642

Supek F, Bosnjak M, Skunca N, Smuc T (2011) REVIGO summarizes and visualizes long lists of gene ontology terms. PLoS One 6(7):e21800. doi:10.1371/journal.pone.0021800

Takahashi H, Kato S, Murata M, Carninci P (2012a) CAGE (cap analysis of gene expression): a protocol for the detection of promoter and transcriptional networks. Methods Mol Biol 786:181–200. doi:10.1007/978-1-61779-292-2_11

Takahashi H, Lassmann T, Murata M, Carninci P (2012b) 5′ end-centered expression profiling using cap-analysis gene expression and next-generation sequencing. Nat Protoc 7(3):542–561. doi:10.1038/nprot.2012.005

Thorvaldsdottir H, Robinson JT, Mesirov JP (2013) Integrative Genomics Viewer (IGV): high-performance genomics data visualization and exploration. Brief Bioinform 14(2):178–192. doi:10.1093/bib/bbs017

Trapnell C, Roberts A, Goff L, Pertea G, Kim D, Kelley DR, Pimentel H, Salzberg SL, Rinn JL, Pachter L (2012) Differential gene and transcript expression analysis of RNA-seq experiments with TopHat and Cufflinks. Nat Protoc 7(3):562–578. doi:10.1038/nprot.2012.016

Wang L, Wang S, Li W (2012) RSeQC: quality control of RNA-seq experiments. Bioinformatics 28(16):2184–2185. doi:10.1093/bioinformatics/bts356

Yendrek CR, Ainsworth EA, Thimmapuram J (2012) The bench scientist's guide to statistical analysis of RNA-Seq data. BMC Res Notes 5:506. doi:10.1186/1756-0500-5-506

Chapter 5
Differential mRNA Alternative Splicing

Albert Lahat and Sushma Nagaraja Grellscheid

5.1 Introduction

Over 90 % of human genes have been reported to be alternatively spliced to produce more than one mRNA isoform from the same gene (Cooper et al. 2009; Pan et al. 2008; Wang et al. 2008). While a proportion of alternative splicing events may be explained by stochasticity in the regulation of individual splicing events, there are many examples of tissue or developmental stage-specific switches in isoform expression as a result of tightly regulated alternative splicing.

When using RNA-seq to measure gene expression, reads mapped within a gene loci directly count towards the gene-level transcript abundance. Most sequencing technologies currently in widespread use yield short reads of less than 250 nucleotides in length making it impossible to unambiguously assign reads to specific isoforms, with the possible exception of some exon-exon junction reads that may be unique. Therefore, when measuring relative transcript-level abundance, read mapping is inferred indirectly, potentially leading to biases in the calculated outcome. Two possible approaches are firstly, to give greater weightage to exon-exon junction reads, which requires a greater sequencing depth compared to gene-level expression analysis.

Secondly, recent approaches developed for full-length cDNA sequencing avoid this problem altogether (Tilgner et al. 2015; Treutlein et al. 2014). However, these approaches are still either more expensive, requiring specialised equipment for library preparation, or have low throughput of the order of just 25,000 transcripts which is at least one order of magnitude less than the total number of expressed transcripts per eukaryotic cell.

A. Lahat, B.Sc. • S.N. Grellscheid, Ph.D. (✉)
School of Biological and Biomedical Sciences, Durham University,
Mountjoy Science Site, Durham DH1 3LE, UK
e-mail: albert.lahat@durham.ac.uk; s.n.grellscheid@durham.ac.uk; sushma@cantab.net

A.M. Aransay, J.L. Lavín Trueba (eds.), *Field Guidelines for Genetic Experimental Designs in High-Throughput Sequencing*,
DOI 10.1007/978-3-319-31350-4_5

This review will focus on approaches available to analyse short-read sequencing datasets. These datasets and approaches to generate them have become widely available and used by numerous laboratories to interrogate gene-level transcript abundance, but existing datasets can frequently be further analysed to infer differential alternative splicing as well. Therefore, the wet-lab methods to generate short-read data for alternative splicing analysis are very similar to standard RNA-seq library preparation approaches (see Chap. 4 for detailed protocols and recommendations). A recent study showed that most methods able to infer both gene-level and transcript-level abundance require 1–3 million reads for accurate quantification of gene-level expression and 10–30 million reads per sample for accurate inference of isoform level transcript abundance (Kanitz et al. 2015).

5.2 Steps in Data Analysis

A few basic steps are common between data processing for gene-level differential expression analysis and for the analysis for alternative splicing. Quality control processes including read trimming to process FASTQ can be carried out according to conventional pipelines. But even though there are many similarities in the data processing required for splicing analysis, there are important differences:

(a) Spliced transcripts don't directly align to a reference genome when mapping as they have components of at least two different noncontinuous loci.
(b) Reads corresponding to different alternatively spliced isoforms often have common exons, causing ambiguous read assignments.
(c) Different types of alternative splicing events have to be modelled.

The early steps are frequently variants to standard RNA-seq analysis such as mapping, assembling and transcript counting.

5.2.1 Read Mapping

Mapping involves uniquely aligning reads onto a reference genome or transcriptome. Mapping outputs are SAM (sequence Alignment/Map) or BAM (binary SAM) files. Spliced exon-exon junction reads do not align directly to the genome due to the read having parts corresponding to two separated loci. Mapping onto a reference transcriptome instead would not allow for de novo splice site discovery. Moreover, aligners to transcriptomes must be able to handle the redundancy resulting from reads mapping to several transcripts of the same gene. Since 2010, several splice-sensitive aligners have been developed such as SpliceMap (Au et al. 2010), HMMsplicer (Dimon et al. 2010), SplitSeek (Ameur et al. 2010), Supersplat (Bryant et al. 2010), MapSplice (Wang et al. 2010), MATS (Shen et al. 2012), TopHat (Trapnell et al. 2009), GSNAP (Wu and Nacu 2010) and STAR (Dobin et al. 2013). This is not an exhaustive list by any means. Alamancos et al. 2014, table 1, list the various splice-sensitive aligners together with a brief summary of attributes such as whether or not they require/use annotation information, can support paired end data, etc.

The relative performance of the various methods has been benchmarked in several recent articles (Engström et al. 2013; Grant et al. 2011; Lindner and Friedel 2012). The benchmarking studies above showed that GSNAP, MapSplice and STAR compared favourably to the other alignment methods. However, every method has pros and cons and must be chosen carefully based on the desired outcome. For example, GSNAP and STAR are highly sensitive, but output a high proportion of false junctions, and require a subsequent step to filter out junctions by the number of supporting alignments. GSNAP and MapSplice require considerable computing time, whereas TopHat2 and STAR were reported to be faster by 3 times and 180 times, respectively (Dobin et al. 2013). Several aligners such as GEM, MapSplice and TopHat carry out splice junction discovery annotation as part of the alignment, while others such as STAR and GSNAP depend on existing annotation, though TopHat output is also improved with annotation. Using genomic annotation, information is clearly beneficial to the latter group of algorithms, but this may not always be available for non-model organisms.

A recently developed method, Sailfish (Patro et al. 2014), uses k-mer statistics on transcripts and disposes of the time-consuming and memory-demanding step of alignment altogether.

The provision of effective mapping strategies for splicing analysis remains an area of much ongoing development. The main future challenges are to address the issue of correct assignment of multi-mapped reads, increasing specificity of exon junctions reported and finally adapting algorithms to accommodate the longer read methods that are beginning to emerge involving higher error rates and multiple exon junctions.

5.2.2 Read Assembly

Instead of mapping all reads onto a reference genome, reads can be overlapped onto each other thus assembling the original transcript. An assembler takes as input reads from a sequencer and attempts to reassemble the original transcript by merging overlapping reads (reads whose ends align to each other). A considerable advantage of this approach is the analysis of sequences from organisms without a good reference genome. Two recent benchmarking studies evaluated the various available algorithms (Li et al. 2014; Steijger et al. 2013; Zhao et al. 2011). There are a variety of assemblers designed to handle alternative splicing events such as Trinity (Grabherr et al. 2011), SOAPdenovo-Trans (Luo et al. 2012) and Trans-ABySS (Robertson et al. 2010). The development of splicing sensitive assemblers is still very much an area of active research. Another class of tools such as PIntron (Bonizzoni et al. 2015) utilises sequencing data together with available EST data to improve exon-intron structure annotation and can be useful for improving annotation files for organisms with good genomic annotation but poor isoform annotation, such as the rat genome.

De novo assembly is a memory-intensive process, requiring either generous memory availability (circa 256–512 G) or multiple nodes to run, and most available software runs exclusively on Linux operating systems. Zhao et al. 2011 showed that in order to assemble 13 billion reads, Trinity used a peak of 57 Gb of RAM and required 150 h. RAM usage and process time increased linearly with increased reads. SOAPdenovo-Trans was more efficient, consuming a peak of 20 Gb for 13 billion reads in 1.5 h. Trans-ABySS being a parallelised assembler running simultaneously on multiple nodes showed the lowest usage of peak RAM (8.2 Gb) during 4 h for assembling 13 billion reads. In all cases, both peak RAM and time increased linearly with read number. Thus, the choice of assembler may depend on the type of high performance computing infrastructure available and the quality of the reference genome annotation. Alamancos et al. 2014, table 6, is a useful compilation of methods for de novo transcriptome assembly, with an indication of whether they are also able to carry out isoform quantification at the same time.

5.2.3 *Isoform Quantification*

In order to analyse and interpret mapped data, a menagerie of analysis tools are available. Most of them employ different models to normalise and quantify the number of reads for each exon/transcript isoform in each sample or sample replicates and compare differential gene expression. Most of these tools take as input BAM/SAM files from a mapping tool. As this is a rapidly evolving field, there are rather few benchmarking studies systematically comparing the commonly used tools. Chandramohan et al. 2013 and Liu et al. 2014 have carried out a limited evaluation of isoform quantification methods and found HTSeq and MATS to perform best, while Kanitz et al. 2015 present a more in-depth comparative benchmarking analysis of several algorithms such as BitSeq, Cufflinks, RSEM and Sailfish among others for determining isoform abundance, but unfortunately do not include the popularly employed HTSeq/DEXSeq or MISO methods used for differential isoform quantification between two biological samples. The mathematical basis of several methods is briefly summarised in the study by Kanitz et al. 2015 and more extensively compared in Pachter 2011. Some of these analysis tools include multivariate statistical comparisons, GO enrichment and other functions, which might be useful depending on the analysis required.

We briefly summarise a few of the many methods available, to enable the reader to choose a suitable analysis tool based on their requirements, data handling expertise and desired output format.

DEXSeq (Anders et al. 2012) takes input of read count data from HTSeq (Anders et al. 2015) and is a widely used method, which is very well annotated with easy to follow vignettes requiring very basic understanding of R to use. It is similar to DESeq2 and uses the same negative binomial distribution model to compare samples but treats exons or smaller variants (such as those defined by alternative 3′ or 5′ splice sites) as units instead of genes. DEXSeq counts the expression of each exon, or smaller variant unit per sample, and normalises this value by the size of the library. When comparing samples it returns the *p*-value, *p*-adjusted value, base means and log2fold changes for each exon. DEXSeq can also generate a fitted expression plot of a gene and its exon usage.

Multivariate Analysis of Transcript Splicing (**MATS**) (Shen et al. 2012) is a command line tool and is another general-purpose analysis tool to study splicing. It can take as an input BAM/SAM files or FASTQ files and align reads directly. It utilises Bayesian statistics to compare splicing between samples and can detect and categorise common splicing events (Skipped exon, alternative 5′ splice site, alternative 3′ splice site, mutually exclusive exons and retained intron).

SplicingCompass (Aschoff et al. 2013) is an R module to quantify changes in isoform abundance regardless of expression-level changes by plotting a gene as an *n* dimensional vector of read counts where *n* is the number of exons. This form of data interpretation enables simple geometry in order to predict alternative isoform expression. It is an R module and it requires R skills to be used. This module can plot normalised exon abundance and normalised junction reads for a given gene.

DiffSplice (Hu et al. 2013) is an analysis tool from the same group that made and maintains MapSplice. DiffSplice is a tool designed to detect ab initio alternative spliced modules (part of genes that are differentially spliced). It produces differentially expressed exon and alternative spliced module tables filtered by desired significance. The significance is calculated using a nonparametric test. It also produces GTF files of the alternative spliced modules and of the isoforms found. Those GTF files can be visualised in genome viewers or used for further analysis. DiffSplice does not require annotation files and takes SAM files as input. This tool can find novel splicing events and splicing categories and is useful for de novo discovery of splicing events and if needed, is one of the few tools capable of producing de novo GTF files.

SpliceR (Vitting-Seerup et al. 2014) is an R bioconductor tool to analyse mapped data. It is designed to work with Cufflinks to find and categorise alternative splicing events (single exon skipping exclusion/inclusion, multiple exon exclusion/inclusion, intron retention/inclusion, alternative 3′/5′ splice sites, alternative transcription start/end site, mutually exclusive exons). It is also capable of discovering de novo splice sites as it does not rely on annotation files to find exons. SpliceR can then generate annotation GTF files, which can be used as input for many other analysis pipelines and also visualised into genome browsers. Splicer can also visualise Venn diagrams comparing splicing events between different samples.

AltAnalyze is a multipurpose tool (Emig et al. 2010). It can be used with microarrays as well as RNA-seq data. It has A GUI which can be used locally or through a server. AltAnalyze provides expression clustering, gene enrichment analysis, pathway visualisation, network analysis and visualisation, alternative exon visualisation, sample classification, Venn diagram creation and ANOVA. It also has full command line usage, so, this program can be automatised or streamlined if needed. This tool is reasonably easy to use and is convenient for multipurpose analysis; it can take as input bed files or microarray files from many suppliers.

CuffDiff is part of the Tuxedo suite which analyses mapped data (SAM or BAM) and compares changes in expression, alternative splicing and promoter use between different treatments and replicates (Trapnell et al. 2012). CuffDiff outputs differential expression files of isoforms, genes, coding sequences and primary transcript in FPKM (fragment per kilobase per million fragments mapped), raw counts and differential expression tests. This test can only be made when comparing two samples (with replicates for each). Like other Tuxedo tools, CuffDiff can be used through usegalaxy.org without command line usage. CuffDiff can use four models; pooled model, a precondition where all condition get modelled independently (needs replicates for each condition), a blind model where all samples are treated as if they were the same condition and a Poisson model. CuffDiff performed rather poorly in the benchmarking study by Kanitz et al. 2015.

Mixture of isoform (**MISO**) (Katz et al. 2010) uses Markov Chain Monte Carlo models to estimate the expression of each isoform. MISO can compare between multiple samples and derive the significance of isoform expression changes. It requires a GFF3 annotation file (a GTF file can be easily converted to a GFF3 file) and cannot detect novel isoforms. It takes as input-sorted and indexed BAM files. It outputs a summary table with one or more samples which can be filtered by MISO itself. MISO can also produce sashimi plots which can also be generated using IGV genome visualiser tool (Robinson et al. 2011).

SwitchSeq (Gonzalez-Porta and Brazma 2014). Considering that 85 % of protein-coding transcripts belong to dominant transcript isoforms (Gonzalez-Porta et al. 2013), which indicates that even though there are many isoforms present of a given gene, only one is dominant. SwitchSeq finds the dominant isoform and exclusively reports on changes in splicing on the dominant isoform (regardless of the gene expression changes). SwitchSeq takes as input normalised counts and outputs an HTML file summarising the results, a table reporting on the switch events found, a list of events discarded if they were not present in the annotation file, distribution plots for events found and star plots for events found.

SplicePlot (Wu et al. 2014) is a command line tool that can analyse mapped data (SAM or BAM files) in the focus of genomic variability. It takes as input BAM files of samples and VCF (variant call format) files, a file that contains information about genetic sequence variation. It can produce plots designed to study genomic variation and splicing such as sashimi plots, hive plots and structure plots. This tool might be useful to visualise the effect of genomic background variation on splicing.

Numerous tools have not been discussed here due to limited space, such as Alt Event Finder (Zhou et al. 2012), ASprofile (Florea et al. 2013), AStalavista (Foissac and Sammeth 2007) and SpliceTrap (Wu et al. 2011) which are similar to others above. SUPPA (Alamancos et al. 2015) and Vast-tools (Langmead et al. 2009) are toolsets for profiling alternative splicing events in RNA-seq data. DSGseq (Wang et al. 2013) compares relative abundance of isoforms between samples, and Spanki (Sturgill et al. 2013) is a flexible tool to analyse alternative splicing events. RSVP (Majoros et al. 2014) is a software package for prediction of alternative isoforms of protein-coding genes, based on both genomic DNA evidence and aligned RNA-seq reads. SAJR (Mazin et al. 2013) calculates the number of the reads that confirm a segment inclusion or exclusion and, then, model these counts by GLM with quasi-binomial distribution to account for biological variability.

5.3 Visualising Alternative Splicing

It is often useful to explore RNA-seq data without needing to generate and analyse numerous plots or tables for each gene. Visualising tools enable to graphically, intuitively and interactively visualise and navigate the genome. Some of them are included in the analysis package and can be used as command line tools (RSEM, SpliceGrapher, DiffSplice, DEXSeq, SplicingCompass) and were covered in the previous section. Others are graphical user interfaces (IGV, IGB, Savant, SpliceSeq). Integrative Genome Viewer (Robinson et al. 2011) from the Broad Institute is a widely used, user-friendly general-purpose genome viewer. Besides interactive navigation of the genome, IGV can also generate sashimi plots for alternative splicing, showing the number of reads at individual exon-exon junctions, with a few mouse clicks. IGV requires sufficient RAM and at least 10 G RAM is recommended. Another popular general-purpose viewer for microarray and RNA-seq data is the Integrated Genome Browser (Nicol et al. 2009). Support to handle splicing was added in 2014 (Gulledge et al. 2014). It can handle a vast variety of file formats and has several plugins available for additional features and file formats. SpliceSEQ (Ryan et al. 2012) is a user-friendly program allowing graphical visualisation of splicing events. SpliceSEQ also categorises reads depending on splicing event type (exon skipping, alternate donor/acceptor site, retained intron, etc.).

5.4 Conclusion

RNA sequencing is now widely used as a method for global expression profiling, due to its large dynamic range, robust reproducibility and most importantly, its ability to detect transcript isoforms. While this technique and its applications have wide impact on understanding gene regulation in health and disease, this is still an area of active development for new tools and benchmarking of existing pipelines for wet-lab as well as analytical methods.

Especially in the case of determination of isoform abundance and relative expression of alternatively spliced isoforms, every step in the pipeline needs to be splicing sensitive and take into account the transcriptomic as well as the genomic view of gene expression. Not unexpectedly, methods developed for calculating isoform-level expression that aggregates the abundances of isoforms are also more accurate estimators of gene-level expression.

There are a variety of tools available, and encouragingly, most of the methods available to estimate transcript isoform abundance produce comparable and reproducible results (Kanitz et al. 2015). In general, isoforms from high-abundance transcripts are more accurately quantified, and as expected this accuracy increases with read depth, up to about 30 million reads after which there is limited improvement in prediction of the presence of transcripts. Higher read depth is likely to continue to have an impact in differential expression of alternatively spliced isoforms, but a large systematic benchmarking study is still lacking. In addition to higher read depth of short-read sequencing output, recent technologies such as from Pacific Biosciences enable full-length transcript readouts and will address many issues surrounding mapping. However, these are still relatively low throughput but expected to improve in the near future. Nevertheless, short-read sequencing methods that have become increasingly cost-effective are highly suitable for isoform abundance and differential alternative splicing analysis.

Thus the main factors influencing choice of analysis method may depend on the availability of computational resources and the researcher level of expertise using informatics. Sailfish is extremely fast and memory efficient as it uses a mapping-free approach. TIGAR2 (Nariai et al. 2014), which is highly accurate, has high memory requirements and takes longer to run. Some tools such as AltAnalyze, DEXSeq and CuffDiff, among others, are easy to use with limited programming knowledge. Newer methods operating on the transcript rather than genomic reference sequence appear to be more accurate but have not yet been adapted for use on a GUI platform and thus require some scripting knowledge.

Annex: Quick Reference Guide

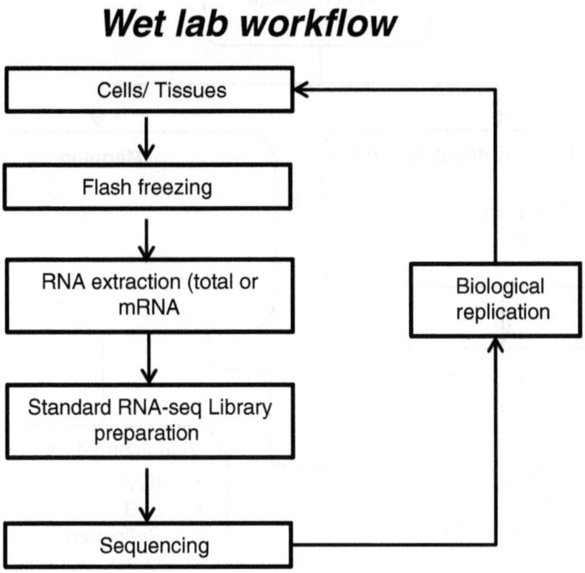

Fig. QG5.1 Representation of the wet-lab procedure workflow

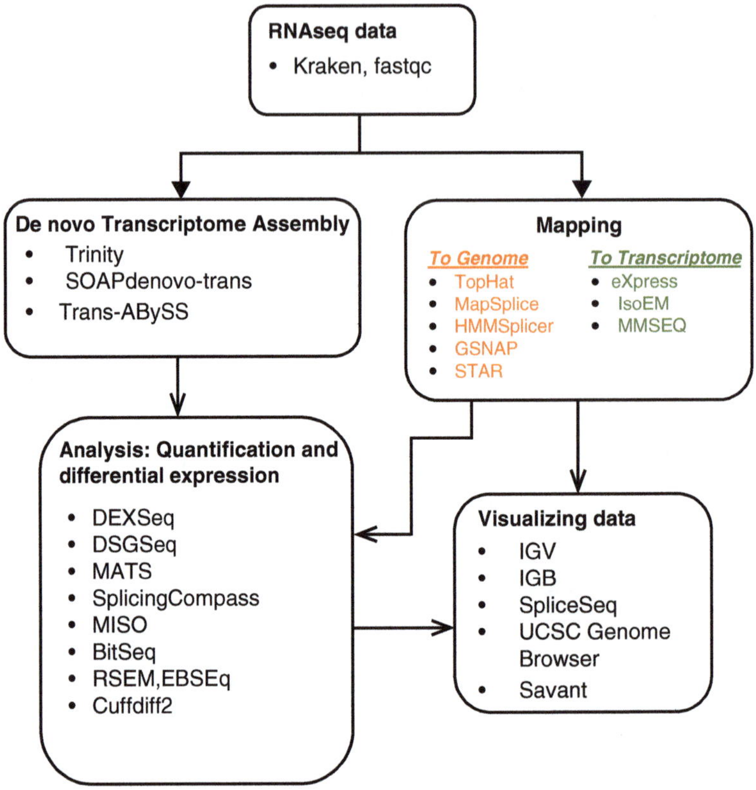

Fig. QG5.2 Main steps of the computational analysis pipeline

Table QG5.1 Experimental design considerations

Technique	Number of replicates	Sequencing depth	Recommended sequencing platforms
RNA-seq for alternative splicing analysis	3 (minimum per condition), 5 recommended	30 million reads uniquely mapped	Illumina HiSeq

Table that comprises relevant experimental design parameters, to carefully consider before applying this methodology

Table QG5.2 Available software recommendations

Application	Software	Reference	Graphical user interface (GUI) or command line (CL) or via Galaxy-UCSC
De novo transcriptome assembly	Trinity	Grabherr et al. (2011)	CL, Galaxy
	SOAPdenovo-trans	Xie et al. (2014)	CL
	Trans-ABySS	Robertson et al. (2010)	CL
Splice-sensitive mapping to genome	TopHat	Trapnell et al. (2009)	CL, Galaxy
	MapSplice	Wang et al. (2010)	CL
	HMMSplicer	Dimon et al. (2010)	CL
	GSNAP	Wu and Nacu (2010)	CL
	STAR	Dobin et al. (2013)	CL
Mapping to transcriptome	eXpress	Roberts and Pachter (2013)	CL
	IsoEM	Nicolae et al. (2011)	CL
	MMSEQ	Turro et al. (2011)	CL
Transcript isoform quantification	BitSeq	Glaus et al. (2012)	CL
	RSEM	Li and Dewey (2011)	CL
	Cufflinks	Trapnell et al. (2010)	CL, Galaxy
Differential expression of isoforms, *isoform-based differential expression*	DEXSEq	Anders et al. (2012)	CL
	MATS	Shen et al. (2012)	CL
	SplicingCompass	Aschoff et al. (2013)	CL
	MISO	Katz et al. (2010)	CL
	Altanalyze	Emig et al. (2010)	GUI, CL
	BitSeq	Glaus et al. (2012)	CL
	EBSeq	Leng et al. (2013)	CL
	Cuffdiff2	Trapnell et al. (2012)	CL, Galaxy
Visualising and reporting results	IGV	Thorvaldsdóttir et al. (2013)	GUI
	IGB	Nicol et al. (2009)	GUI
	UCSC Genome Browser	Raney et al. (2014)	GUI
	SpliceSeq	Ryan et al. (2012)	GUI
	SwitchSeq	Gonzalez-Porta and Brazma (2014)	CL

Table displaying a selection of the recommended software available for the computational analysis of data yielded by this technique

References

Alamancos GP, Agirre E, Eyras E (2014) Methods to study splicing from high-throughput RNA Sequencing data. Methods Mol Biol 1126:357

Alamancos GP, Pagès A, Trincado JL, Bellora N, Eyras E (2015) Leveraging transcript quantification for fast computation of alternative splicing profiles. RNA 21:1521. doi:10.1101/008763

Ameur A, Wetterbom A, Feuk L, Gyllensten U (2010) Global and unbiased detection of splice junctions from RNA-seq data. Genome Biol 11(3):R34. doi:10.1186/gb-2010-11-3-r34

Anders S, Reyes A, Huber W (2012) Detecting differential usage of exons from RNA-seq data. Genome Res 22(10):2008–2017. doi:10.1101/gr.133744.111

Anders S, Pyl PT, Huber W (2015) HTSeq--a Python framework to work with high-throughput sequencing data. Bioinformatics 31(2):166–169

Aschoff M, Hotz-Wagenblatt A, Glatting K-H, Fischer M, Eils R, König R (2013) SplicingCompass: differential splicing detection using RNA-seq data. Bioinformatics 29(9):1141–1148. doi:10.1093/bioinformatics/btt101

Au KF, Jiang H, Lin L, Xing Y, Wong WH (2010) Detection of splice junctions from paired-end RNA-seq data by SpliceMap. Nucleic Acids Res 38:4570–4578. doi:10.1093/nar/gkq211

Bonizzoni P, Della Vedova G, Pesole G, Picardi E, Pirola Y, Rizzi R (2015) Transcriptome assembly and alternative splicing analysis. Methods Mol Biol 1269:173–188

Bryant DW, Shen R, Priest HD, Wong W-K, Mockler TC (2010) Supersplat--spliced RNA-seq alignment. Bioinformatics 26(12):1500–1505. doi:10.1093/bioinformatics/btq206

Chandramohan R, Wu PY, Phan JH, Wang MD (2013) Benchmarking RNA-seq quantification tools. Conf Proc IEEE Eng Med Biol Soc 2013:647–650

Cooper TA, Wan L, Dreyfuss G (2009) RNA and disease. Cell 136(4):777–793. doi:10.1016/j.cell.2009.02.011

Dimon MT, Sorber K, DeRisi JL (2010) HMMSplicer: a tool for efficient and sensitive discovery of known and novel splice junctions in RNA-seq data. PLoS One 5:e13875. doi:10.1371/journal.pone.0013875

Dobin A, Davis CA, Schlesinger F, Drenkow J, Zaleski C, Jha S et al (2013) STAR: ultrafast universal RNA-seq aligner. Bioinformatics 29(1):15–21. doi:10.1093/bioinformatics/bts635

Emig D, Salomonis N, Baumbach J, Lengauer T, Conklin BR, Albrecht M (2010) AltAnalyze and DomainGraph: analyzing and visualizing exon expression data. Nucleic Acids Res 38(Web Server issue):W755–W762. doi:10.1093/nar/gkq405

Engström PG, Steijger T, Sipos B, Grant GR, Kahles A, Rätsch G et al (2013) Systematic evaluation of spliced alignment programs for RNA-seq data. Nat Methods 10(12):1185–1191. doi:10.1038/nmeth.2722

Florea L, Song L, Salzberg SL (2013) Thousands of exon skipping events differentiate among splicing patterns in sixteen human tissues. F1000Research 2:188, doi: 10.12688/f1000research.2-188.v1

Foissac S, Sammeth M (2007) ASTALAVISTA: dynamic and flexible analysis of alternative splicing events in custom gene datasets. Nucleic Acids Res 35(Web Server issue):W297–W299. doi:10.1093/nar/gkm311

Glaus P, Honkela A, Rattray M (2012) Identifying differentially expressed transcripts from RNA-seq data with biological variation. Bioinformatics 28(13):1721–1728. doi:10.1093/bioinformatics/bts260

Gonzalez-Porta, M., & Brazma, A. (2014). Identification, annotation and visualisation of extreme changes in splicing from RNA-seq experiments with SwitchSeq. bioRxiv. Cold Spring Harbor Labs Journals. doi:10.1101/005967

Gonzalez-Porta M, Frankish A, Rung J, Harrow J, Brazma A (2013) Transcriptome analysis of human tissues and cell lines reveals one dominant transcript per gene. Genome Biol 14(7):R70. doi:10.1186/gb-2013-14-7-r70

Grabherr MG, Haas BJ, Yassour M, Levin JZ, Thompson DA, Amit I, Adiconis X, Fan L, Raychowdhury R, Zeng Q, Chen Z, Mauceli E, Hacohen N, Gnirke A, Rhind N, di Palma F, Birren BW, Nusbaum C, Lindblad-Toh K, Friedman N, Regev A (2011) Full-length transcriptome assembly from RNA-seq data without a reference genome. Nat Biotechnol 29(7):644–652. doi:10.1038/nbt.1883

Grant GR, Farkas MH, Pizarro AD, Lahens NF, Schug J, Brunk BP et al (2011) Comparative analysis of RNA-seq alignment algorithms and the RNA-seq unified mapper (RUM). Bioinformatics 27:2518–2528. doi:10.1093/bioinformatics/btr427

Gulledge AA, Vora H, Patel K, Loraine AE (2014) A protocol for visual analysis of alternative splicing in RNA-seq data using integrated genome browser. Methods Mol Biol 1158:123–137. doi:10.1007/978-1-4939-0700-7_8

Hu Y, Huang Y, Du Y, Orellana CF, Singh D, Johnson AR et al (2013) DiffSplice: the genome-wide detection of differential splicing events with RNA-seq. Nucleic Acids Res 41(2):e39. doi:10.1093/nar/gks1026

Kanitz A, Gypas F, Gruber AJ, Gruber AR, Martin G, Zavolan M (2015) Comparative assessment of methods for the computational inference of transcript isoform abundance from RNA-seq data. Genome Biol 16(1):150. doi:10.1186/s13059-015-0702-5

Katz Y, Wang ET, Airoldi EM, Burge CB (2010) Analysis and design of RNA sequencing experiments for identifying isoform regulation. Nat Methods 7(12):1009–1015. doi:10.1038/nmeth.1528

Langmead B, Trapnell C, Pop M, Salzberg SL (2009) Ultrafast and memory-efficient alignment of short DNA sequences to the human genome. Genome Biol 10:R25. doi:10.1186/gb-2009-10-3-r25

Leng N, Dawson JA, Thomson JA, Ruotti V, Rissman AI, Smits BM, Haag JD, Gould MN, Stewart RM, Kendziorski C (2013) EBSeq: an empirical Bayes hierarchical model for inference in RNA-seq experiments. Bioinformatics 29(8):1035–1043. doi:10.1093/bioinformatics/btt087

Li B, Dewey CN (2011) RSEM: accurate transcript quantification from RNA-seq data with or without a reference genome. BMC Bioinformatics 12:323. doi:10.1186/1471-2105-12-323

Li B, Fillmore N, Bai Y, Collins M, Thomson JA, Stewart R, Dewey CN (2014) Evaluation of *de novo* transcriptome assemblies from RNA-seq data. Genome Biol 15(12):553. doi:10.1186/s13059-014-0553-5

Lindner R, Friedel CC (2012) A comprehensive evaluation of alignment algorithms in the context of RNA-seq. PLoS One 7(12):e52403. doi:10.1371/journal.pone.0052403

Liu R, Loraine AE, Dickerson JA (2014) Comparisons of computational methods for differential alternative splicing detection using RNA-seq in plant systems. BMC Bioinformatics 15(1):364

Luo R, Liu B, Xie Y, Li Z, Huang W, Yuan J, He G, Chen Y, Pan Q, Liu Y, Tang J, Wu G, Zhang H, Shi Y, Liu Y, Yu C, Wang B, Lu Y, Han C, Cheung DW, Yiu SM, Peng S, Xiaoqian Z, Liu G, Liao X, Li Y, Yang H, Wang J, Lam TW, Wang J (2012) SOAPdenovo2: an empirically improved memory-efficient short-read *de novo* assembler. Gigascience 1(1):18. doi:10.1186/2047-217X-1-18

Majoros WH, Lebeck N, Ohler U, Li S (2014) Improved transcript isoform discovery using ORF graphs. Bioinformatics 30(14):1958–1964. doi:10.1093/bioinformatics/btu160

Mazin P, Xiong J, Liu X, Yan Z, Zhang X, Li M et al (2013) Widespread splicing changes in human brain development and aging. Mol Syst Biol 9(1):633. doi:10.1038/msb.2012.67

Nariai N, Kojima K, Mimori T, Sato Y, Kawai Y, Yamaguchi-Kabata Y, Nagasaki M (2014) TIGAR2: sensitive and accurate estimation of transcript isoform expression with longer RNA-Seq reads. BMC Genomics 15(Suppl 10):S5. doi:10.1186/1471-2164-15-S10-S5

Nicol JW, Helt GA, Blanchard SG, Raja A, Loraine AE (2009) The Integrated Genome Browser: free software for distribution and exploration of genome-scale datasets. Bioinformatics 25(20):2730–2731. doi:10.1093/bioinformatics/btp472

Nicolae M, Mangul S, Măndoiu II, Zelikovsky A (2011) Estimation of alternative splicing isoform frequencies from RNA-seq data. Algorithms Mol Biol 6(1):9. doi:10.1186/1748-7188-6-9

Pachter L (2011) Models for transcript quantification from RNA-seq. Genomics; Methodology. Available from http://arxiv.org/abs/1104.3889

Pan Q, Shai O, Lee LJ, Frey BJ, Blencowe BJ (2008) Deep surveying of alternative splicing complexity in the human transcriptome by high-throughput sequencing. Nat Genet 40(12):1413–1415. doi:10.1038/ng.259

Patro R, Mount SM, Kingsford C (2014) Sailfish enables alignment-free isoform quantification from RNA-seq reads using lightweight algorithms. Nat Biotechnol 32(5):462–464. doi:10.1038/nbt.2862

Raney BJ, Dreszer TR, Barber GP, Clawson H, Fujita PA, Wang T, Nguyen N, Paten B, Zweig AS, Karolchik D, Kent WJ (2014) Track data hubs enable visualization of user-defined genome-wide annotations on the UCSC Genome Browser. Bioinformatics 30(7):1003–1005. doi:10.1093/bioinformatics/btt637

Roberts A, Pachter L (2013) Streaming fragment assignment for real-time analysis of sequencing experiments. Nat Methods 10(1):71–73. doi:10.1038/nmeth.2251

Robertson G, Schein J, Chiu R, Corbett R, Field M, Jackman SD, Mungall K, Lee S, Okada HM, Qian JQ, Griffith M, Raymond A, Thiessen N, Cezard T, Butterfield YS, Newsome R, Chan SK, She R, Varhol R, Kamoh B, Prabhu AL, Tam A, Zhao Y, Moore RA, Hirst M, Marra MA, Jones SJ, Hoodless PA, Birol I (2010) *De novo* assembly and analysis of RNA-seq data. Nat Methods 7(11):909–912. doi:10.1038/nmeth.1517

Robinson JT, Thorvaldsdóttir H, Winckler W, Guttman M, Lander ES, Getz G, Mesirov JP (2011) Integrative genomics viewer. Nat Biotechnol 29:24. doi:10.1038/nbt.1754

Ryan MC, Cleland J, Kim R, Wong WC, Weinstein JN (2012) SpliceSeq: a resource for analysis and visualization of RNA-seq data on alternative splicing and its functional impacts. Bioinformatics 28(18):2385–2387. doi:10.1093/bioinformatics/bts452

Shen S, Park JW, Huang J, Dittmar KA, Lu Z, Zhou Q et al (2012) MATS: a Bayesian framework for flexible detection of differential alternative splicing from RNA-seq data. Nucleic Acids Res 40(8):e61. doi:10.1093/nar/gkr1291

Steijger T, Abril JF, Engström PG, Kokocinski F, Hubbard TJ, Guigó R, Harrow J, Bertone P, RGASP Consortium (2013) Assessment of transcript reconstruction methods for RNA-seq. Nat Methods 10(12):1177–1184

Sturgill D, Malone JH, Sun X, Smith HE, Rabinow L, Samson M-L, Oliver B (2013) Design of RNA splicing analysis null models for post hoc filtering of Drosophila head RNA-seq data with the splicing analysis kit (Spanki). BMC Bioinformatics 14(1):320. doi:10.1186/1471-2105-14-320

Thorvaldsdóttir H, Robinson JT, Mesirov JP (2013) Integrative Genomics Viewer (IGV): high-performance genomics data visualization and exploration. Brief Bioinform 14(2):178–192. doi:10.1093/bib/bbs017

Tilgner H, Jahanbani F, Blauwkamp T, Moshrefi A, Jaeger E, Chen F et al (2015) Comprehensive transcriptome analysis using synthetic long-read sequencing reveals molecular co-association of distant splicing events. Nat Biotechnol 33(7):736–742. doi:10.1038/nbt.3242

Trapnell C, Pachter L, Salzberg SL (2009) TopHat: discovering splice junctions with RNA-seq. Bioinformatics 25:1105–1111. doi:10.1093/bioinformatics/btp120

Trapnell C, Williams BA, Pertea G, Mortazavi A, Kwan G, van Baren MJ, Salzberg SL, Wold BJ, Pachter L (2010) Transcript assembly and quantification by RNA-seq reveals unannotated transcripts and isoform switching during cell differentiation. Nat Biotechnol 28(5):511–515. doi:10.1038/nbt.1621

Trapnell C, Roberts A, Goff L, Pertea G, Kim D, Kelley DR, Pimentel H, Salzberg SL, Rinn JL, Pachter L (2012) Differential gene and transcript expression analysis of RNA-seq experiments with TopHat and Cufflinks. Nat Protoc 7(3):562–578. doi:10.1038/nprot.2012.016

Treutlein B, Gokce O, Quake SR, Südhof TC (2014) Cartography of neurexin alternative splicing mapped by single-molecule long-read mRNA sequencing. Proc Natl Acad Sci U S A 111(13):E1291–E1299. doi:10.1073/pnas.1403244111

Turro E, Su SY, Gonçalves Â, Coin LJ, Richardson S, Lewin A (2011) Haplotype and isoform specific expression estimation using multi-mapping RNA-seq reads. Genome Biol 12(2):R13. doi:10.1186/gb-2011-12-2-r13

Vitting-Seerup K, Porse BT, Sandelin A, Waage J (2014) spliceR: an R package for classification of alternative splicing and prediction of coding potential from RNA-seq data. BMC Bioinformatics 15(1):81. doi:10.1186/1471-2105-15-81

Wang ET, Sandberg R, Luo S, Khrebtukova I, Zhang L, Mayr C et al (2008) Alternative isoform regulation in human tissue transcriptomes. Nature 456(7221):470–476. doi:10.1038/nature07509

Wang K, Singh D, Zeng Z, Coleman SJ, Huang Y, Savich GL et al (2010) MapSplice: accurate mapping of RNA-seq reads for splice junction discovery. Nucleic Acids Res 38:e178. doi:10.1093/nar/gkq622

Wang W, Qin Z, Feng Z, Wang X, Zhang X (2013) Identifying differentially spliced genes from two groups of RNA-seq samples. Gene 518(1):164–170. doi:10.1016/j.gene.2012.11.045

Wu TD, Nacu S (2010) Fast and SNP-tolerant detection of complex variants and splicing in short reads. Bioinformatics 26(7):873–881. doi:10.1093/bioinformatics/btq057

Wu J, Akerman M, Sun S, McCombie WR, Krainer AR, Zhang MQ (2011) SpliceTrap: a method to quantify alternative splicing under single cellular conditions. Bioinformatics 27(21):3010–3016. doi:10.1093/bioinformatics/btr508

Wu E, Nance T, Montgomery SB (2014) SplicePlot: a utility for visualizing splicing quantitative trait loci. Bioinformatics 30:1025–1026. doi:10.1093/bioinformatics/btt733

Xie Y, Wu G, Tang J, Luo R, Patterson J, Liu S, Huang W, He G, Gu S, Li S, Zhou X, Lam TW, Li Y, Xu X, Wong GK, Wang J (2014) SOAPdenovo-Trans: *de novo* transcriptome assembly with short RNA-seq reads. Bioinformatics 30(12):1660–1666. doi:10.1093/bioinformatics/btu077

Zhao Q-Y, Wang Y, Kong Y-M, Luo D, Li X, Hao P (2011) Optimizing *de novo* transcriptome assembly from short-read RNA-seq data: a comparative study. BMC Bioinformatics 12(Suppl 14):S2. doi:10.1186/1471-2105-12-S14-S2

Zhou A, Breese MR, Hao Y, Edenberg HJ, Li L, Skaar TC, Liu Y (2012) Alt Event Finder: a tool for extracting alternative splicing events from RNA-seq data. BMC Genomics 13(Suppl 8):S10. doi:10.1186/1471-2164-13-S8-S10

Chapter 6
microRNA Discovery and Expression Analysis in Animals

Bastian Fromm

6.1 Introduction

When the first miRNA Lin-4 was discovered in *Caenorhabditis elegans* in 1993 (Lee et al. 1993), the importance of the discovery was underestimated, and only few imagined that these small noncoding molecules could represent a completely new and major class of gene regulators in worms or even further (Wickens and Takayama 1994). And indeed it took another 7 years until the second miRNA Let-7 was discovered (Pasquinelli et al. 2000). Soon after several more miRNAs and their wide distribution across, the majority of animal groups was simultaneously published by three groups (Lee and Ambros 2001; Lau et al. 2001; Lagos-Quintana et al. 2001). However, not the fact that they are evolutionary highly conserved across the animal tree of life, but that they represented a novel way of gene regulation in all animals, triggered a vast range of studies and a new field of molecular biology: the small noncoding RNA field.

In this chapter I will give an overview of miRNAs in animals and the challenges of miRNA discovery, annotation, and expression analysis using high-throughput sequencing.

MiRNAs in animals are single-stranded, 20–26 nucleotide long small RNAs that derive from hairpin precursor and regulate gene expression by negative posttranscriptional regulation of messenger RNAs (mRNAs) (Fromm et al. 2015a). In the canonical pathway, a pri-miRNA is transcribed by RNA polymerase II and processed by the RNase Drosha to the pre-miRNAs. Via Exportin 5 channel proteins, the pre-miRNA is exported from the nucleus into the cytosol where another RNase, Dicer, removes the remaining loop sequence. The miRNA/miRNA* RNA duplex is

B. Fromm, Ph.D. (✉)
Department of Tumor Biology, Institute for Cancer Research, Norwegian Radium Hospital, Oslo University Hospital, PO Box 4950, Nydalen, 0424 Oslo, Norway
e-mail: BastianFromm@gmail.com

A.M. Aransay, J.L. Lavín Trueba (eds.), *Field Guidelines for Genetic Experimental Designs in High-Throughput Sequencing*,
DOI 10.1007/978-3-319-31350-4_6

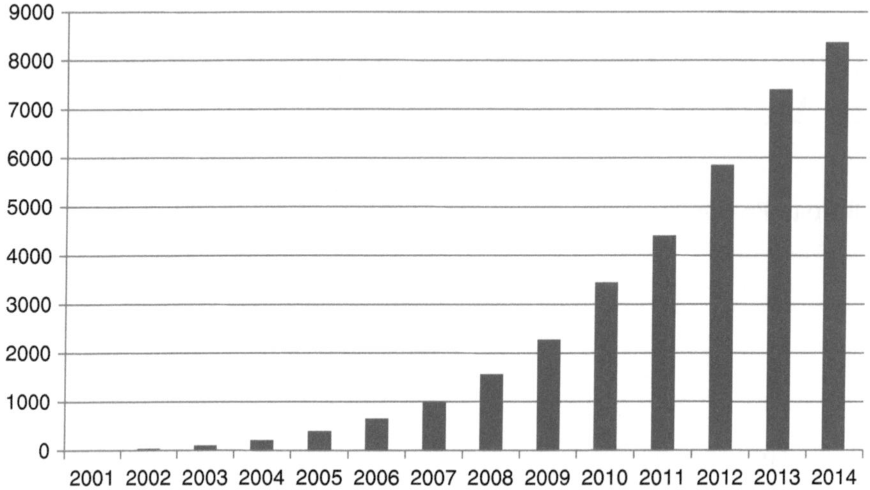

Fig. 6.1 PubMed entries with "microRNA" in title per year 2001–2014

typically transferred to AGO2 that removes the passenger strand, retains the mature miRNA, and exposes positions 2–8 of the sequence (miRNA seed) via conformational change (Schirle et al. 2014). Subsequently the RISC-complex, a complex of several proteins, is assembled and modulates the seed-sequence-directed binding to the 3′ UTR of a target mRNA (but see (La Rocca et al. 2015)). This interaction leads to inhibition of translation and degradation of the respective mRNA that negatively affects the protein levels of the corresponding genes (for more details see for instance (Pasquinelli 2012; Krol et al. 2010; Berezikov 2011)). Today it is known that miRNAs play key roles in a broad variety of biological processes, such as, e.g., cell proliferation and metabolism (Brennecke and Cohen 2003), tissue identity (Christodoulou et al. 2010), developmental timing (Reinhart et al. 2000), cell death (Baehrecke 2003), hematopoiesis (Chen et al. 2004), neuron development (Johnston and Hobert 2003), tumorigenesis (Esquela-Kerscher and Slack 2006), DNA methylation, and chromatin modification (Bao et al. 2004), as well as in immune defense against viruses (Sarnow et al. 2006). Recently it has also been shown that miRNAs can mediate interspecies cross talk and immune regulation via extracellular vesicles (Buck et al. 2014; Fromm et al. 2015b).

While in the early days the presence of mature products was confirmed by cloning and subsequent sequencing (Lau et al. 2001), it was impossible to assess relative expression accurately or if the retained sequences were derived from appropriately folding hairpin precursors. This rapidly changed when the number and quality of available genome sequences increased and modern sequencing methods became available also for short RNAs (Lu et al. 2005). Consequently the number of published miRNAs virtually exploded (Fig. 6.1).

However, studies on the human genome uncovered massive numbers of putative miRNA hairpins with a high probability for false-positives (Bentwich 2005).

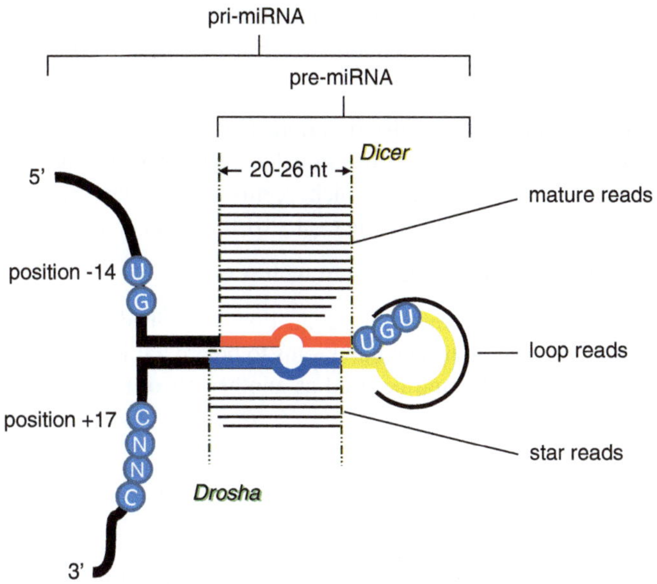

Fig. 6.2 Simplified structure of a canonical human miRNA and its read representation (based on (Fromm et al. 2015a; Nguyen et al. 2015; Auyeung et al. 2013)). Note the 5′ read homogeneity and the 2 nt overlap between the mature and passenger strand

This and the flood of steadily increasing amounts of small RNA sequencing data exposed the need for bioinformatic tools and more accurate criteria for the annotation of miRNAs. Increasing knowledge about the details of miRNA processing and their structural features is used to improve miRNA predictions.

6.2 Unique Structural Features of miRNAs

While in the last two decades, the small noncoding RNA field significantly expanded leading to a vast number of new small RNA families like piRNAs (Lau et al. 2006), siRNAs (Hamilton and Baulcombe 1999), snRNAs and snoRNAs (Matera et al. 2007), and novel small RNA functions of, e.g., tRNA (Goodarzi et al. 2015) and rRNA fragments (Chak et al. 2015), no other group like (canonical) miRNAs has so many unique structural features that clearly separate them from any other RNA family (summarized in Fromm et al. 2015a; see Fig. 6.2):

1. miRNAs are between 20 and 26 nt long.
2. They are genome encoded.
3. They derive from hairpin precursor that shows imperfect complementarity (~16 nt).
4. Mature products of both hairpin arms are expressed (mature, co-mature, or star sequence).

5. They show a 5′ read homogeneity in 90 % of the reads.
6. They show a 2 nt offset on both ends, which is a consequence of Drosha/Dicer processing.

Besides those features, it was also described that mature miRNAs show significantly higher ratios of As or Us at position 1 or, alternatively, show mismatched hairpin sequences at this position which seems to facilitate arm selection by Argonaute at least in mammals (Schirle et al. 2014; Suzuki et al. 2015).

7. Mature miRNA sequences usually start with A or U.

Additionally, recently, a number of publications described a set of motifs in the flanking regions and loop sequence that seems to be crucial for processing of most miRNAs in mammals and could be used as another set of criteria (Nguyen et al. 2015; Auyeung et al. 2013).

8. The flanking region upstream shows UG motif at position 14, loop shows UGU motif at the 3′ end of the 5′ arm, and flanking region downstream shows CNNC motif at positions 17–18.

One of the key features of miRNA complements of most metazoan animals is that they usually consist of species-specific miRNAs that are evolutionary novelties (novel miRNAs) and miRNAs that are representatives of evolutionary conserved miRNA families that are found in other animal groups, too, and reflect their hierarchical acquisitions over evolutionary time (conserved miRNAs). The latter have been shown to be highly conserved across all bilaterian animals (Hertel et al. 2006; Sempere et al. 2006; Heimberg et al. 2008; Sperling and Peterson 2009; Wheeler et al. 2009), and in some cases identical mature miRNA sequences are found, for instance, in human and fly (Hsa-Let-7-P1-5p, Dme-Let-7-5p) underlining their evolutionary age and critical importance for some key processes in animals.

9. At least some miRNAs of any higher animal taxon are representatives of phylogenetically conserved miRNA families and show very high sequence similarities.

If all these features would be carefully implemented in a bioinformatic pipeline that scores predictions based on them, I argue that miRNA prediction could be relatively simple. However, some of the listed features are very recent findings that were validated only for specific animal groups so far (feature 7 A or U as mature start), and others are known to be specific for such groups only (e.g., feature 8 motifs in mammals) and are therefore not found in any existing pipeline. For many of the early-days studies, several of the listed features that would allow checking for many of the required miRNA annotation criteria from above weren't available at the time either (read coverage, genome availability). As a consequence, many miRNAs have been described that lack falsifiable data for virtually all the structural features listed above, except for feature 9: phylogenetic conservation that is very often used to recover the expected set of miRNAs based on existing databases. Hitherto, this is not only historically relevant: in particular for studies on "non-genome organisms," miRNA complements are published continuously that often lack crucial parts of the actual miRNA complement (Xu et al. 2012) and require additional work (Fromm et al. 2015b). Unfortunately, it is not the absence of miRNAs from published complements that is the main issue, but false-positive miRNAs, hundreds and thousands of putative miRNAs that do not meet the simplest of the mentioned structural features.

6.3 Generating miRNA Next-Generation Sequencing Data from Biological Samples

To introduce the generation of miRNA data, I would like to mention common steps, challenges, and pitfalls that can influence the quality and the significance of the data generated. I hereby only focus on next-generation sequencing (NGS) technology because it is the only method that is not restricted to known sequences and isoforms. Detailed comparisons between available NGS methods and array or qPCR-based detection methods have been reported elsewhere (Leshkowitz et al. 2013; Git et al. 2010; Baker 2010; Willenbrock et al. 2009; Chen et al. 2009; Mestdagh et al. 2014; Knutsen et al. 2013, 2015). For NGS approaches, four steps are distinguishable in the process from sample to miRNA sequences:

1. Sampling
2. RNA extraction
3. Library preparation
4. Sequencing

1. Sampling of material for the generation of miRNA data is a crucial step as this step can have a big impact on the quality of the RNA produced and the amount of small RNAs in it. To assess RNA quality, usually a RIN (RNA Integrity Number) value will be determined (Schroeder et al. 2006). While a RIN value of 10 denotes the highest possible overall RNA quality, an RNA sample where even high molecular weight RNA molecules like 28S and many mRNAs are not degraded, RIN values below ten denote stepwise degradation of them. Consequently, it has been argued that the RIN value has an important function to distinguish representative samples from biased ones, which is obviously true for RNAseq studies where the focus lays on relatively long mRNA sequences. However, it has been argued that the RIN value does reflect degradation of miRNAs and thus does not affect miRNA studies (Jung et al. 2010). Unfortunately, this has not been shown for NGS studies, where short fragments of high molecular RNAs would at least "dilute" the amount of small RNAs in a given library. More importantly, recent research shows that even mature miRNAs can degrade and do this heterogeneously based on the sequential composition; thus, the introduction of bias in samples with low RINs has to be expected (pers. communication Francesco Nicassio, IIT). Essentially three possible routes can be followed that have been used successfully before. The first and obviously best method is to sample fresh material and immediately proceed to #2 RNA extraction. Alternatively, often used in clinical studies, fresh material is sampled and stored

on −80 °C where it can be stored for long times until #2 RNA extraction will be done. Similarly fresh samples can be stored in RNA stabilizing solution that is available from many suppliers (e.g., RNAlater). Dependent on the size of organisms of interest, it might be required to think about the design of the study more carefully—e.g., in cases where organisms are very small, several individuals might have to be pooled in order to arrive at a big enough number RNA molecules that allow for NGS sequencing (often around 1 μg total RNA required). Or in other cases, particular tissue types have to be collected. Anyway, clean and quick work on ice is recommended.

2. Historically most labs had own RNA extraction protocols; however, today many commercially available protocols exist that promise high-quality and high-yield total RNA samples that include miRNAs from a range of different tissue, cell, and organism types with acceptable price/value ratios. Nevertheless, when they are compared for, e.g., very little sample inputs, particular sample, and tissue types or in comparison to other nucleic acids, substantial differences were observed (Fromm et al. 2011; Bergallo et al. 2015; Grabmuller et al. 2015; Hantzsch et al. 2014; Monleau et al. 2014; Guo et al. 2014). In conclusion, there is no optimal method that satisfies all demands and it appears that each method has specific merits and flaws. I recommend to carefully choose from the available methods, based on literature and centered on the experimental needs. RNA quality (RIN) and purity (DNA or protein contamination) should be carefully assessed.
3. Library preparation is highly dependent on the sequencing strategy chosen, but many different methods exist here, too, and it has been suggested that this step—which is usually done as a service, but can also been done by anyone that purchases the respective kits or chemicals—is the most crucial step in the generation of miRNA data (Knutsen et al. 2015; van Dijk et al. 2014; Toedling et al. 2012; Jackson et al. 2014; Jayaprakash et al. 2011; Hafner et al. 2011). Common methods of library preparation are laborious and require training and a significant amount of RNA (between 500 ng and 1 μg totalRNA). Often, yields in read number and miRNA content cannot be predicted accurately and much depends on individual skills and personal experience. Recently, new library preparation kits are emerging that claim an improved sensitivity and accuracy in generating high-quality libraries from as little as 100 pg RNA (see www.trilinkbiotech.com/cleantag/ligation-kit.asp). Realistically you might not be able to choose a library preparation method if you are within a, e.g., clinical setting, but it is important to be aware of the potential pitfalls and could be advisable to talk to sequencing facilities before you conduct an experiment.
4. During the years different sequencing platforms emerged with ever-increasing read counts and different techniques that all perform in their own ways (Knutsen et al. 2013; Toedling et al. 2012; Raabe et al. 2014). Today, however, Illumina technology is clearly leading the small RNA sequencing field and used the most. As for the relatively short size of miRNAs, usually short reads will be sequenced in order to maximize the amount of biological sequences as opposed to adapters or primers that are ligated to them, may it be in the preferable single-end or paired-end sequencing strategy. The desired number of reads that should be

aimed for depends significantly on the organism (i.e., how many miRNAs are expected), the biological question (characterization of few vs many miRNAs or expression analysis of, e.g., top 100 miRNAs), the aimed for resolution (e.g., detection of rare miRNAs in pooled samples requires sometimes extreme measures like the generation of hundreds of millions of reads for individual samples), and the yield of miRNA sequences of the library preparation protocol used (also dependent on who uses it, variations of 10–80 % miRNA content of the reads can occur between facilities; personal observation). In my opinion, single-end sequencing of as short as possible read length is the best option.

6.4 Status of miRNA Repositories

Given the historical development and technical challenges miRNA predictions face, it is not surprising that the overall quality and completeness of published and deposited miRNA complements are very heterogenic. An miRNA reference however is very important not only to understand transcription or biology of a known organism, but crucial to make accurate predictions in novel—previously not sequenced—organisms. Until recently, miRBase was the only online repository for miRNAs, and it regulates new entries by accepting published miRNAs only, assuming that peer review would eliminate incorrect calls. In the latest version (Release 21) of the database, it contains 28,645 hairpin precursor miRNAs and 35,828 mature miRNA products of 223 organisms, roughly half of them from animals (Kozomara and Griffiths-Jones 2014). While annotation and nomenclature of miRNAs have admittedly been problematic, and, therefore, many representatives of conserved miRNA families are named redundantly, which is partly covered by miRBase (but see table 2 in (Tarver et al. 2013)), it was recently found that almost half of all animal entries in miRBase are not derived from *bona fide* miRNA genes (Fromm et al. 2015a), supporting earlier doubts (Castellano and Stebbing 2013; Chiang et al. 2010; Jones-Rhoades 2012; Langenberger et al. 2011; Meng et al. 2012; Tarver et al. 2012; Taylor et al. 2014; Wang and Liu 2011) and questioning a system of accepting miRNAs based on their publication alone. Consequently, a database of manually curated miRNA genes—MirGeneDB.org—was erected that aims at providing high-confidence miRNA complement with low false-positive and low false-negative rates at the same time (Fromm et al. 2015a). A uniform system for the annotation of miRNAs based on a set of consistent criteria (see above: structural features) was used to decide whether or not a given putative miRNA entry in miRBase is likely to be derived from a *bona fide* miRNA gene or not. Additionally, a new consistent nomenclature was put in place (while keeping old names for relocation) that is simple and stable over time, comprehensible especially between species, and is predictive in evolutionary terms, so it reveals the expected number of miRNAs in any species and can expose instances of miRNA loss or absence (see (Fromm et al. 2015a) for details). Currently MirGeneDB.org consists of 1421 fully annotated miRNA genes from human, mouse, chicken, and zebrafish and will significantly increase in numbers of species (invertebrates and vertebrates) in the next years.

6.5 Prediction of miRNAs in Genome and Non-genome Organisms

Today the prediction of miRNAs is a bioinformatic task and highly reliant on programmers and their biological knowledge of structural features of miRNAs. However, the prediction depends much on the quality and purity of the samples (degradation or contamination of RNA), the depth of sequencing (i.e., enough reads to detect passenger reads of lower expressed miRNAs, too), the quality of miRNA references (number of false-positives), and reference genomes (completeness, redundancy). While lately more and more web-interface-based program limit the required computer literacy for miRNA work for some organisms and questions (Rueda et al. 2015), it is advisable to be able to work on command line interfaces and know basic UNIX commands for text manipulations, especially if nonstandard organisms, i.e., without genome found within available genomes. However, programming skills are usually not required as many pipelines have been developed that include all required steps. It is of importance to remark that all currently available pipelines require manual curation of predictions. Among the most influential pipelines for animal miRNAs are (alphabetically) DARIO (Fasold et al. 2011), miRanalyzer (Hackenberg et al. 2009, 2011), miRDeep2 (Friedländer et al. 2008, 2012), miRDeep* (An et al. 2013), mirTools2 (Wu et al. 2013; Zhu et al. 2010), miRTRAP (Hendrix et al. 2010), and UEA sRNA workbench (Rueda et al. 2015; Stocks et al. 2012). Although the differences between the pipelines in approach, requirements, performance, and ease of use are substantial (Table 6.1 and for detailed review of some of the listed programs, see (Kang and Friedlander 2015; Li et al. 2012)), many of them share a common set of steps (see Fig. 6.3 for generalized workflow):

1. Preprocessing of reads

   Usually miRNA sequencing data (Illumina Inc.) will be provided in unprocessed (raw) format that requires preprocessing like adaptor handling, quality and length filtering, and setting required number of unique reads (depending on library and sequencing protocol and sequencing depth). While some of the programs provide these steps as options, stand-alone tools like fastx-toolkit (http://hannonlab.cshl.edu/fastx_toolkit/index.html), Galaxy implementations (https://usegalaxy.org), or custom scripts are regularly used for this purpose, too. Eventually reads between 20 and 26 nt in length should be used for further analysis (see structural feature 1).
2. Profiling of known miRNAs

   Not necessarily the second step in each pipeline, it is common to screen the read data of small RNA sequencing experiments for known miRNAs. This is useful for getting a first impression of the quality of the sequencing run (e.g., depth), and, in cases where samples of organisms with known miRNA complement are sequenced or where no reference genome is available, this step might already be sufficient to identify differences in expression of known miRNAs of interests given the biological samples or scientific question (e.g., normal vs. tumor

Table 6.1 Overview over the most influential miRNA prediction pipelines

Pipeline	Preprocessing of reads	Mapping to genome	Expression profiling	Target prediction	User interface	Reference
miRTRAP	No	Not included	No	No	No graphics	Hendrix et al. (2010)
DARIO	No	Not included	Yes	No	Graphics, webserver	Fasold et al. (2011)
miRDeep2	Yes	Bowtie	Yes	No	Graphics, standalone	(Friedländer et al. 2008, 2012)
miRanalyzer	Yes	Bowtie	Differential expression	TargetSpy	Graphics, webserver, and standalone	Hackenberg et al. (2009, 2011)
mirTools2	Yes	Bowtie	Differential expression	TargetSpy	Graphics, webserver	Wu et al. (2013), Zhu et al. (2010)
				miRanda		
				MirRNAMap		
				microTv4.0		
				MicroCosm		
				MirTarget2		
UEA sRNA workbench	Yes	Bowtie	Differential expression	TargetSpy	Graphics, webserver, and standalone	Rueda et al. (2015), Stocks et al. (2012)
				miRanda		
				PITA		
				psRobot		
				TAPIR FASTA engine		
				TAPIR RNA hybrid engine		

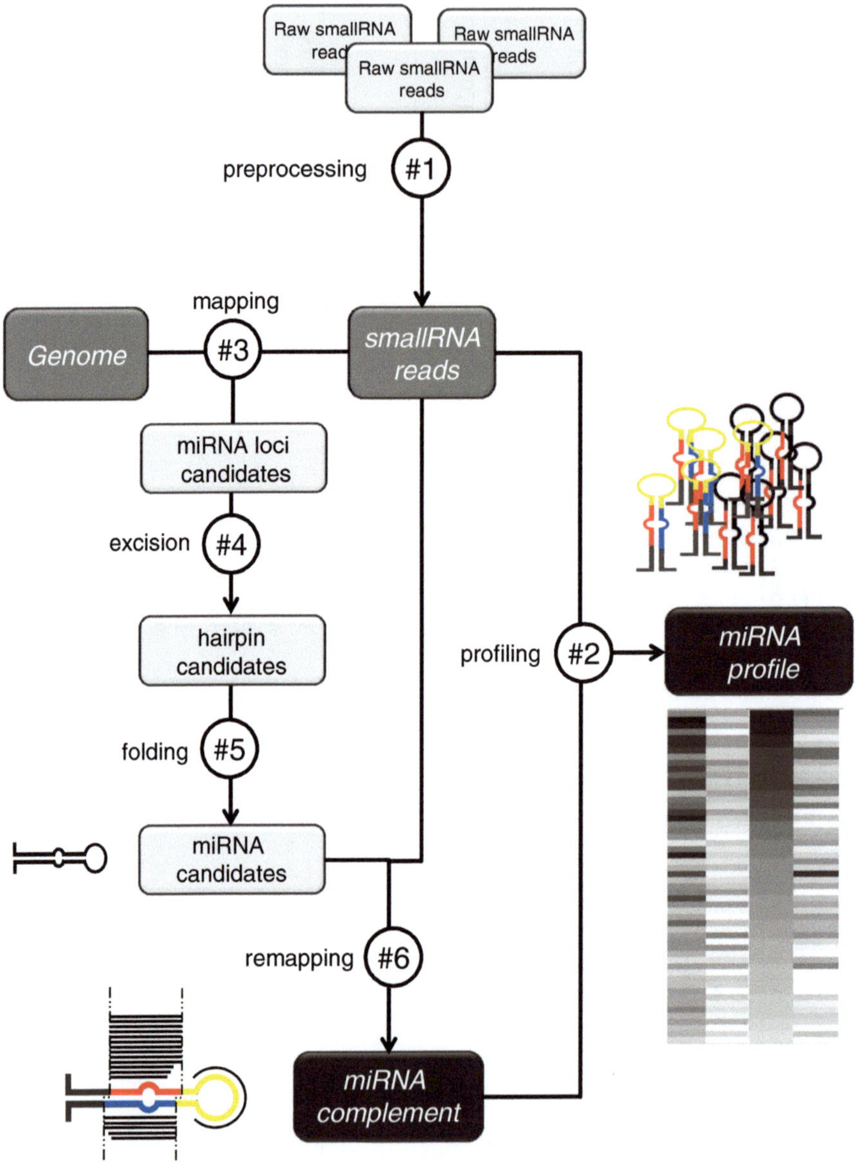

Fig. 6.3 Workflow for miRNA prediction. #1: Many pipelines offer preprocessing of raw sequencing reads in order to get them in the right format and to filter putative miRNA reads based on quality and length only. #2: Based on a given reference (usually miRBase but often a custom reference is possible, too), known miRNAs are identified and profiled (e.g., in heat map changed from (Fromm et al. 2015b)) within or between samples. #3: Small RNA reads are further filtered by mapping them to reference genome. #4: Regions where read map in so-called read stacks with clear boundaries rather than smears are excised for #5: Folding of putative miRNA loci (usually mfold or similar). #6: Small RNA reads are remapped to the approved hairpins in order to be able to bioinformatically assess structural features and score candidates accordingly

samples in human cancer studies). For other studies on, e.g., novel organisms, it is possible to compare the identity of the conserved miRNAs with the expected complement given the taxonomic position and draw preliminary conclusions that would need further confirmations (Tarver et al. 2013). All of the abovementioned pipelines profile known and novel (putative) miRNAs.

3. Mapping of reads to genome

   In order to be able to confirm criteria feature 2 (genomic origin), only reads that map to the required (!) reference genome sequence can be considered. Here, somewhat relaxed features can be used that allow for some mismatches as it is known that posttranscriptionally modified miRNAs occur and can play important roles, too (Neilsen et al. 2012). This step is also a condition for the identification of genomic regions where reads map in stacks that can trigger the selection or precursor candidates (for details see Box 1 in (Berezikov 2011)). Recently a method (MirCandRef) was described that uses small RNAseq data and genomic sequencing data without prior need for genome assembly to create so-called crystal contigs that can be used as a reference genome (Fromm et al. 2013).

4 & 5. Identifying hairpins

   After regions around read stacks are identified, they can be excised and separately folded (e.g., Mfold (Zuker 2003), RNAfold (Gruber et al. 2015; Lorenz et al. 2011)) for assessing the folding energy and structure of a possible hairpin that is required as a structural feature (feature 3). At this step, the size of the excised region for folding the hairpin is of course important to assess corresponding minimal folding energies (the longer the higher the energy cutoff). Nevertheless, it has also been shown that hairpin size can differ dramatically among certain groups of invertebrates and indeed can exceed the average value of 59 nt (Fromm et al. 2015a) by sometimes hundreds of nucleotides (Fromm et al. 2013, 2015a). A variable size option is clearly desirable but, to my knowledge, not available at the moment. As a sidenote, hairpin size is one of the main differences between the structural features of animals and plants.

6. Remapping reads to putative hairpins

   After a set of putative hairpins has been identified, reads are often mapped back to them in order to assess structural features 4, 5, and 6. If both arms are expressed (feature 4—at least if the miRNA is not already known), reads show 5′ homogeneity (feature 5), and a 2 nt offset is obvious (feature 6), candidates should be accepted as miRNAs or are given a high score in the assessment of candidates.

Usually, a list of predicted miRNAs is created and sorted for expression levels of known and novel miRNA predictions that are given scores according to their structure and structural features (e.g., DARIO, miRDeep2). In some cases, multiple samples can be analyzed, differential expression profiles are produced, and target prediction software is included for downstream analyses, too (miRanalyzer, mirTools2, UEA sRNA workbench). In this case, normalization of expression values across samples is a crucial and controversially discussed issue that is best dealt with by sequencing samples one wants to compare with exactly the same method and ideally on the same platform (Leshkowitz et al. 2013; Mestdagh et al. 2014; Knutsen et al. 2013, 2015; Bergallo et al. 2015; Monleau et al. 2014; Guo et al. 2014; Li et al. 2012; Sauer et al. 2014a, b).

6.6 Discussion and Outlook

The discovery of miRNAs in the last 20 years has triggered the development of new fields in biological research and further accelerated the understanding of human diseases like cancer. With the advent of novel sequencing technologies and availability of ever-increasing datasets, great bioinformatic efforts were undertaken to catch up with the demand of the field. However, the current surplus of bioinformatic pipelines is not necessarily reflecting a progress of the field as all have high numbers of false-positive (erroneously identified miRNAs) and false-negative rates (miRNAs that are present in the data but not detected) despite the known structural features of miRNAs and lead to many incorrectly predicted miRNAs in public databases. It seems to be a recent trend—at least in human miRNA research—to rather describe more putative miRNAs than accurate predictions (Friedländer et al. 2014; Londin et al. 2015a, b; Backes and Keller 2015). A throughout comparison of all available pipelines that would focus on false-discovery rates and the accuracy of annotations is clearly missing. More so, the current status of the main online repository miRBase is questionable as about 50 % of all entries seem to be incorrect putting miRNA predictions based on this database on jeopardy of being biased toward false-positives (Fromm et al. 2015a).

Besides noncanonical miRNAs like Mir-451 (Yang et al. 2010), where assessing all the described features is impossible, also several cases of extra-long hairpins have been described for animal species that are currently not detectable in an automatized fashion (see (Fromm et al. 2013; Grimson et al. 2008) and others).

While most of the presented pipelines show non-template reads of miRNAs (isomiRs) in their results (i.e., miRNAs that can be assigned to a particular miRNA locus but differ, for example, in a few additional nucleotides or show other polymorphisms), a detailed analysis of this subspecies of miRNAs is currently not available in a systematic and detailed fashion (sRNA bench gives at least an overall count of most prominent additions), although their potential role in some biological setting such as cancer has been proposed (Koppers-Lalic et al. 2014).

Much is known about the structural features of miRNAs, and new findings of distinct sequence motifs in the flanking regions (Nguyen et al. 2015; Auyeung et al. 2013), or by-products of miRNA processing like miRNA-offset RNAs (moRs) (Shi et al. 2009; Langenberger et al. 2009; Asikainen et al. 2015; Bortoluzzi et al. 2011; Babiarz et al. 2008), are very interesting, but it has yet to be demonstrated that they represent features that can be implemented in miRNA prediction pipelines, also for nonmammalian species. However, currently not a single pipeline exists that uses all the confirmed structural features of miRNAs. It is therefore not surprising that the knowledge about the phylogenetic distribution of miRNA families that could be used to predict miRNA complements of any given animal is nowhere implemented either (Fromm et al. 2013, 2015a; Tarver et al. 2013).

To summarize, miRNA prediction is possible with many different programs, but their performance is not only different but far from being optimal, as high rates of false-positives and false-negatives persist. Today, bioinformatic knowledge is less needed than before, because more comprehensive, web-based pipelines are available. Nevertheless, a certain level is still advisable because manual curation of miRNA candidates is continuously required given the high rates of incorrect or incomplete identifications. When asking more complex biological questions or looking for many samples in parallel, or from "unusual" reference genomes that are not part of the pipeline, all programs lack at least some important features. This is generally true for accurate screening for structural features (especially for the most recently discovered ones) and for high resolution of isomiRs distribution.

Acknowledgment The work was supported by South-Eastern Norway Regional Health Authority grant #2014041. I would like to express my gratitude to M. Hackenberg, J. Ramalho Carvalho, E. Høye, J. Quintana-Alcalá, and K.J. Peterson for useful comments on the manuscript.

Annex: Quick Reference Guide

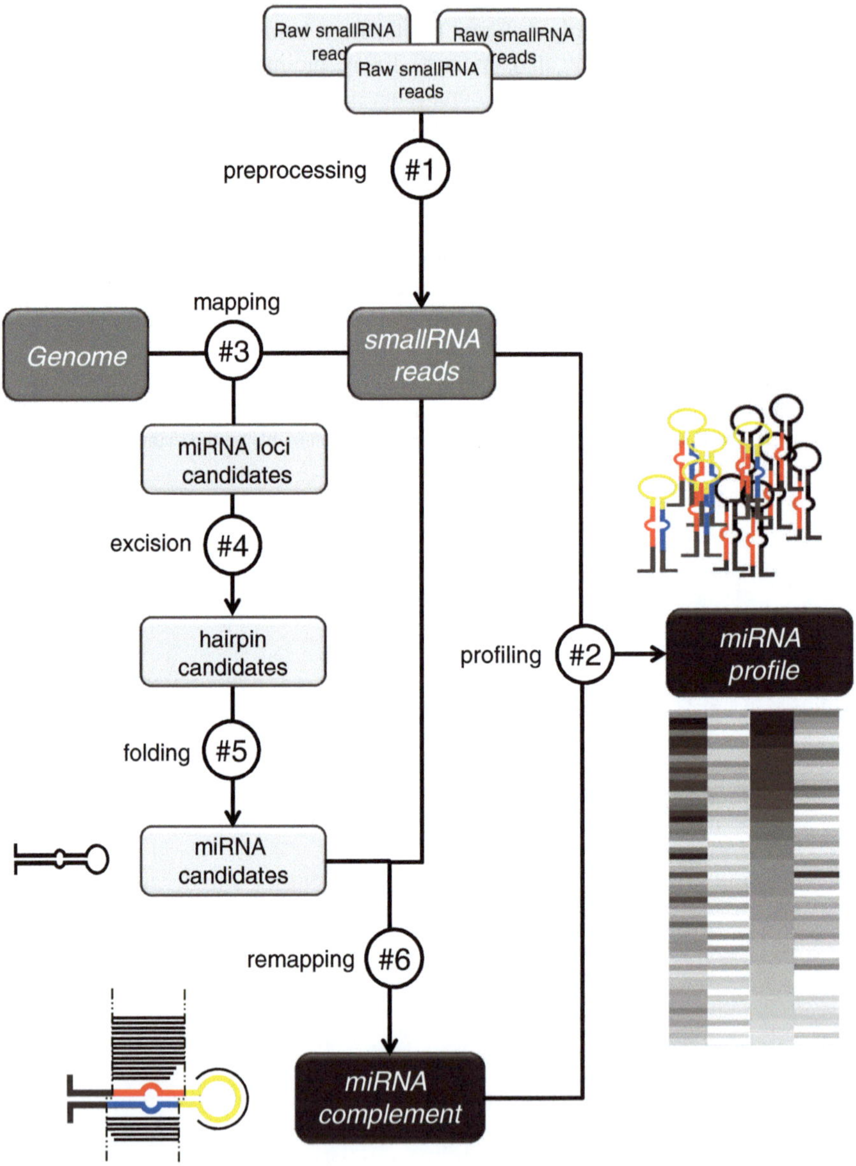

Fig. QG6.2 Main steps of the computational analysis pipeline

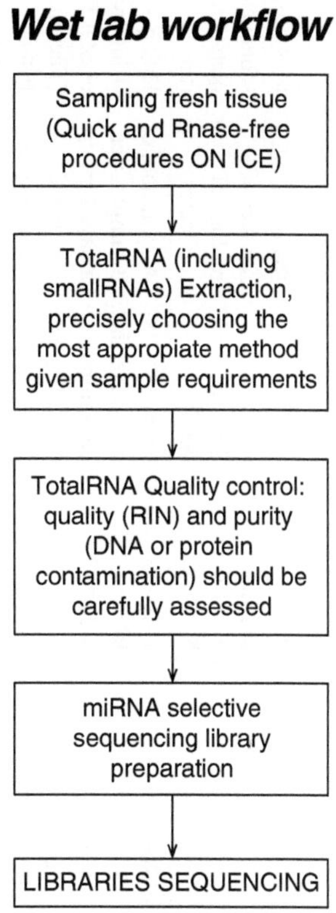

Fig. QG6.1 Representation of the wet-lab procedure workflow

Table QG6.1 Experimental design considerations

Technique	Protocol	References	Recommended starting material (see Note 1)	Sequencing depth	Recommended number of biological replicates	Recommended sequencing platform and run
Good-quality RNA, miRNA library preparation	NEXTflex™ Small RNA Sequencing kit (Bioo Scientific) with modified adaptors [size selection by acrylamide gel]	http://www.biooscientific.com/Next-Gen-Sequencing/NEXTflex-Illumina-Small-RNA-Seq-Library-Prep-Kit-v2	1–10 mg total RNA (RIN[a] > 7)	From 3 to 200 million reads per sample, depending on the projects aims	A minimum of 3 samples per assayed group for animal/plant models or	HiSeq (Illumina Inc.) or SOLID (Thermo-Fisher-Scientific)
		Nakashe et al. (2011) OMICS Res 1(1): 6–11				
		Sorefan et al. (2012) Silence 3:4				
	TruSeq® Small RNA Library Prep (Illumina Inc.) [size selection by acrylamide gel]	http://www.illumina.com/applications/sequencing/rna/small_rna.html	1–10 mg TotalRNA (RIN[a] > 7)			
		http://www.google.com/patents/US20120108440				
		Lopez et al. (2015) BMC Medical Genomics 8:35				
	SOLID™ Small RNA Expression Kit (Thermo-Fisher-Scientific) [size selection by acrylamide gel]	https://www.thermofisher.com/order/catalog/product/4463013?ICID=search-product	0.25–5 mg total RNA (RIN[a] > 7)		A minimum of 6 samples per assayed group for human clinical samples or wild animal/plant collections	Single read ×50 nt
		Toedling et al. (2012) PLoS ONE 7(2): e32724				
	NEBNext® Multiplex Small RNA Library Preparation Set for Illumina® (BioLabs Inc.) [optional size selection by acrylamide gel or gel-free]	https://www.neb.com/products/e7300-nebnext-multiplex-small-rna-library-prep-set-for-illumina-set-1	0.1–1 μg total RNA (RIN[a] > 7)			
	CleanTag™ Ligation Kit for Small RNA Library Preparation (TriLink Biotechnologies) plus recommended reagents [gel-free]	http://www.trilinkbiotech.com/cleantag/ligation-kit.asp	0.001–1 μg total RNA (RIN[a] > 7)			
		Vigneault et al. (2012) chapter 11: Unit–11.1210, current protocols in Human Genetics/Editorial Board, Jonathan L. Haines … [et al.]				

FFPE[b] RNA, miRNA Library Preparation	small RNA population enrichment (See NOTE 2) followed by any of the previously mentioned kits	Meng et al. (2013) PLoS ONE 8(5): e64393 Buitrago et al. (2015) PLoS ONE 10(3): e0121521	0.1–10 mg total RNA (RIN[a] < 7)	From 6 to 300 million reads per sample, depending on the total RNA degradation level and the project's aims		

Table that comprises relevant experimental design parameters, to carefully consider before applying this methodology
This table has been generated by the editors for the quick reference guide corresponding to this chapter
The protocol (kit) described within the gray cell is the mostly recommended, since this protocol reduces the ligation biases observed for the other methods
Note 1. At small RNA sequencing projects, be aware of choosing Total RNA extraction kits that preserve the small RNA molecules (i.e., keep all RNAs > 17 nt). In addition, dedicated kits should be used for Total RNA or miRNA extraction from FFPE samples
Note 2. Small RNA populations can be enriched by exclusive extraction of RNAs between 17 and 200 nt from the FFPE tissues with specific kits or by size selection of Total RNA on acrylamide gel
[a]RIN: RNA Integrity Number, which indicates the quality of the Total RNA as estimated by Bioanalyzer (Agilent Technologies) RNA nano- or pico-chips
[b]FFPE: Formalin-fixed paraffin-embedded samples

Table QG6.2 Available software recommendations

Pipeline	Preprocessing of reads	Mapping to genome	Expression profiling	Target prediction	User interface	Reference
miRTRAP	No	Not included	No	No	No graphics	Hendrix et al. (2010)
DARIO	No	Not included	Yes	No	Graphics, webserver	Fasold et al. (2011)
miRDeep2	Yes	Bowtie	Yes	No	Graphics, standalone	(Friedländer et al. 2008, 2012)
miRanalyzer	Yes	Bowtie	Differential expression	TargetSpy	Graphics, webserver, and standalone	Hackenberg et al. (2009, 2011)
mirTools2	Yes	Bowtie	Differential expression	TargetSpy	Graphics, webserver	Wu et al. (2013), Zhu et al. (2010)
				miRanda		
				MirRNAMap		
				microTv4.0		
				MicroCosm		
				MirTarget2		
UEA sRNA workbench	Yes	Bowtie	Differential expression	TargetSpy	Graphics, webserver, and standalone	(Rueda et al. (2015), Stocks et al. (2012)
				miRanda		
				PITA		
				psRobot		
				TAPIR FASTA engine		
				TAPIR RNA hybrid engine		

Table displaying a selection of the recommended software available for the computational analysis of data yielded by this technique

References

An J, Lai J, Lehman ML, Nelson CC (2013) miRDeep*: an integrated application tool for miRNA identification from RNA sequencing data. Nucleic Acids Res 41(2):727–737

Asikainen S et al (2015) Selective microRNA-Offset RNA expression in human embryonic stem cells. PLoS One 10(3):e0116668

Auyeung VC, Ulitsky I, McGeary SE, Bartel DP (2013) Beyond secondary structure: primary-sequence determinants license pri-miRNA hairpins for processing. Cell 152(4):844–858

Babiarz JE, Ruby JG, Wang Y, Bartel DP, Blelloch R (2008) Mouse ES cells express endogenous shRNAs, siRNAs, and other Microprocessor-independent, Dicer-dependent small RNAs. Genes Dev 22(20):2773–2785

Backes C, Keller A (2015) Reanalysis of 3,707 novel human microRNA candidates. Proc Natl Acad Sci U S A 112(22):E2849–E2850

Baehrecke EH (2003) miRNAs: micro managers of programmed cell death. Curr Biol 13(12): R473–R475

Baker M (2010) MicroRNA profiling: separating signal from noise. Nat Methods 7(9):687–692

Bao N, Lye KW, Barton MK (2004) MicroRNA binding sites in Arabidopsis class III HD-ZIP mRNAs are required for methylation of the template chromosome. Dev Cell 7(5):653–662

Bentwich I (2005) Identification of hundreds of conserved and nonconserved human microRNAs. Nat Genet 37:766–770

Berezikov E (2011) Evolution of microRNA diversity and regulation in animals. Nat Rev Genet 12(12):846–860

Bergallo M et al. (2015) Comparison of two available RNA extraction protocols for microRNA amplification in serum samples. J Clin Lab Anal 00:1–7. http://onlinelibrary.wiley.com/doi/10.1002/jcla.21848/epdf

Bortoluzzi S, Biasiolo M, Bisognin A (2011) MicroRNA-offset RNAs (moRNAs): by-product spectators or functional players? Trends Mol Med 17(9):473–474

Brennecke J, Cohen SM (2003) Towards a complete description of the microRNA complement of animal genomes. Genome Biol 4(9):228

Buck AH et al (2014) Exosomes secreted by nematode parasites transfer small RNAs to mammalian cells and modulate innate immunity. Nat Commun 5:5488

Castellano L, Stebbing J (2013) Deep sequencing of small RNAs identifies canonical and non-canonical miRNA and endogenous siRNAs in mammalian somatic tissues. Nucleic Acids Res 41(5):3339–3351

Chak LL, Mohammed J, Lai EC, Tucker-Kellogg G, Okamura K (2015) A deeply conserved, non-canonical miRNA hosted by ribosomal DNA. RNA 21(3):375–384

Chen J, Li WX, Xie D, Peng JR, Ding SW (2004) Viral virulence protein suppresses RNA silencing-mediated defense but upregulates the role of microrna in host gene expression. Plant Cell 16(5):1302–1313

Chen Y, Gelfond JA, McManus LM, Shireman PK (2009) Reproducibility of quantitative RT-PCR array in miRNA expression profiling and comparison with microarray analysis. BMC Genomics 10:407

Chiang HR et al (2010) Mammalian microRNAs: experimental evaluation of novel and previously annotated genes. Genes Dev 24(10):992–1009

Christodoulou F et al (2010) Ancient animal microRNAs and the evolution of tissue identity. Nature 463(7284):1084–1088

Esquela-Kerscher A, Slack FJ (2006) Oncomirs - microRNAs with a role in cancer. Nat Rev Cancer 6(4):259–269

Fasold M, Langenberger D, Binder H, Stadler PF, Hoffmann S (2011) DARIO: a ncRNA detection and analysis tool for next-generation sequencing experiments. Nucleic Acids Res 39(Web Server issue):W112–W117

Friedlander MR et al (2008) Discovering microRNAs from deep sequencing data using miRDeep. Nat Biotech 26(4):407–415

Friedländer MR, Mackowiak SD, Li N, Chen W, Rajewsky N (2012) miRDeep2 accurately identifies known and hundreds of novel microRNA genes in seven animal clades. Nucleic Acids Res 40:37

Friedländer M et al (2014) Evidence for the biogenesis of more than 1,000 novel human microRNAs. Genome Biol 15(R57):1–17

Fromm B, Harris PD, Bachmann L (2011) MicroRNA preparations from individual monogenean Gyrodactylus salaris-a comparison of six commercially available totalRNA extraction kits. BMC Res Notes 4:217

Fromm B, Worren MM, Hahn C, Hovig E, Bachmann L (2013) Substantial loss of conserved and gain of novel MicroRNA families in flatworms. Mol Biol Evol 30(12):2619–2628

Fromm B et al (2015a) A uniform system for the annotation of human microRNA genes and the evolution of the human microRNAome. Annu Rev Genet 49(1):213

Fromm B et al (2015b) The revised microRNA complement of Fasciola hepatica reveals a plethora of overlooked microRNAs and evidence for enrichment of immuno-regulatory microRNAs in extracellular vesicles. Int J Parasitol 45:697

Git A et al (2010) Systematic comparison of microarray profiling, real-time PCR, and next-generation sequencing technologies for measuring differential microRNA expression. RNA 16(5):991–1006

Goodarzi H et al (2015) Endogenous tRNA-derived fragments suppress breast cancer progression via YBX1 displacement. Cell 161(4):790–802

Grabmuller M, Madea B, Courts C (2015) Comparative evaluation of different extraction and quantification methods for forensic RNA analysis. Forensic Sci Int Genet 16:195–202

Grimson A et al (2008) Early origins and evolution of microRNAs and Piwi-interacting RNAs in animals. Nature 455(7217):1193–1197

Gruber AR, Bernhart SH, Lorenz R (2015) The Vienna RNA web services. Methods Mol Biol 1269:307–326

Guo Y et al (2014) A comparison of microRNA sequencing reproducibility and noise reduction using mirVana and TRIzol isolation methods. Int J Comput Biol Drug Des 7(2-3):102–112

Hackenberg M, Sturm M, Langenberger D, Falcon-Perez JM, Aransay AM (2009) miRanalyzer: a microRNA detection and analysis tool for next-generation sequencing experiments. Nucleic Acids Res 37(Web Server issue):W68–W76

Hackenberg M, Rodriguez-Ezpeleta N, Aransay AM (2011) miRanalyzer: an update on the detection and analysis of microRNAs in high-throughput sequencing experiments. Nucleic Acids Res 39(Web Server issue):W132–W138

Hafner M et al (2011) RNA-ligase-dependent biases in miRNA representation in deep-sequenced small RNA cDNA libraries. RNA 17(9):1697–1712

Hamilton AJ, Baulcombe DC (1999) A species of small antisense RNA in posttranscriptional gene silencing in plants. Science 286(5441):950–952

Hantzsch M et al (2014) Comparison of whole blood RNA preservation tubes and novel generation RNA extraction kits for analysis of mRNA and MiRNA profiles. PLoS One 9(12):e113298

Heimberg AM, Sempere LF, Moy VN, Donoghue PC, Peterson KJ (2008) MicroRNAs and the advent of vertebrate morphological complexity. Proc Natl Acad Sci U S A 105(8):2946–2950

Hendrix D, Levine M, Shi W (2010) MiRTRAP, a computational method for the systematic identification of miRNAs from high throughput sequencing data. Genome Biol 11(4):R39

Hertel J et al (2006) The expansion of the metazoan microRNA repertoire. BMC Genomics 7:25

Jackson TJ, Spriggs RV, Burgoyne NJ, Jones C, Willis AE (2014) Evaluating bias-reducing protocols for RNA sequencing library preparation. BMC Genomics 15:569

Jayaprakash AD, Jabado O, Brown BD, Sachidanandam R (2011) Identification and remediation of biases in the activity of RNA ligases in small-RNA deep sequencing. Nucleic Acids Res 39(21):e141

Johnston RJ, Hobert O (2003) A microRNA controlling left/right neuronal asymmetry in Caenorhabditis elegans. Nature 426(6968):845–849

Jones-Rhoades MW (2012) Conservation and divergence in plant microRNAs. Plant Mol Biol 80(1):3–16

Jung M et al (2010) Robust microRNA stability in degraded RNA preparations from human tissue and cell samples. Clin Chem 56(6):998–1006

Kang W, Friedlander MR (2015) Computational prediction of miRNA genes from small RNA sequencing data. Front Bioeng Biotechnol 3:7

Knutsen E et al (2013) Performance comparison of digital microRNA profiling technologies applied on human breast cancer cell lines. PLoS One 8(10):e75813

Knutsen E, Perander M, Fiskaa T, Johansen S (2015) Performance comparison and data analysis strategies for MicroRNA profiling in cancer research. In: Wu W, Choudhry H (eds) Next generation sequencing in cancer research, vol 2. Springer International Publishing, New York, NY, pp 239–265

Koppers-Lalic D et al (2014) Nontemplated nucleotide additions distinguish the small RNA composition in cells from exosomes. Cell Rep 8(6):1649–1658

Kozomara A, Griffiths-Jones S (2014) miRBase: annotating high confidence microRNAs using deep sequencing data. Nucleic Acids Res 42(D1):D68–D73

Krol J, Loedige I, Filipowicz W (2010) The widespread regulation of microRNA biogenesis, function and decay. Nat Rev Genet 11(9):597–610

La Rocca G et al (2015) In vivo, Argonaute-bound microRNAs exist predominantly in a reservoir of low molecular weight complexes not associated with mRNA. Proc Natl Acad Sci U S A 112(3):767–772

Lagos-Quintana M, Rauhut R, Lendeckel W, Tuschl T (2001) Identification of novel genes coding for small expressed RNAs. Science 294:853–858

Langenberger D et al (2009) Evidence for human microRNA-offset RNAs in small RNA sequencing data. Bioinformatics 25(18):2298–2301

Langenberger D et al (2011) MicroRNA or not microRNA? Advances in Bioinformatics and Computational Biology. In: de Souza ON, Telles GP, Palakal MJ (eds) Vol 6th Brazilian symposium on bioinformatics. Springer, Brasilia, pp 1–9

Lau NC, Lim LP, Weinstein EG, Bartel DP (2001) An abundant class of tiny RNAs with probable regulatory roles in *Caenorhabditis elegans*. Science 294(5543):858–862

Lau NC et al (2006) Characterization of the piRNA complex from rat testes. Science 313(5785):363–367

Lee RC, Ambros V (2001) An extensive class of small RNAs in *Caenorhabditis elegans*. Science 294:862–864

Lee RC, Feinbaum RL, Ambros V (1993) The C. elegans heterochronic gene lin-4 encodes small RNAs with antisense complementarity to lin-14. Cell 75:843–854

Leshkowitz D, Horn-Saban S, Parmet Y, Feldmesser E (2013) Differences in microRNA detection levels are technology and sequence dependent. RNA 19(4):527–538

Li Y et al (2012) Performance comparison and evaluation of software tools for microRNA deep-sequencing data analysis. Nucleic Acids Res 40(10):4298–4305

Londin E et al (2015a) Analysis of 13 cell types reveals evidence for the expression of numerous novel primate- and tissue-specific microRNAs. Proc Natl Acad Sci U S A 112(10): E1106–E1115

Londin E, Loher P, Rigoutsos I (2015b) Reply to Backes and Keller: identification of novel tissue-specific and primate-specific human microRNAs. Proc Natl Acad Sci U S A 112(22):e2851

Lorenz R et al (2011) ViennaRNA Package 2.0. Algorithms Mol Biol 6:26

Lu C et al (2005) Elucidation of the small RNA component of the transcriptome. Science 309(5740):1567–1569

Matera AG, Terns RM, Terns MP (2007) Non-coding RNAs: lessons from the small nuclear and small nucleolar RNAs. Nat Rev Mol Cell Biol 8(3):209–220

Meng Y, Shao C, Wang H, Chen M (2012) Are all the miRBase-registered microRNAs true? A structure- and expression-based re-examination in plants. RNA Biol 9(3):249–253

Mestdagh P et al (2014) Evaluation of quantitative miRNA expression platforms in the microRNA quality control (miRQC) study. Nat Methods 11(8):809–815

Monleau M et al (2014) Comparison of different extraction techniques to profile microRNAs from human sera and peripheral blood mononuclear cells. BMC Genomics 15:395

Neilsen CT, Goodall GJ, Bracken CP (2012) IsomiRs--the overlooked repertoire in the dynamic microRNAome. Trends Genet 28(11):544–549

Nguyen TA et al (2015) Functional anatomy of the human microprocessor. Cell 161(6):1374–1387

Pasquinelli AE (2012) MicroRNAs and their targets: recognition, regulation and an emerging reciprocal relationship. Nat Rev Genet 13(4):271–282

Pasquinelli AE et al (2000) Conservation of the sequence and temporal expression of *let-7* heterochronic regulatory RNA. Nature 408:86–89

Raabe CA, Tang TH, Brosius J, Rozhdestvensky TS (2014) Biases in small RNA deep sequencing data. Nucleic Acids Res 42(3):1414–1426

Reinhart B et al (2000) The 21-nucleotide let-7 RNA regulates developmental timing in Caenorhabditis elegans. Nature 403:901–906
Rueda A et al (2015) sRNA toolbox: an integrated collection of small RNA research tools. Nucleic Acids Res 43(W1):W467–W473
Sarnow P, Jopling CL, Norman KL, Schutz S, Wehner KA (2006) MicroRNAs: expression, avoidance and subversion by vertebrate viruses. Nat Rev Microbiol 4(9):651–659
Sauer E, Babion I, Madea B, Courts C (2014a) An evidence based strategy for normalization of quantitative PCR data from miRNA expression analysis in forensic organ tissue identification. Forensic Sci Int Genet 13:217–223
Sauer E, Madea B, Courts C (2014b) An evidence based strategy for normalization of quantitative PCR data from miRNA expression analysis in forensically relevant body fluids. Forensic Sci Int Genet 11:174–181
Schirle NT, Sheu-Gruttadauria J, MacRae IJ (2014) Gene regulation. Structural basis for microRNA targeting. Science 346(6209):608–613
Schroeder A et al (2006) The RIN: an RNA integrity number for assigning integrity values to RNA measurements. BMC Mol Biol 7:3
Sempere LF, Cole CN, McPeek MA, Peterson KJ (2006) The phylogenetic distribution of metazoan microRNAs: insights into evolutionary complexity and constraint. J Exp Zool B 306B(6):575–588
Shi W, Hendrix D, Levine M, Haley B (2009) A distinct class of small RNAs arises from pre-miRNA-proximal regions in a simple chordate. Nat Struct Mol Biol 16(2):183–189
Sperling EA, Peterson KJ (2009) microRNAs and metazoan phylogeny: big trees from little genes. In: Telford MJ, Littlewood DTJ (eds) Animal evolution—genomes, trees and fossils. Oxford University Press, Oxford
Stocks MB et al (2012) The UEA sRNA workbench: a suite of tools for analysing and visualizing next generation sequencing microRNA and small RNA datasets. Bioinformatics 28(15):2059–2061
Suzuki HI et al (2015) Small-RNA asymmetry is directly driven by mammalian Argonautes. Nat Struct Mol Biol 22(7):512–521
Tarver JE, Donoghue PC, Peterson KJ (2012) Do miRNAs have a deep evolutionary history? BioEssays 34(10):857–866
Tarver JE et al (2013) miRNAs: small genes with big potential in metazoan phylogenetics. Mol Biol Evol 30:2369
Taylor RS, Tarver JE, Hiscock SJ, Donoghue PC (2014) Evolutionary history of plant microRNAs. Trends Plant Sci 19(3):175–182
Toedling J et al (2012) Deep-sequencing protocols influence the results obtained in small-RNA sequencing. PLoS One 7(2):e32724
van Dijk EL, Jaszczyszyn Y, Thermes C (2014) Library preparation methods for next-generation sequencing: tone down the bias. Exp Cell Res 322(1):12–20
Wang X, Liu XS (2011) Systematic curation of miRBase annotation using integrated small RNA high-throughput sequencing data for *C. elegans* and *Drosophila*. Front Genet 2(25):1–15
Wheeler BM et al (2009) The deep evolution of metazoan microRNAs. Evol Dev 11(1):50–68
Wickens M, Takayama K (1994) RNA. Deviants--or emissaries. Nature 367(6458):17–18
Willenbrock H et al (2009) Quantitative miRNA expression analysis: comparing microarrays with next-generation sequencing. RNA 15(11):2028–2034
Wu J et al (2013) mirTools 2.0 for non-coding RNA discovery, profiling, and functional annotation based on high-throughput sequencing. RNA Biol 10(7):1087–1092
Xu M-J et al (2012) Comparative characterization of MicroRNAs from the liver flukes Fasciola gigantica and F. hepatica. PLoS One 7(12):e53387
Yang JS et al (2010) Conserved vertebrate mir-451 provides a platform for Dicer-independent, Ago2-mediated microRNA biogenesis. Proc Natl Acad Sci U S A 107(34):15163–15168
Zhu E et al (2010) mirTools: microRNA profiling and discovery based on high-throughput sequencing. Nucleic Acids Res 38(Web Server issue):W392–W397
Zuker M (2003) Mfold web server for nucleic acid folding and hybridization prediction. Nucleic Acids Res 31(13):3406–3415

Chapter 7
Analysis of Long Noncoding RNAs in RNA-Seq Data

Farshad Niazi and Saba Valadkhan

7.1 Introduction

One of the most exciting outcomes of the high-throughput analysis of the transcriptome of higher eukaryotes has been the discovery of thousands of novel transcripts that do not seem to have any protein-coding capacity. These RNAs, collectively named the long noncoding RNAs (lncRNAs), are found in both prokaryotes and eukaryotes; however, they seem to be particularly abundant in higher eukaryotes including both animals and plants (Rinn and Chang 2012; Morris and Mattick 2014). Some lncRNAs can be tens of thousands of nucleotides long, and while an originally proposed arbitrary lower length limit of 200 nucleotides should not be applied too strictly, it serves to distinguish this class of RNAs from the small noncoding classes of RNAs such as small nuclear RNAs (snRNAs), small nucleolar RNAs (snoRNAs), microRNAs (miRNAs), etc. (Clark and Mattick 2011; Rinn and Chang 2012; Mattick and Rinn 2015). Due to their relatively recent discovery, lncRNAs remain poorly characterized, and every RNA-seq experiment of sufficient depth will yield several novel lncRNAs that are not present in the existing reference annotations. Further, many protein-coding RNAs have alternatively processed isoforms that do not have protein-coding capacity and fall into the category of lncRNAs (Carninci et al. 2005; Djebali et al. 2012; Bernstein et al. with ENCODE Project Consortium

F. Niazi, M.D.
Department of Molecular Biology and Microbiology, Case Western Reserve University School of Medicine, 10900 Euclid Avenue, W210b, Cleveland, OH 44106, USA
e-mail: farshad.niazi@gmail.com

S. Valadkhan, M.D., Ph.D. (✉)
Department of Molecular Biology and Microbiology, Case Western Reserve University School of Medicine, 10900 Euclid Avenue, W210a, Cleveland, OH 44106, USA
e-mail: saba.valadkhan@case.edu

A.M. Aransay, J.L. Lavín Trueba (eds.), *Field Guidelines for Genetic Experimental Designs in High-Throughput Sequencing*,
DOI 10.1007/978-3-319-31350-4_7

et al. 2012). Thus, discovery of novel lncRNAs and noncoding isoforms of protein-coding genes is quickly becoming a major aspect of analysis of every RNA-seq experiment.

While the computational steps involved in detection and analysis of the noncoding transcriptome are largely identical to the workflow of a typical RNA-seq analysis aiming at study of the protein-coding genes (reviewed by Ramsköld et al. 2012a; Trapnell et al. 2012; Anders et al. 2013, and Chaps. 4 and 5 in this volume), some aspects of the biology of lncRNAs require fine-tuning of several steps of the workflow. The low expression level of many lncRNAs, the abundance of transposon-derived sequences in them, and frequent genomic overlap with other protein-coding and noncoding genes (Carninci et al. 2005; Djebali et al. 2012; Bernstein et al. with ENCODE Project Consortium et al. 2012) present unique challenges in identification and accurate quantification of expression of these transcripts. Further, the lack of poly(A) tails in many lncRNAs and monoexonic genomic architecture requires changes in the sample preparation steps for RNA-seq experiments. Finally, many lncRNAs are expressed in a cell type- and state-specific manner, and thus, provided there is sufficient sequencing depth, almost every RNA-seq experiment can potentially yield novel lncRNAs that are specific to the cell type and state being studied. Identification of such novel transcripts requires the additional computational steps of transcriptome assembly and sequence-based analysis of protein-coding potential.

Even among the currently annotated lncRNAs, the vast majority remain unstudied, providing an exciting opportunity for uncovering novel aspects of biological processes. Functional analysis of a very small fraction of lncRNAs suggests their involvement in virtually every aspect of cellular function, with regulation of nuclear events including epigenetic state of chromatin and transcription emerging as major themes in lncRNA function (Rinn and Chang 2012; Amaral et al. 2013; Rinn 2014). Although computational analyses cannot replace functional "wet bench" studies for defining the cellular role of lncRNAs, they can provide clues that can guide the wet bench studies and help select the most exciting candidates for further analysis. In this review, we will discuss the specific requirements and considerations needed for a successful analysis of the long noncoding transcriptome in RNA-seq experiments, followed by some computational analysis steps that will serve as a first step toward functional characterization of the lncRNAs identified in the RNA-seq computational analysis steps.

7.2 Practical Considerations in Defining the Long Noncoding Transcriptome by RNA-Seq

As previously mentioned, the majority of the computational analysis steps used for characterization of the long noncoding transcriptome are similar to those used for the study of protein-coding genes. However, considering the challenges discussed above, some changes to the protocols will help improve the detection and characterization of the lncRNAs, as detailed below. The following recommendations are written with the higher eukaryotic transcriptome analysis using short-read RNA-seq (Illumina technology) in mind. However, with minor modifications, they can also be applicable to other sequencing platforms and organisms.

7.2.1 Design of the Study and the Need for Replicates

An interesting feature of lncRNAs, which distinguishes them from protein-coding RNAs, is that their expression can be highly specific to a certain cell type, a certain developmental stage, or a certain cellular state (Mercer et al. 2008; Djebali et al. 2012; Bernstein et al. with ENCODE Project Consortium et al. 2012). In a large-scale study across 15 human cell types, over half of protein-coding genes showed ubiquitous expression, with 7 % being cell type specific. For lncRNAs, the ubiquitous and cell type-specific RNA comprised 10 % and 29 % of all expressed lncRNAs, respectively (Djebali et al. 2012). Thus, the use of a heterogeneous population of cells comprising different cell types may result in loss of signal for lncRNAs that are expressed in a small subset of cells and an overall lncRNA expression profile that is very difficult or impossible to deconvolute without resorting to additional studies. The issue is compounded by the low expression level of many lncRNAs, which makes the detection of their expression difficult even in highly homogeneous samples. If profiling the gene expression pattern of complex tissues or highly heterogeneous samples are of interest, single-cell RNA-seq (see below) is likely to be a better choice than the commonly used population-level RNA-seq. The use of more-or-less homogeneous cell populations such as cell lines or primary cultured cells of high purity, when possible, will provide the cleanest and most informative lncRNA profiles when the population-level RNA-seq is employed.

Another important aspect of lncRNA biology is the responsiveness of the promoter of many lncRNAs to cellular stress. This becomes important during the preparation of cells for RNA extraction, as any preparation or processing step that results in cellular stress can result in an expression pattern that does not reflect the condition being studied, but rather the impact of the cellular stress caused during the processing steps. Thus, gentle handling of cells prior to the harvest of cellular RNA for high-throughput sequencing is strongly recommended.

7.2.1.1 Determining the Number of Needed Replicates

Since lncRNAs are expressed at levels lower than protein-coding genes, a higher depth of sequencing and/or higher replicate number will be required for obtaining sufficient statistical power in the differential expression analysis. Calculation of statistical power for lncRNA analysis (Ching et al. 2014) and general RNA-seq analysis (Hart et al. 2013) has been previously discussed. In addition, the studies cited above have provided simple tools that allow investigators to define the number of needed replicates and depth of sequencing required to achieve an acceptable statistical power.

7.2.2 Preparation of RNA for the Use in Deep Sequencing

A significant percentage of lncRNAs function in regulation of nuclear events and are therefore predominantly or even exclusively nuclear in their subcellular localization, although the exact ratio of nuclear to cytoplasmic copies of RNA may depend on the cell type being studied (Djebali et al. 2012). Thus, if the analysis of the noncoding transcriptome is one of the goals of the RNA-seq study, care must be taken to ensure the inclusion of nuclear RNAs during the RNA-extraction process. Importantly, many lncRNAs are chromatin associated and may be discarded along with the chromatin fraction during the extraction processes. To prevent this from occurring, DNase treatment of the chromatin fraction and re-extraction for RNA are recommended. If a deeper analysis of lncRNAs is of interest, cellular fractionation into cytoplasmic, nucleoplasmic, and chromatin fractions followed by RNA extraction and sequencing will provide key insights into the potential functional category that the identified lncRNAs may belong to. For example, the absence of a novel identified RNA in the cytoplasmic fraction combined with computational evidence of lack of protein-coding capacity (see below) is strongly indicative of the noncoding nature of the RNA. Similarly, if RNA is predominantly found in the chromatin-associated fraction, it is likely to function in regulation of an aspect of chromatin function, such as epigenetic events or transcription.

Large-scale studies have shown that many lncRNAs are found in both polyadenylated and non-polyadenylated cellular RNA fractions, and a small percentage (~5 % in human) of the annotated lncRNAs are exclusively found in the non-polyadenylated fraction (Djebali et al. 2012; Bernstein et al. with ENCODE Project Consortium et al. 2012). The percentage of non-polyadenylated RNAs is likely to increase with technical improvements in RNA-seq, as many such RNAs are expressed at low levels or have short half-lives. Thus, in order to capture the entire complexity of the long noncoding transcriptome, it is important to refrain from the use of RNA preparation strategies that select for polyadenylated RNAs. Another common feature of lncRNAs is their low expression level compared to protein-coding genes (Derrien et al. 2012; Djebali et al. 2012; Iyer et al. 2015). This makes them particularly vulnerable to loss as a result of even minute amounts of degradation, so for studies involving the analysis of lncRNAs, extra care must be taken during the RNA-extraction process. In our own experience, the use of RNA-extraction strategies that involve a column purification step results in loss of a fraction of low copy number RNAs, and it's best to avoid them.

7.2.3 Preparation of Sequencing Libraries

During the library preparation step for RNA-seq, cellular RNAs are converted to cDNA fragments which will hopefully reflect the RNA population within the sample, and additional sequences are added to the ends of the cDNA fragments to assist in the sequencing step. As mentioned above, in addition to the small percentage of lncRNAs that are exclusively found in the non-polyadenylated fraction, many other lncRNAs exist in both polyadenylated and non-polyadenylated forms (Djebali et al. 2012; Bernstein et al. with ENCODE Project Consortium et al. 2012). Thus, to define their cellular abundance both polyadenylated and non-polyadenylated transcripts should be included in the RNA-seq experiment. Several library preparation kits have been developed that lack a poly(A) selection step and, thus, include all cellular RNAs in the library. However, since ribosomal RNAs and other abundant housekeeping RNAs can make up over 90 % of cellular RNA content, it is essential to exclude them from the library preparation step. All "total RNA" library preparation kits have a ribosomal RNA depletion step that should be carefully followed to ensure elimination of these abundant transcripts from the resulting library. However, it should be mentioned that when total cellular RNA is used in library preparation, nascent transcripts will also be included in the library, and this should be taken into consideration in the downstream computational analysis steps, especially if quantitation of the level of fully processed RNAs is desired (Sultan et al. 2014).

Another consideration regarding the library preparation step for lncRNA analysis is related to the diversity of genomic loci from which lncRNAs originate. The large-scale transcriptome analyses have revealed a complex and overlapping pattern of transcription in higher eukaryotes in which many transcribed units overlap each other in sense or antisense orientations (Djebali et al. 2012; ENCODE Project Consortium et al. 2012). This is particularly the case with lncRNAs, which originate from genomic loci both within and outside of other transcribed units. As can be seen in Fig. 7.1, lncRNAs can overlap protein-coding or other noncoding RNAs in the sense or antisense orientation by originating from a promoter within an exon or intron of the overlapped gene or from a promoter located in its 3′ UTR or further downstream. Another commonly observed conformation of transcribed units in the higher eukaryotes is the "twin" transcripts originating from the so-called bidirectional promoters (Fig. 7.1) (Adachi and Lieber 2002; Wakano et al. 2012; Uesaka et al. 2014). While promoters inherently lack directionality, it has been shown that their transcription is often limited to one direction due to the sequence context (Almada et al. 2013; Ntini et al. 2013). However, it has been shown that about 11 % of human genes have a detectable transcript originating from the same promoter

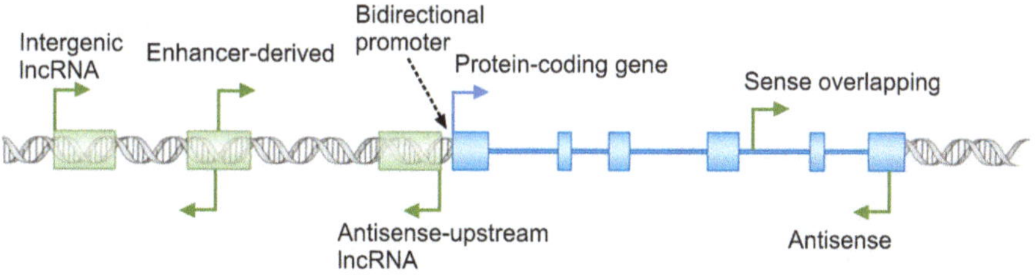

Fig. 7.1 Diverse genomic origins of lncRNAs. The broken arrows mark the location of transcription start sites and direction of transcription

region in the opposite orientation (Adachi and Lieber 2002; Trinklein et al. 2004). In many cases, these "twin" transcripts are noncoding/protein-coding pairs, and in some studied examples, one of the twins regulates the expression of their promoter-sharing RNA (Wei et al. 2011; Uesaka et al. 2014). Thus, inclusion of lncRNAs originating from bidirectional promoters in RNA-seq analysis is needed for obtaining a complete picture of the cellular regulatory networks. In addition, many lncRNAs originate from bidirectional promoters in enhancer loci and are thought to be needed for the function of the enhancer in which they originate (Fig. 7.1) (Lam et al. 2014). Other lncRNAs arise from genomic loci that don't overlap with other genes or enhancer elements and fall into the "intergenic" lncRNA class (Fig. 7.1). From the above discussion, it is clear that knowledge of the directionality of transcription at the lncRNA loci is essential for detecting their presence, especially in the case of lncRNAs overlapping another gene in the antisense manner and those originating from bidirectional promoters. For example, in the absence of directional data, a lncRNA originating from a bidirectional promoter that also gives rise to a protein-coding gene can be mistaken for an alternative isoform of the protein-coding gene resulting from the use of an alternative upstream promoter. Further, as mentioned above, RNA-seq experiments often lead to discovery of novel intergenic transcripts for which no directionality data is available in public databases. While some transcript assembly packages such as Cufflinks (see below) (Trapnell et al. 2012) can predict directionality from canonical splice site information in some cases, this is often not feasible due to lack of splice sites which is observed in many lncRNAs, the presence of non-canonical splice sites, and at complex loci. Thus, preservation of the directionality of the cellular RNAs during the library preparation step is essential for analysis of the lncRNAs. Many library preparation kits preserve strandedness information either through the use of distinct RT and PCR primers or via incorporation of chemically distinct nucleotides such as dUTP during the cDNA synthesis step. While both methods can successfully capture directionality, in our experience and that of others, the second method has provided cleaner data (Levin et al. 2010).

As mentioned above, lncRNAs are often expressed at much lower levels compared to protein-coding genes. Estimation of the abundance of low expression level genes by RNA-seq is often plagued by the higher level of technical "noise" which partly results from the library preparation step. The use of primers carrying random sequence tags (a.k.a. unique molecular identifiers, UMI) that will allow identification of the PCR-amplified fragments in the downstream analyses has been shown to improve reproducibility especially for low-abundance genes (Islam et al. 2014; Grün et al. 2014) and, thus, should be used in RNA-seq studies aiming at analysis of lncRNAs.

7.2.4 Sequencing

While there are several sequencing platforms available, the majority of RNA-seq results are obtained using the Illumina technology, with tens of millions of shorter (~100 nucleotide long) reads generated. However, other platforms that generate longer reads such as the 454 technology and Pacific Bio instruments have also been successfully used for the detection of lncRNAs (Tilgner et al. 2013). With the use of either technology, obtaining a sufficiently high number of reads is essential to detect the expression of lncRNAs, as many of them are expressed in lower copy numbers compared to protein-coding genes (Derrien et al. 2012; Djebali et al. 2012; Iyer et al. 2015). With the use of Illumina platform, a sequencing depth of 60–100 million reads seems to yield a good coverage of most lncRNAs, although deeper sequencing will provide more detailed information. Also, the paired-end sequencing option is strongly recommended for RNA-seq studies aiming to analyze the noncoding transcriptome. Since many lncRNA are rich in retroelement-derived sequences (Kelley and Rinn 2012; Kapusta et al. 2013), paired-end sequencing improves the mappability of the reads originating from lncRNAs and, thus, enhances their detection. Even for more abundant RNAs, although gene-level abundance can be determined using single-end RNA-seq, paired-end reads greatly improve the detection of splicing patterns and isoform-level expression quantitation (Li and Dewey 2011).

7.2.5 *Quality Control and Preprocessing*

Once the sequencing results are accessible, a quality control step should be performed to ensure that sufficient high-quality reads are obtained for the downstream computational steps. Several quality control packages are available; perhaps the most commonly used ones are FastQC (http://www.bioinformatics.babraham.ac.uk/projects/fastqc/) and ShortRead (Morgan et al. 2009; Anders et al. 2013). Removal of adaptor-derived sequences may improve the downstream alignment step, and commonly used packages include trim galore (http://www.bioinformatics.babraham.ac.uk/projects/trim_galore/), Trimmomatic (Bolger et al. 2014) and ShortRead (Morgan et al. 2009). Importantly, removal of reads that map to ribosomal RNAs or other abundant cellular RNAs such as tRNAs and snRNAs can improve the accuracy of gene expression quantitation for low-abundance transcripts such as lncRNAs. In addition to the removal of ribosomal RNAs during the library preparation step, the reads originating from the remaining ribosomal RNAs can be removed at the preprocessing step using the sortmerna package (Kopylova et al. 2012). Sortmerna eliminates ribosomal RNAs from the fastq files using fasta-formatted reference files that contain the sequence of ribosomal RNAs, tRNAs, snRNAs, and other abundant cellular RNAs, which can be obtained from Rfam (Daub et al. 2015) or similar RNA databases. This step is particularly helpful if the quality control step indicates enrichment for sequences with a high (>50 %) GC content, which is found in ribosomal RNAs from most organisms. It is also possible to replace this step with masking or removal of ribosomal RNAs during the alignment and quantification steps, as detailed below.

7.2.6 Alignment of RNA-Seq Reads to the Genome

The next step in RNA-seq pipelines is the mapping or alignment (or pseudoalignment in the case of newer quantification tools, see below) of the reads in fastq files to the target genome (or transcriptome). As is the case for all RNA-seq experiments, genome mapping should be performed with an aligner that is able to map splice sites such as TopHat (Trapnell et al. 2012), STAR (Dobin et al. 2013), GSNAP (Wu and Nacu 2010), MapSplice (Wang et al. 2010), or RUM (Grant et al. 2011), among others. Many of the available aligner tools have been subjected to detailed side-by-side comparison in benchmarking studies (Hatem et al. 2013; Engström et al. 2013; Benjamin et al. 2014). Overall, most aligners were found to perform well especially when transcriptome annotations were provided; however, an important consideration regarding the use of RNA-seq for lncRNA discovery is the impact of the chosen aligner on the transcript assembly step. An analysis of the influence of aligners on transcript assembly indicated that when transcript assembly was performed with Cufflinks, the best results were obtained when TopHat was used in the alignment step along with a reference annotation file (Palmieri et al. 2012; Engström et al. 2013; Hayer et al. 2015). Thus, the combination of TopHat-Cufflinks is likely to be the pipeline of choice in RNA-seq studies aiming at lncRNA analysis and discovery.

If ribosomal RNA-derived reads have not been removed prior to the alignment step, it is possible to remove them by performing a preliminary alignment of the reads to a fasta file containing the sequence of ribosomal RNAs and other abundant cellular RNAs such as tRNAs and snRNAs. The reads that are rejected from this preliminary alignment can then be aligned to the entire genome of choice. A commonly used alternative approach, the use of GTF files lacking rRNAs in the alignment and subsequent steps, is not recommended when the study includes a novel gene discovery step, as the "masked" genes will be "discovered" in that step.

In every RNA-seq experiment, it is essential to perform frequent reality checks on the data as it is processed through the pipeline. For example, after the alignment step, the integrated genome viewer (IGV) or a similar genome viewer can be used to visualize the alignment of the reads to the genome and the transcriptome in order to ensure that the reads are split at the predicted splice junctions. Also, if there are genes which are highly likely to show a change in expression, they should be checked to ensure that there is a detectable change in the number of mapped reads.

7.2.7 Transcript Assembly

The long noncoding transcriptome is at present very poorly characterized, and in every RNA sequencing effort, several novel lncRNAs and novel isoforms of known lncRNAs are discovered. This is partly due to the novelty of this class of RNAs and partly due to the fact that they show a much higher level of cell type and state specificity than protein-coding genes. Very large numbers of such transcripts are discovered in large-scale transcriptome studies (Iyer et al. 2015), and it is likely that the number of lncRNAs is strongly underestimated in current reference annotation databases. The transcript assembly step should be part of every RNA-seq analysis even when the focus is on the protein-coding transcriptome, as even for protein-coding genes, novel isoforms may be observed. Importantly, some less commonly observed isoforms of the protein-coding genes seem to lack protein-coding capacity. In the absence of transcript assembly and transcript-level expression analysis, changes in the expression of a noncoding isoform may be erroneously attributed to an increase in the expression of the main protein-coding isoforms of the gene.

Transcript assemblers, which use the mapped reads to assemble a gene model that explains the observed mapping of the reads and splice junctions, are thus critical for accurate interpretation of RNA-seq data. These fall into two general categories, one category which includes Cufflinks (Trapnell et al. 2012), IsoLasso (Li et al. 2011), and Scripture (Guttman et al. 2010) requires the use of a reference annotation file for optimal function. However, when RNA-seq is performed on a non-model organism with no reference genome and transcriptome available, or when the transcriptome under study contains many structural rearrangements such as those arising within complex cancer genomes, there is a need for transcript assembly platforms that can operate without reference transcriptomes. To address this requirement, several de novo transcript assemblers have been developed including Trinity (Grabherr et al. 2011), SOAPdenovo (Li et al. 2009), transAbyss (Robertson et al. 2010), and Oases (Schulz et al. 2012). However, de novo assemblers cannot effectively assemble low expression level transcripts and thus are only able to reconstruct the most abundant lncRNAs and miss the low copy number RNAs which include the majority of lncRNAs (Schulz et al. 2012). Further, at least in some independent benchmarking studies, they seem to have a higher error rate compared to Cufflinks, which performed better than all the other assemblers studied especially when paired with TopHat as the read aligner (Hayer et al. 2015). Importantly, Cufflinks can also perform novel transcript discovery which should be attempted if the study of lncRNA expression is of interest. Taken together, if a reference transcriptome is available, it is good practice to use reference-based transcript assemblers such as Cufflinks. The use of de novo assemblers is warranted only if a large number of structural rearrangements are suspected. However, even with Cufflinks, the transcript assembly is far from perfect in complex loci (Hayer et al. 2015), and thus, visual inspection of the functionally important subset of transcripts should be performed. A newer transcript assembler, Astroid, has been developed in an attempt to improve the accuracy of the assembly process (Huang et al. 2014). However, it has not yet been independently benchmarked against other available packages.

There are a number of reference annotations available for use in the assembly step and other RNA-seq computational processes including the reference transcriptomes from Gencode (Harrow et al. 2012), UCSC genome browser (Rosenbloom et al. 2015), and RefSeq (Pruitt et al. 2014), all of which include both lncRNA and protein-coding RNA annotations. In addition, there are lncRNA-centric databases and annotations such as lncRNAdb (Quek et al. 2015), NONCODE (Xie et al. 2014),

MiTranscriptome (Iyer et al. 2015), LNCipedia (Volders et al. 2015), and RNAcentral (RNAcentral Consortium 2015) which can be used as additional guides for transcript assembly when the analysis of lncRNAs is required. These annotations are frequently updated, so it is important to use the latest annotation version.

In every RNA-seq experiment of sufficient depth, transcript assembly tools will yield a number of transcript models that originate from loci which are not annotated as genes in the reference transcriptome provided to the transcript assembly tool. These transcripts are likely to be specific to the condition and cell type being studied and hence absent in reference annotations. These are usually flagged as novel RNAs in the output of the transcript assembly step. Further, there are usually a number of novel isoforms for annotated genes, including protein-coding genes. In addition, many novel monoexonic transcripts found in large-scale transcriptome efforts, a large percentage of which are likely to be long noncoding RNAs (Derrien et al. 2012; Djebali et al. 2012), can be very difficult to computationally distinguish from sequencing artifacts resulting from low levels of genomic DNA contamination and are therefore largely excluded from the reference annotations (Cabili et al. 2011; Harrow et al. 2012). Thus, it should be expected that the transcriptome produced during the transcript assembly step of the RNA-seq analysis will contain a significant number of predicted transcripts that are not found in the reference databases. However, it should also be considered that the reference annotations are not necessarily in full agreement with each other even for protein-coding RNAs and much less so for lncRNAs (Frankish et al. 2015), so an RNA that is flagged as novel with the use of one reference transcriptome may already be annotated in another. Once it is confirmed that the transcripts flagged as novel in the assembled transcriptome are indeed not found in any of the reference transcriptomes, they should be analyzed for protein-coding capacity as discussed below.

7.2.8 *Differential Expression Analysis*

7.2.8.1 Quantification of Read Mapping to Desired Features

Once an experiment-specific transcriptome is assembled and the reads are aligned to the genome/transcriptome, the number of reads that map to each gene or transcript should be calculated and used in the differential expression analysis. This quantification step can be performed at the gene level with htseq-count tool from the HTSeq python library (Anders et al. 2015), which is the most commonly used package for determining the number of raw read counts that map to a gene. Alternatively, one of the packages developed for transcript-level quantification of reads can be used. In addition to several packages that perform combined quantification and differential expression analysis at transcript level (see below), a number of newer tools for transcript-level quantification of reads have been developed. RSEM (Li and Dewey 2011) can determine transcript-level abundance from aligned reads in a highly accurate manner (Bray et al. 2015) and can also work in the absence of a reference genome. Sailfish (Patro et al. 2014) and kallisto (Bray et al. 2015) are alignment-free algorithms that offer very fast transcript-level quantification. A newer package from the Sailfish team, Salmon (http://salmon.readthedocs.org/en/latest/), can perform both alignment-based and alignment-free transcript-level quantification and has improved accuracy compared to Sailfish. Although no independent comparison of these tools has been performed, it is likely that RSEM, kallisto, and Salmon compare favorably to the older tools available for transcript-level expression quantification.

7.2.8.2 Removal of Ribosomal RNAs

If ribosomal RNAs and other abundant small housekeeping RNAs were not removed in preprocessing or alignment steps, they should be masked in the differential expression analysis step. Masking the reads that map to ribosomal RNA and abundant small RNA species will help improve quantitation for the low expression level RNAs such as lncRNAs. Depending on the differential expression platform, this can be achieved by using a masking option included in the package (e.g., the -M option in Cuffquant) or manual removal (e.g., using the grep command on a Linux system) of the ribosomal RNAs, tRNAs, and snRNAs from the annotation files provided to the quantification software.

7.2.8.3 Filtering Low Read Count Genes/Transcripts

Since many lncRNAs are expressed at lower levels compared to protein-coding genes (Djebali et al. 2012; Bernstein et al. with ENCODE Project Consortium et al. 2012; Iyer et al. 2015), it is important to perform the filtering of low-abundance transcripts in a very conservative manner when analysis of lncRNAs is one of the goals of the RNA-seq experiment. In general, the use of transcript length- and depth-adjusted abundance indicators such as RPKM/FPKM/TPM (Hebenstreit et al. 2011; Wagner et al. 2012, 2013) is preferable to raw read counts for defining the filtering threshold. Determining the exact value of the threshold must be guided by the empirical examination of several lncRNA loci to determine what would be a good threshold in the particular set of samples being studied. In general, the threshold needed for studies of the noncoding transcriptome is much lower than what is commonly used for the analysis of protein-coding genes (Hebenstreit et al. 2011; Wagner et al. 2012, 2013).

7.2.8.4 Choice of a Differential Expression Analysis Platform

As discussed above, to obtain a picture of the entire long noncoding transcriptome, differential expression analysis should be performed at both gene and transcript levels. While the gene-level quantification algorithms show a high level of accuracy, until recently, the transcript-level quantification tools had not reached the desired level of robustness. Since the concept of "gene" is likely to be functionally outdated soon and replaced by transcripts as the functionally relevant entities, it is likely that transcript-level analyses will become the dominant method of differential expression analysis in near future. However, since many downstream packages such as pathway analysis tools have been developed with gene-level analysis in mind, for the moment, both gene-level and transcript-level differential expression analyses should be attempted. A number of packages have been developed for performing the differential expression analysis, and based on the type of quantitation data provided (gene- or transcript-level quantitation of reads), they can perform gene-level or transcript-level differential expression tests. There are also a number of "combined" packages that perform both transcript-level quantification and differential expression analysis. It is advisable to use more than one analysis platform to improve quantitation accuracy for low-abundance transcripts which include most lncRNAs. The following paragraphs contain a discussion of the strengths and weaknesses of the most commonly used packages.

7.2.8.5 Differential Expression Analysis Tools

Similar to the other steps in the RNA-seq analysis pipeline, there are several special considerations that must be taken for optimal detection and analysis of lncRNAs. During the differential expression analysis in particular, two aspects of the biology of lncRNAs should be taken into account, namely, their generally low expression level compared to protein-coding genes and their highly state-specific expression pattern resulting in a high rate of binary (all or none) expression changes.

A benchmarking study by Rapaport and colleagues has compared several commonly used differential expression analysis packages for their ability to accurately determine differential expression of the low expression level genes/transcripts (Rapaport et al. 2013). The results of this study suggest that poissonSeq (Li et al. 2012) and edgeR (Robinson et al. 2010) packages perform the most robust differential expression analysis overall for low expression genes especially when at least three replicates were analyzed from each study group. Even with these two packages, the depth of sequencing is very important to the extent of detection (sensitivity) for low expression genes, as is the number of replicates (Rapaport et al. 2013). However, specificity of the analysis did not seem to be affected with low sequencing depth. On the other hand, for high expression genes, the impact of sequencing depth on sensitivity decreased as the level of expression of genes increased. In both low and highly expressed genes, the number of replicates made a stronger contribution to the accuracy of differential expression analysis than sequencing depth. Thus, when budget is limited, it's better to divide the same read number into 3–4 replicates of moderate depth (e.g., 60 million reads each) rather than having fewer replicates with higher read numbers (Rapaport et al. 2013). Other benchmarking studies (Seyednasrollah et al. 2015; Ching et al. 2014) also found edgeR to show one of the best performances in overall sensitivity and specificity, although these analyses did not specifically assess the accuracy of detection of low expression level genes. Also, poissonSeq was not included in these benchmarking studies.

In cases when the expression level of a gene is zero in one of the conditions, poissonSeq, Limma (Ritchie et al. 2015), and Bayseq (Hardcastle and Kelly 2010) packages performed best in accurately calculating the significance of the change (Rapaport et al. 2013). The commonly used package Cuffdiff did not perform strongly in gene-level analysis in benchmarking studies (Rapaport et al. 2013). Thus, edgeR and poissonSeq packages seem to be the best differential expression analysis tools for the study of lncRNAs.

7.2.8.6 Transcript-Level Differential Expression

Several packages have been developed for performing both read quantitation and differential expression analysis at transcript level. The Cufflinks package (Trapnell et al. 2012) is perhaps the most commonly used one. Additional packages include DEXseq (Anders et al. 2012), which can determine differential expression at exon level; rMATS (Shen et al. 2014), which can perform differential splicing analysis; EBseq (Leng et al. 2013); BitSeq (Glaus et al. 2012); and rnaSeqMap (Okoniewski et al. 2012), among others. As mentioned above in Sect. 7.2.8.1, a number of new tools for transcript-level quantitation of reads have been developed, and their output can be used with most differential expression analysis packages discussed in the previous subsection. While no independent benchmarking studies have done a side-by-side comparison of the performance of the transcript-level differential expression packages, it is likely that the newer quantification tools will be good candidates to try, especially considering their significantly shorter processing time.

Finally, although many of the existing software packages for differential expression analysis provide a robust assessment of changes in expression, it is important to tailor the approach to the particular requirements of the study in hand. For example, a large-scale study on the long noncoding transcriptome of cancer (Iyer et al. 2015) used a custom-made nonparametric differential expression method which allowed sensitive detection of differential expression in the highly heterogeneous samples such as tumor subtypes.

7.2.8.7 Quality Control Check

As in all high-throughput studies, it is important to do frequent reality checks to ensure the differential expression analysis conforms to the visual inspection of the read density on genes/transcripts. This can be done for some of the genes/transcripts identified as the top differentially expressed species using IGV or a similar genome viewer to ensure that the computationally defined level of differential expression is commensurate with the difference in the number of reads mapping to the gene or transcript of interest.

7.2.9 Data Sharing

In addition to the deposition of the RNA-seq raw data into the public sequencing data repositories, the assembled transcriptomes that include lncRNAs should be deposited into GenBank (http://www.ncbi.nlm.nih.gov/genbank/), RNAcentral (http://rnacentral.org/), or similar warehouse-type websites.

7.3 Discovery of Novel lncRNAs: Defining the Protein-Coding Potential

As discussed in the transcript assembly section, every RNA-seq experiment will yield a number of assembled transcripts that are not found in reference databases. These will include transcripts originating from novel genic regions and novel isoforms of known genes. These novel isoforms are functionally important to characterize, as large-scale experiments indicate that many protein-coding RNAs have noncoding isoforms (Carninci et al. 2005; Bernstein et al. with ENCODE Project Consortium et al. 2012). Such isoforms may not only complicate the interpretation of the differential expression data for the genes they originate from but also may regulate the protein-coding function of the coding isoforms. For example, if they share 3′ UTR sequences with the protein-coding isoforms, they can act as sinks for miRNAs and other regulatory factors, underscoring the importance of their detection and analysis. The same is true with expressed pseudogenes, which fall under the broad category of lncRNAs. The first step in the characterization of these transcripts and identifying novel lncRNAs among them is a rigorous computational study of their protein-coding potential (Dinger et al. 2008; Ilott and Ponting 2013; Mattick and Rinn 2015).

Effort should be made to ensure the accurate identification of the beginning and ends of the transcripts and their splicing architecture before the analysis of their coding potential. Although many lncRNAs are multiexonic, mono-exonic RNAs seem to constitute a much larger proportion of lncRNAs compared to protein-coding genes (Niazi and Valadkhan 2012; Derrien et al. 2012; Djebali et al. 2012). Detection of unspliced transcripts can also result from artifacts caused by a low level of genomic DNA contamination or due to currently incomplete information on transcript structures. This, in turn, can result from the low expression level of the RNA which prevents the accurate assignment of its 5′ and 3′ ends or splicing patterns. In such cases, the use of publicly available datasets from relevant tissues or cellular states that may have better coverage of this region is recommended.

7.3.1 ORF Analysis

The first step in defining the protein-coding capacity of a novel transcript is ORF analysis, which can be performed with the NCBI ORF finder or similar tools. It has been shown that the vast majority of protein-coding genes in mammals are over 100 amino acid long; thus, the length of predicted ORFs in a novel transcript can provide clues into its protein-coding capacity. However, in the absence of additional information, ORF length is not very useful, as on the one hand, short peptides can be functional, and, on the other, longer ORFs can occur fortuitously in longer RNAs without being in a context conducive to translation. To further characterize the protein-coding potential of an ORF, a number of tools have been developed including PhyloCSF (Lin et al. 2011), which relies on phylogenetic codon substitution frequency among other parameters; CONC (Liu et al. 2006) and CPC (Kong et al. 2007), both of which rely on support vector machine (SVM)-mediated classification of RNAs based on the robustness of ORFs and their protein-coding features; CNCI (Sun et al. 2013b) and PLEK (Li et al. 2014), both of which use k-mer frequencies and a SVM algorithm to separate lncRNAs and protein-coding RNAs; and CPAT (Wang et al. 2013), which uses ORF size and coverage, Fickett statistics, and hexamer nucleotide usage. Although no independent benchmarking studies have been performed to compare the robustness of the predictions made by these tools, the developing team of PLEK compared the sensitivity and specificity of their tool with CPC, phyloCS, and CNCI (Li et al. 2014). Their results, overall, suggested that each package has its own strengths and weaknesses, and thus, the use of more than one tool may be necessary to ensure a more robust prediction. Also, the details of the RNA-seq experiment may determine which tool will be most useful. For example, PhyloCSF strongly relies on interspecies alignments, which makes it restricted to well-aligned regions. The alignment-free tools that use the intrinsic sequence features of the transcript are more powerful when annotations are not complete, but are sensitive to errors caused by indels that occur during sequencing, which are common with 454 and Pacific Biosciences sequencing platforms (Quail et al. 2012; Loman et al. 2012).

7.3.2 Comparison with Existing Protein Databases

Another helpful approach in distinguishing novel protein-coding genes from noncoding genes is comparison of their sequence and potential coded peptides with protein domain databases such as PFAM (Finn et al. 2014) and databanks of large-scale proteomics efforts. For example, one can search large proteomics datasets for peptides that uniquely map to transcripts of unknown coding potential, as implemented in Pinstripe suite of programs (Gascoigne et al. 2012) and also performed in a large-scale effort at identification of cancer-related lncRNAs (Iyer et al. 2015). A partial list of suitable peptide databases includes the EBI Proteomics Identifications Database (PRIDE) peptide database (Vizcaíno et al. 2013), the Human Proteome Map (Kim et al. 2014), UniProt/Swiss-Prot (The UniProt Consortium 2015), PFAM (Finn et al. 2014), and more recent databases explicitly aiming at defining the coding potential of novel transcripts in the genome (Khatun et al. 2013). This approach, while certainly valuable, has two major shortcomings. First, lack of a match does not mean lack of protein-coding capacity, as a novel RNA may code for a protein that is not similar to any previously analyzed protein. On the other hand, the presence of a domain similar to a known protein domain does not necessarily mean that the RNA containing it is translated, as this sequence may be in a sequence or location context within the transcript that is not amenable to translation. For example, an RNA that is strictly nuclear in its localization is unlikely to code for a peptide even if it contains a short ORF that can potentially code for a peptide containing a known protein motif. Further, some lncRNAs and transcribed pseudogenes have evolved from protein-coding genes or overlap them and, thus, do carry sequences corresponding to those coding for known protein motifs. Further, many proteome databases such as PFAM contain endogenous retroviral protein sequences which will match many transposon-derived sequences found in lncRNAs, leading to erroneous identification of lncRNAs as coding transcripts.

A number of platforms have been developed for quick and user-friendly annotation of the genes found in the output of RNA-seq experiments. A recent example is Annocript (Musacchia et al. 2015), which uses several annotation databases, BLASTX and BLASTP searches, and two packages (dna2pep (Wernersson 2006) and Protrait (Arrial et al. 2009)) for determining the protein-coding capacity of putative lncRNAs in a transcriptome. Additional examples of such packages have been developed in recent years (Sun et al. 2013a).

7.3.3 Ribosome Profiling

Determining whether the novel identified transcripts associate with the polysomes, or ribosome profiling, is also a helpful approach (Ingolia et al. 2011). It has been shown that a small number of transcripts that were assigned to the lncRNA category do associate with the ribosomes (Guttman et al. 2013); however, whether this results in the formation of any functional peptides or has any other functional consequences remains to be determined. Nonetheless, analysis of the polysome-associated RNAs under conditions that match the ones used to obtain the RNA for the sequencing experiment will be very helpful. Computational analysis of the existing RNA-seq experiments performed on ribosome-bound RNAs can also be insightful, although the association of RNAs with polysomes may depend on cell type and cellular state being studied.

Based on the outcome of the above analyses, it should be possible to categorize the novel RNAs into most likely coding, most likely noncoding, and transcripts of unknown coding potential, this latter group being the novel RNAs that seem to have some coding potential, but it's not clear enough to assign them to the protein-coding category (Cabili et al. 2011). While not all novel transcripts identified in an RNA-seq study will be noncoding, large-scale studies to date suggest that the majority of such transcripts indeed do not code for peptides (Bánfai et al. 2012; Djebali et al. 2012; Khatun et al. 2013).

7.4 Computational Characterization of the lncRNAs

In some cases, a novel but potentially interesting transcript is expressed at such a low level that distinguishing it from sequencing artifacts becomes necessary. In such cases, analysis of the publicly available ChIP-seq and DNaseI hypersensitivity data mapping to the genomic locus of the novel RNA can be very helpful. For example, the presence of histone 3 lysine 4 trimethylation (H3K4me3) marks and DNase I hypersensitivity sites close to its transcription start site and the presence of RNA polymerase II (Pol II) and H3K36me3 broad peaks over the body of the putative transcript are strong evidence for the presence of a transcript in this region. Further, chromatin marks can provide insights into the potential function of lncRNAs. For example, the presence of H3K4me1 marks which are associated with active or poised enhancers at the locus of lncRNA can point to an enhancer-associated function for the RNA (Lam et al. 2014). The genomic locus of an RNA can provide additional clues to its function, for example, 11 % of human genes are thought to originate from bidirectional promoters (Adachi and Lieber 2002; Trinklein et al. 2004) and many such promoters give rise to lncRNA/protein-coding RNA pairs which often affect the expression of each other (Wei et al. 2011; Uesaka et al. 2014).

Although it has been shown that lack of conservation of the primary sequence of lncRNAs does not indicate lack of a conserved function (Pang et al. 2006; Ulitsky et al. 2011), the presence of a high level of conservation can strengthen the likelihood that the lncRNA plays an important cellular role. A number of packages, including phyloP (Pollard et al. 2010) and phaseCons (Siepel et al. 2005), can be used for defining the extent of conservation of lncRNAs. Finally, predicting the general area of function of lncRNAs through identification of protein-coding genes with similar expression patterns or via more sophisticated, weighted gene co-expression network analyses (Langfelder and Horvath 2008) has been attempted (Liao et al. 2011; Guo et al. 2013; Jiang et al. 2015; Xiao et al. 2015; Bergmann et al. 2015). However, the usefulness of such approaches remains to be determined.

7.4.1 RNA Editing in lncRNAs

A large fraction of higher eukaryotic cellular RNAs are subjected to posttranscriptional modifications. Most frequently observed RNA editing events involve deamination of A residues to inosine, but other modifications including methylation are also abundantly found in cellular RNAs. While editing occurs both in coding and noncoding RNAs, it seems to be most abundant in noncoding RNAs and noncoding regions of protein-coding genes. Since these changes affect the structure and function of the RNAs (Nishikura 2010; Li and Mason 2014), defining the location of such changes is of interest. The use of RNA-seq for this purpose has been very fruitful (Picardi et al. 2010; Ramaswami et al. 2012; Ramaswami and Li 2014), and there are a number of tools that have been developed to simplify the detection of RNA editing sites in deep sequencing data (Picardi et al. 2014).

7.4.2 Detection of Circular RNAs

Another interesting class of cellular RNAs, circular RNAs, have been implicated in regulation of transcription and miRNA function by acting as miRNA sponges, which places them within the broader category of regulatory long noncoding RNAs (Guo et al. 2014; Lasda and Parker 2014; Chen and Yang 2015). A number of tools have been developed for discovery of circular RNAs in transcriptomic studies (Zhang et al. 2014; Gao et al. 2015; Pan and Xiong 2015).

7.5 Special Considerations for Post-analysis Validation Steps on lncRNAs

Similar to other high-throughput studies, RNA-seq results should also be validated using low-throughput approaches. RT-PCR-based validation experiments are the simplest and most commonly used validation experiments. In the case of lncRNAs, it is important to perform strand-specific RT-PCR, especially for low expression level lncRNAs, and to use primers that flank predicted exon-exon junctions to ensure that the obtained signal is not affected by genomic DNA contamination. Using both gel-based and qPCR-based approaches is recommended, as for low expression level lncRNAs, RT-qPCR may result in artifactual results. We recommend the use of radioactively labeled primers for low expression level lncRNAs followed by visualization on PAGE in order to obtain a clear signal without the need for too many PCR amplification cycles.

For downstream functional studies, it is important to appreciate that unlike protein-coding RNAs, addition of sequences to the beginning and ends of lncRNAs is not appropriate, as the RNA itself is the functional molecule, and thus, it should not be modified. Similarly, "fusion" to GFP or similar protein tags is completely unacceptable. If cloning of the lncRNA into a plasmid is to be performed, the annotated 5′ end of the lncRNA should be placed at the transcription start site of the promoter used in the plasmid, and the 3′ end of the lncRNA should be placed at the cleavage site of the plasmid for polyadenylated lncRNAs and transcription stop site for non-polyadenylated ones. Clearly, a non-polyadenylated RNA should not be expressed with a poly(A) tail, as it changes the localization and proteome of the RNA. In our experience with peer review of the literature, such mistakes are unfortunately common and result in artifactual data being reported in literature. Additional guidelines for experimental analysis and manipulation of lncRNAs and lncRNA genes are discussed in a recent review (Bassett et al. 2014).

7.6 Naming Novel lncRNAs

Guidelines by the HUGO Genome Nomenclature Committee (HGNC) (Wright 2014) provide a helpful framework for naming the newly discovered putative lncRNAs. The suggested guidelines propose including some information about the genomic context of the lncRNA locus in its name and promote the inclusion of functional information when available. For example, a novel putative lncRNA of unknown function that overlaps the *MET* gene in the antisense orientation should be named MET-AS1, if it were encoded within an intron of the *MET* gene, MET-IT1 (for intronic), and if it originated from a bidirectional promoter that also gives rise to the *MET* gene, MET-AU1 (for antisense upstream). However, once there is functional information available on a gene, it should be named based on that function (Wright 2014).

7.7 Capturing the Full Complexity of the Noncoding Transcriptome

7.7.1 Capture-Seq

As mentioned above, the low expression level of many lncRNAs and their highly state-dependent expression pattern, together with their relatively recent emergence as a functionally important class of transcripts, have limited our current knowledge of the extent and pattern of expression of long noncoding RNAs, even in highly studied organisms such as human and mouse. With improvements in sequencing technology resulting in increase of sequencing depth and reduced cost, and wider interest in discovery of the function of the noncoding transcriptome, many of the existing gaps in our knowledge of this class of RNAs are likely to be addressed in the coming years. In addition, creative use of existing technologies can make a significant contribution to our understanding of the complexity of the long noncoding transcriptome. In an exciting step in this direction, Mercer and colleagues have made clever use of tiling arrays to select and enrich RNAs transcribed from a targeted region of the genome, followed by RNA-seq analysis (Mercer et al. 2012, 2014). This enrichment-sequencing approach (capture-seq) has yielded unprecedented insight into the extent of intergenic transcription and the enormous complexity of the nonprotein-coding (and protein-coding) transcriptome. The use of this approach can provide invaluable clues into the expression pattern of rare transcripts or rare isoforms of more abundant transcripts that may have important functions or act as biomarkers.

7.7.2 Single-Cell RNA-Seq

The use of conventional RNA-seq, which is performed on RNA extracted from a large number of cells, will likely continue to yield pivotal information on the long noncoding transcriptome for many years to come. However, studies focusing on protein-coding genes have shown a significant level of cell-to-cell variation in expression level and splicing pattern of many genes within the same cell population, both under basal conditions and in response to external stimuli (Shalek et al. 2013). Further, many cellular processes such as reprogramming are stochastic, and thus, cell population-level RNA-seq is of very limited use in elucidating the processes and pathways involved. The emergence of single-cell RNA-seq technologies (Saliba et al. 2014; Stegle et al. 2015) has provided the means to address these shortcomings of cell population-based RNA-seq through the analysis of the individual transcriptomes of a large number of cells. In addition to shedding light on the cell-to-cell heterogeneity of gene expression, the increased resolution provided by this technology has the potential to identify novel pathways which could not have been detected using the traditional, cell population-level RNA-seq (Trapnell et al. 2014).

Although very few studies have used this technique to analyze lncRNA expression (Yan et al. 2013; Kim et al. 2015), this technology clearly has the potential to provide a much more in-depth look into the regulation and expression pattern of the lncRNAs at the cellular level. Current data from in situ hybridization studies on a subset of lncRNAs using well-established cell lines suggest that at least for this subset, lncRNAs show a cell-to-cell expression heterogeneity similar to that of protein-coding genes (Cabili et al. 2015). However, considering the cell type- and developmental stage-specific expression pattern of many lncRNAs, the use of single-cell techniques will be required to interrogate the lncRNA expression pattern in samples more complex than cell lines, such as patient-derived tissues or during development. Also, it is possible that even in cell lines, some lncRNAs may show a very high level of heterogeneity in their basal expression level or in their response to extrinsic stimuli. Finally, many lncRNAs have allele-specific expression patterns and are involved in modulation of the expression of genes nearby, which can be best studied using single-cell sequencing techniques (Stegle et al. 2015).

Although the use of single-cell RNA-seq for lncRNA expression analysis is an exciting prospect, there are several limitations and caveats that should be considered (Saliba et al. 2014). The number of lncRNAs that can be detected by current single-cell techniques such as smartSeq (Ramsköld et al. 2012b) is very small (Marinov et al. 2014). Without the use of targeted primers during the library preparation step (Armour et al. 2009), a large proportion of cellular RNAs remain undetected or minimally covered, especially for low-abundance RNAs such as lncRNAs. Even when represented in the sequenced population, the high level of technical noise makes the reliable detection of differential expression difficult for low-abundance transcripts such as lncRNAs (Shalek et al. 2013; Islam et al. 2014; Grün et al. 2014). Fortunately, this concern can be addressed by the use of bar-coded oligonucleotides (UMIs) during the library preparation steps, which can improve transcript quantification and cell-to-cell reproducibility (Islam et al. 2014; Grün et al. 2014), especially for low-abundance transcripts including lncRNAs. Further, existing methodologies for preparation of cellular RNA for single-cell sequencing are limited to polyadenylated RNAs, which will eliminate a large proportion of lncRNA-derived transcripts from analysis. Finally, there are also technical limitations for simultaneous detection of the transcript isoforms and maintenance of strand-specific information when short reads are used in the sequencing step, which reduces the complexity and accuracy of detection of lncRNAs. However, it is likely that with further development of single-cell techniques, many of these concerns will be appropriately addressed.

7.7.3 *In situ RNA-Seq*

An exciting addition to the available techniques for analysis of transcriptome is in situ sequencing or sequencing of RNAs without displacing them from their cellular location (Ke et al. 2013; Avital et al. 2014; Lovatt et al. 2014; Lee et al. 2014). In this technique, the transcripts are chemically linked to the cellular protein matrix and are converted to cDNA using either gene-specific or random primers. Amplification of the cDNAs by PCR is followed by sequencing using a fluorescent microscope, allowing the detection of the location of cellular RNAs in cellular compartments, changes in localization in response to stimuli, and co-localization of RNAs. Although this technique is still in its infancy, further development and adaption of this technique for detection of lncRNAs, for example, through the use of lncRNA-specific primers for cDNA synthesis step, will yield a plethora of information about both abundance and subcellular localization of all cellular lncRNAs. This, in turn, will provide novel clues into the function of this class of RNAs and regulation of their expression level and subcellular localization in a population of cells or within tissues in a high-throughput manner.

7.8 Concluding Remarks

Despite existing shortcomings, analysis of lncRNAs in RNA-seq experiments continues to provide key insights into the extent of noncoding transcription and the regulation and function of this class of transcripts. Improvements on existing technologies will likely make single-cell sequencing a viable method for analysis of the lncRNAs, and development of new sequencing platforms such as nanopore-based sequencing will eliminate many of the current bottlenecks in analysis of cellular lncRNAs in near future. However, developing algorithms for extraction of knowledge from the high volume of emerging sequencing data is likely to remain a challenge for the years to come.

Acknowledgment This work is supported by subawards from grant number 5P30AR039750-22-23 from NIH NIAMS and CFAR grant number P30-AI036219 to Saba Valadkhan.

Annex: Quick Reference Guide

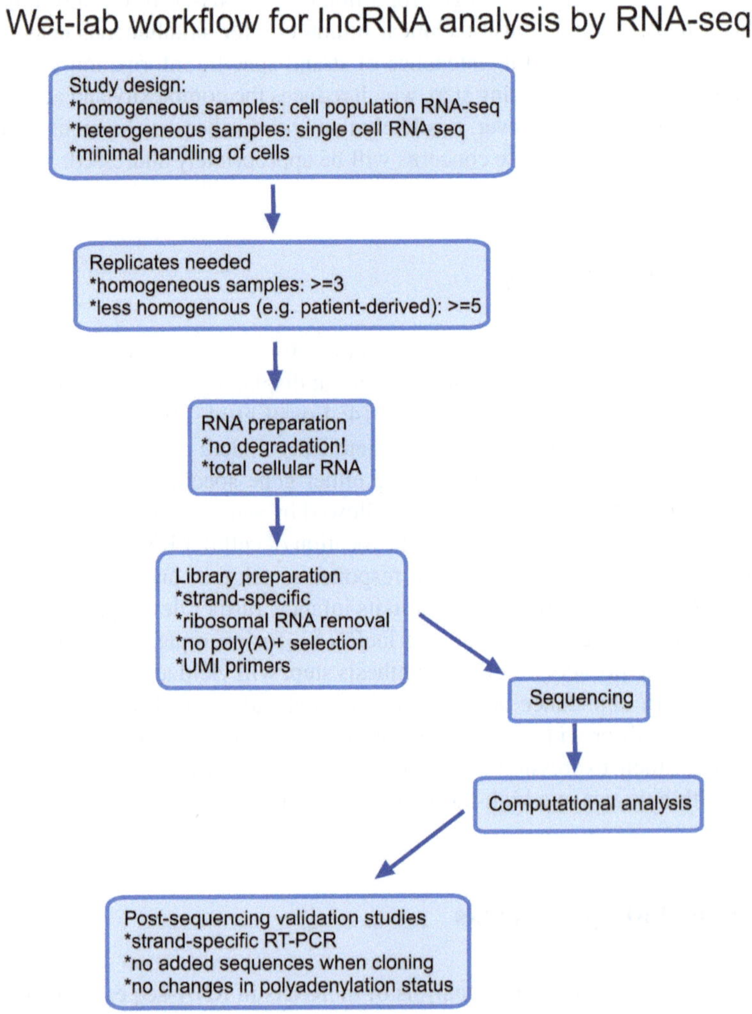

Fig. QG7.1 Representation of the wet-lab procedure workflow

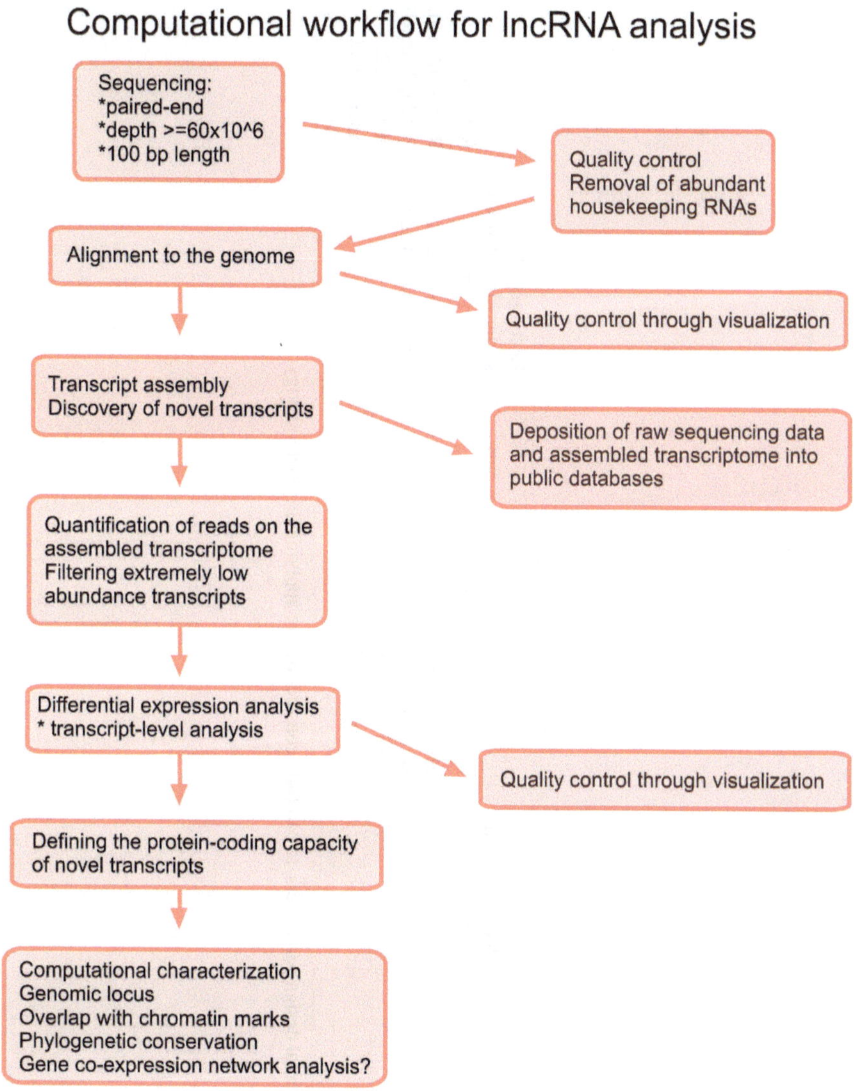

Fig. QG7.2 Main steps of the computational analysis pipeline

Table QG7.1 Experimental design considerations

Technique	Protocol	Control library (if applicable)	Recommended starting material	Number of replicates	Sequencing depth	Recommended sequencing platform and run type	Reference
RNA-seq	Total RNA extraction	Untreated sample, when appropriate	1 microgram of total RNA	Minimum of 3 if homogeneous samples, otherwise minimum of 5	60–100 million reads	Illumina Hiseq 100 bp paired end	Ramsköld et al. 2012a; Trapnell et al. 2012; Anders et al. 2013

Table that comprises relevant experimental design parameters, to carefully consider before applying this methodology

Table QG7.2 Available software recommendations

Software	Function	Input	Reference	Results output	Results format
FastQC	Quality control	Fastq files	N/A	Quality control tables/text	text
TopHat	Genome aligner	Fastq files	Trapnell et al. 2012	Alignment files	Bam
Cufflinks	Transcript assembler	Bam files	Trapnell et al. 2012	Transcriptome	GTF
RSEM/kallisto/Salmon	Transcript quantifier	Fastq/Bam files	Li and Dewey 2011; Bray et al. 2015, unpublished	Quantification files	Text
EdgeR/poissonSeq	Differential expression analyzer	Quantification files	Li et al. 2012; Robinson et al. 2010	Differential expression lists	Text

Table displaying a selection of the recommended software available for the computational analysis of data yielded by this technique

References

Adachi N, Lieber MR (2002) Bidirectional gene organization: a common architectural feature of the human genome. Cell 109:807–809

Almada AE, Wu X, Kriz AJ et al (2013) Promoter directionality is controlled by U1 snRNP and polyadenylation signals. Nature 499:360–363. doi:10.1038/nature12349

Amaral PP, Dinger ME, Mattick JS (2013) Non-coding RNAs in homeostasis, disease and stress responses: an evolutionary perspective. Brief Funct Genomics 12:254–278. doi:10.1093/bfgp/elt016

Anders S, Reyes A, Huber W (2012) Detecting differential usage of exons from RNA-seq data. Genome Res 22:2008–2017. doi:10.1101/gr.133744.111

Anders S, McCarthy DJ, Chen Y et al (2013) Count-based differential expression analysis of RNA sequencing data using R and Bioconductor. Nat Protoc 8:1765–1786. doi:10.1038/nprot.2013.099

Anders S, Pyl PT, Huber W (2015) HTSeq—a Python framework to work with high-throughput sequencing data. Bioinformatics 31:166–169. doi:10.1093/bioinformatics/btu638

Armour CD, Castle JC, Chen R et al (2009) Digital transcriptome profiling using selective hexamer priming for cDNA synthesis. Nat Methods 6:647–649. doi:10.1038/nmeth.1360

Arrial RT, Togawa RC, Brigido Mde M (2009) Screening non-coding RNAs in transcriptomes from neglected species using PORTRAIT: case study of the pathogenic fungus Paracoccidioides brasiliensis. BMC Bioinformatics 10:239. doi:10.1186/1471-2105-10-239

Avital G, Hashimshony T, Yanai I (2014) Seeing is believing: new methods for in situ single-cell transcriptomics. Genome Biol 15:110. doi:10.1186/gb4169

Bánfai B, Jia H, Khatun J et al (2012) Long noncoding RNAs are rarely translated in two human cell lines. Genome Res 22:1646–1657. doi:10.1101/gr.134767.111

Bassett AR, Akhtar A, Barlow DP et al (2014) Considerations when investigating lncRNA function in vivo. eLife Sci 3:e03058. doi:10.7554/eLife.03058

Benjamin AM, Nichols M, Burke TW et al (2014) Comparing reference-based RNA-Seq mapping methods for non-human primate data. BMC Genomics 15:570. doi:10.1186/1471-2164-15-570

Bergmann JH, Li J, Eckersley-Maslin MA et al (2015) Regulation of the ESC transcriptome by nuclear long non-coding RNAs. Genome Res 25:1336. doi:10.1101/gr.189027.114

Bernstein BE, Birney E, Dunham I, et al with ENCODE Project Consortium (2012) An integrated encyclopedia of DNA elements in the human genome. Nature 489:57–74. doi: 10.1038/nature11247

Bolger AM, Lohse M, Usadel B (2014) Trimmomatic: a flexible trimmer for Illumina Sequence Data. Bioinformatics 30:2114. doi:10.1093/bioinformatics/btu170

Bray N, Pimentel H, Melsted P, Pachter L (2015) Near-optimal RNA-Seq quantification.

Cabili MN, Trapnell C, Goff L et al (2011) Integrative annotation of human large intergenic noncoding RNAs reveals global properties and specific subclasses. Genes Dev 25:1915–1927. doi:10.1101/gad.17446611

Cabili MN, Dunagin MC, McClanahan PD et al (2015) Localization and abundance analysis of human lncRNAs at single-cell and single-molecule resolution. Genome Biol 16:20. doi:10.1186/s13059-015-0586-4

Carninci P, Kasukawa T, Katayama S, et al, FANTOM Consortium, RIKEN Genome Exploration Research Group and Genome Science Group (Genome Network Project Core Group) (2005) The transcriptional landscape of the mammalian genome. Science 309:1559–1563. doi:10.1126/science.1112014

Chen L-L, Yang L (2015) Gear up in circles. Mol Cell 58:715–717. doi:10.1016/j.molcel.2015.05.027

Ching T, Huang S, Garmire LX (2014) Power analysis and sample size estimation for RNA-Seq differential expression. RNA 20:1684. doi:10.1261/rna.046011.114

Clark MB, Mattick JS (2011) Long noncoding RNAs in cell biology. Semin Cell Dev Biol 22:366–376. doi:10.1016/j.semcdb.2011.01.001

Daub J, Eberhardt RY, Tate JG, Burge SW (2015) Rfam: annotating families of non-coding RNA sequences. Methods Mol Biol 1269:349–363. doi:10.1007/978-1-4939-2291-8_22

Derrien T, Johnson R, Bussotti G et al (2012) The GENCODE v7 catalog of human long noncoding RNAs: analysis of their gene structure, evolution, and expression. Genome Res 22:1775–1789. doi:10.1101/gr.132159.111

Dinger ME, Pang KC, Mercer TR, Mattick JS (2008) Differentiating protein-coding and noncoding RNA: challenges and ambiguities. PLoS Comput Biol 4:e1000176. doi:10.1371/journal.pcbi.1000176

Djebali S, Davis CA, Merkel A et al (2012) Landscape of transcription in human cells. Nature 489:101–108. doi:10.1038/nature11233

Dobin A, Davis CA, Schlesinger F et al (2013) STAR: ultrafast universal RNA-seq aligner. Bioinformatics 29:15–21. doi:10.1093/bioinformatics/bts635

Engström PG, Steijger T, Sipos B et al (2013) Systematic evaluation of spliced alignment programs for RNA-seq data. Nat Methods 10:1185–1191. doi:10.1038/nmeth.2722

Finn RD, Bateman A, Clements J et al (2014) Pfam: the protein families database. Nucleic Acids Res 42:D222–D230. doi:10.1093/nar/gkt1223

Frankish A, Uszczynska B, Ritchie GR et al (2015) Comparison of GENCODE and RefSeq gene annotation and the impact of reference geneset on variant effect prediction. BMC Genomics 16:S2. doi:10.1186/1471-2164-16-S8-S2

Gao Y, Wang J, Zhao F (2015) CIRI: an efficient and unbiased algorithm for *de novo* circular RNA identification. Genome Biol 16:4. doi:10.1186/s13059-014-0571-3

Gascoigne DK, Cheetham SW, Cattenoz PB et al (2012) Pinstripe: a suite of programs for integrating transcriptomic and proteomic datasets identifies novel proteins and improves differentiation of protein-coding and non-coding genes. Bioinformatics 28:3042–3050. doi:10.1093/bioinformatics/bts582

Glaus P, Honkela A, Rattray M (2012) Identifying differentially expressed transcripts from RNA-seq data with biological variation. Bioinformatics 28:1721–1728. doi:10.1093/bioinformatics/bts260

Grabherr MG, Haas BJ, Yassour M et al (2011) Full-length transcriptome assembly from RNA-Seq data without a reference genome. Nat Biotechnol 29:644–652. doi:10.1038/nbt.1883

Grant GR, Farkas MH, Pizarro AD et al (2011) Comparative analysis of RNA-Seq alignment algorithms and the RNA-Seq unified mapper (RUM). Bioinformatics 27:2518–2528. doi:10.1093/bioinformatics/btr427

Grün D, Kester L, van Oudenaarden A (2014) Validation of noise models for single-cell transcriptomics. Nat Methods 11:637–640. doi:10.1038/nmeth.2930

Guo X, Gao L, Liao Q et al (2013) Long non-coding RNAs function annotation: a global prediction method based on bi-colored networks. Nucleic Acids Res 41:e35. doi:10.1093/nar/gks967

Guo JU, Agarwal V, Guo H, Bartel DP (2014) Expanded identification and characterization of mammalian circular RNAs. Genome Biol 15:409. doi:10.1186/s13059-014-0409-z

Guttman M, Garber M, Levin JZ et al (2010) Ab initio reconstruction of cell type-specific transcriptomes in mouse reveals the conserved multi-exonic structure of lincRNAs. Nat Biotechnol 28:503–510. doi:10.1038/nbt.1633

Guttman M, Russell P, Ingolia NT et al (2013) Ribosome profiling provides evidence that large noncoding RNAs do not encode proteins. Cell 154:240–251. doi:10.1016/j.cell.2013.06.009

Hardcastle TJ, Kelly KA (2010) baySeq: empirical Bayesian methods for identifying differential expression in sequence count data. BMC Bioinformatics 11:422. doi:10.1186/1471-2105-11-422

Harrow J, Frankish A, Gonzalez JM et al (2012) GENCODE: the reference human genome annotation for The ENCODE Project. Genome Res 22:1760–1774. doi:10.1101/gr.135350.111

Hart SN, Therneau TM, Zhang Y et al (2013) Calculating sample size estimates for RNA sequencing data. J Comput Biol 20:970–978. doi:10.1089/cmb.2012.0283

Hatem A, Bozdağ D, Toland AE, Çatalyürek ÜV (2013) Benchmarking short sequence mapping tools. BMC Bioinformatics 14:184. doi:10.1186/1471-2105-14-184

Hayer K, Pizzaro A, Lahens NL et al (2015) Benchmark analysis of algorithms for determining and quantifying full-length mRNA splice forms from RNA-Seq data. Bioinformatics 31:3938. doi:10.1101/007088

Hebenstreit D, Fang M, Gu M et al (2011) RNA sequencing reveals two major classes of gene expression levels in metazoan cells. Mol Syst Biol 7:497. doi:10.1038/msb.2011.28

Huang Y, Hu Y, Liu J (2014) Piecing the puzzle together: a revisit to transcript reconstruction problem in RNA-seq. BMC Bioinformatics 15:S3. doi:10.1186/1471-2105-15-S9-S3

Ilott NE, Ponting CP (2013) Predicting long non-coding RNAs using RNA sequencing. Methods 63:50–59. doi:10.1016/j.ymeth.2013.03.019

Ingolia NT, Lareau LF, Weissman JS (2011) Ribosome profiling of mouse embryonic stem cells reveals the complexity and dynamics of mammalian proteomes. Cell 147:789–802. doi:10.1016/j.cell.2011.10.002

Islam S, Zeisel A, Joost S et al (2014) Quantitative single-cell RNA-seq with unique molecular identifiers. Nat Methods 11:163–166. doi:10.1038/nmeth.2772

Iyer MK, Niknafs YS, Malik R et al (2015) The landscape of long noncoding RNAs in the human transcriptome. Nat Genet 47:199–208. doi:10.1038/ng.3192

Jiang Q, Ma R, Wang J et al (2015) LncRNA2Function: a comprehensive resource for functional investigation of human lncRNAs based on RNA-seq data. BMC Genomics 16(Suppl 3):S2. doi:10.1186/1471-2164-16-S3-S2

Kapusta A, Kronenberg Z, Lynch VJ et al (2013) Transposable elements are major contributors to the origin, diversification, and regulation of vertebrate long noncoding RNAs. PLoS Genet 9:e1003470. doi:10.1371/journal.pgen.1003470

Ke R, Mignardi M, Pacureanu A et al (2013) In situ sequencing for RNA analysis in preserved tissue and cells. Nat Methods 10:857–860. doi:10.1038/nmeth.2563

Kelley D, Rinn J (2012) Transposable elements reveal a stem cell-specific class of long noncoding RNAs. Genome Biol 13:R107. doi:10.1186/gb-2012-13-11-r107

Khatun J, Yu Y, Wrobel JA et al (2013) Whole human genome proteogenomic mapping for ENCODE cell line data: identifying protein-coding regions. BMC Genomics 14:141. doi:10.1186/1471-2164-14-141

Kim M-S, Pinto SM, Getnet D et al (2014) A draft map of the human proteome. Nature 509:575–581. doi:10.1038/nature13302

Kim DH, Marinov GK, Pepke S et al (2015) Single-cell transcriptome analysis reveals dynamic changes in lncRNA expression during reprogramming. Cell Stem Cell 16:88–101. doi:10.1016/j.stem.2014.11.005

Kong L, Zhang Y, Ye Z-Q et al (2007) CPC: assess the protein-coding potential of transcripts using sequence features and support vector machine. Nucleic Acids Res 35:W345–W349. doi:10.1093/nar/gkm391

Kopylova E, Noé L, Touzet H (2012) SortMeRNA: fast and accurate filtering of ribosomal RNAs in metatranscriptomic data. Bioinformatics 28:3211–3217. doi:10.1093/bioinformatics/bts611

Lam MTY, Li W, Rosenfeld MG, Glass CK (2014) Enhancer RNAs and regulated transcriptional programs. Trends Biochem Sci 39:170–182. doi:10.1016/j.tibs.2014.02.007

Langfelder P, Horvath S (2008) WGCNA: an R package for weighted correlation network analysis. BMC Bioinformatics 9:559. doi:10.1186/1471-2105-9-559

Lasda E, Parker R (2014) Circular RNAs: diversity of form and function. RNA 20:1829–1842. doi:10.1261/rna.047126.114

Lee JH, Daugharthy ER, Scheiman J et al (2014) Highly multiplexed subcellular RNA sequencing in situ. Science 343:1360–1363. doi:10.1126/science.1250212

Leng N, Dawson JA, Thomson JA et al (2013) EBSeq: an empirical Bayes hierarchical model for inference in RNA-seq experiments. Bioinformatics 29:1035–1043. doi:10.1093/bioinformatics/btt087

Levin JZ, Yassour M, Adiconis X et al (2010) Comprehensive comparative analysis of strand-specific RNA sequencing methods. Nat Methods 7:709–715. doi:10.1038/nmeth.1491

Li B, Dewey CN (2011) RSEM: accurate transcript quantification from RNA-Seq data with or without a reference genome. BMC Bioinformatics 12:323. doi:10.1186/1471-2105-12-323

Li S, Mason CE (2014) The pivotal regulatory landscape of RNA modifications. Annu Rev Genomics Hum Genet 15:127–150. doi:10.1146/annurev-genom-090413-025405

Li R, Yu C, Li Y et al (2009) SOAP2: an improved ultrafast tool for short read alignment. Bioinformatics 25:1966–1967. doi:10.1093/bioinformatics/btp336

Li W, Feng J, Jiang T (2011) IsoLasso: a LASSO regression approach to RNA-Seq based transcriptome assembly. J Comput Biol 18:1693–1707. doi:10.1089/cmb.2011.0171

Li J, Witten DM, Johnstone IM, Tibshirani R (2012) Normalization, testing, and false discovery rate estimation for RNA-sequencing data. Biostatistics 13:523–538. doi:10.1093/biostatistics/kxr031

Li A, Zhang J, Zhou Z (2014) PLEK: a tool for predicting long non-coding RNAs and messenger RNAs based on an improved k-mer scheme. BMC Bioinformatics 15:311. doi:10.1186/1471-2105-15-311

Liao Q, Liu C, Yuan X et al (2011) Large-scale prediction of long non-coding RNA functions in a coding–non-coding gene co-expression network. Nucleic Acids Res 39:3864–3878. doi:10.1093/nar/gkq1348

Lin MF, Jungreis I, Kellis M (2011) PhyloCSF: a comparative genomics method to distinguish protein coding and non-coding regions. Bioinformatics 27:i275–i282. doi:10.1093/bioinformatics/btr209

Liu J, Gough J, Rost B (2006) Distinguishing protein-coding from non-coding RNAs through support vector machines. PLoS Genet 2:e29. doi:10.1371/journal.pgen.0020029

Loman NJ, Misra RV, Dallman TJ et al (2012) Performance comparison of benchtop high-throughput sequencing platforms. Nat Biotechnol 30:434–439. doi:10.1038/nbt.2198

Lovatt D, Ruble BK, Lee J et al (2014) Transcriptome in vivo analysis (TIVA) of spatially defined single cells in live tissue. Nat Methods 11:190–196. doi:10.1038/nmeth.2804

Marinov GK, Williams BA, McCue K et al (2014) From single-cell to cell-pool transcriptomes: stochasticity in gene expression and RNA splicing. Genome Res 24:496–510. doi:10.1101/gr.161034.113

Mattick JS, Rinn JL (2015) Discovery and annotation of long noncoding RNAs. Nat Struct Mol Biol 22:5–7. doi:10.1038/nsmb.2942

Mercer TR, Dinger ME, Sunkin SM et al (2008) Specific expression of long noncoding RNAs in the mouse brain. Proc Natl Acad Sci U S A 105:716–721. doi:10.1073/pnas.0706729105

Mercer TR, Gerhardt DJ, Dinger ME et al (2012) Targeted RNA sequencing reveals the deep complexity of the human transcriptome. Nat Biotechnol 30:99–104. doi:10.1038/nbt.2024

Mercer TR, Clark MB, Crawford J et al (2014) Targeted sequencing for gene discovery and quantification using RNA CaptureSeq. Nat Protoc 9:989–1009. doi:10.1038/nprot.2014.058

Morgan M, Anders S, Lawrence M et al (2009) ShortRead: a bioconductor package for input, quality assessment and exploration of high-throughput sequence data. Bioinformatics 25:2607–2608. doi:10.1093/bioinformatics/btp450

Morris KV, Mattick JS (2014) The rise of regulatory RNA. Nat Rev Genet 15:423–437. doi:10.1038/nrg3722

Musacchia F, Basu S, Petrosino G et al (2015) Annocript: a flexible pipeline for the annotation of transcriptomes able to identify putative long noncoding RNAs. Bioinformatics 31:2199. doi:10.1093/bioinformatics/btv106

Niazi F, Valadkhan S (2012) Computational analysis of functional long noncoding RNAs reveals lack of peptide-coding capacity and parallels with 3′ UTRs. RNA 18:825–843. doi:10.1261/rna.029520.111

Nishikura K (2010) Functions and regulation of RNA editing by ADAR deaminases. Annu Rev Biochem 79:321–349. doi:10.1146/annurev-biochem-060208-105251

Ntini E, Järvelin AI, Bornholdt J et al (2013) Polyadenylation site-induced decay of upstream transcripts enforces promoter directionality. Nat Struct Mol Biol 20:923–928. doi:10.1038/nsmb.2640

Okoniewski MJ, Leśniewska A, Szabelska A et al (2012) Preferred analysis methods for single genomic regions in RNA sequencing revealed by processing the shape of coverage. Nucleic Acids Res 40:e63. doi:10.1093/nar/gkr1249

Palmieri N, Nolte V, Suvorov A et al (2012) Evaluation of different reference based annotation strategies using RNA-Seq - a case study in Drososphila pseudoobscura. PLoS One 7:e46415. doi:10.1371/journal.pone.0046415
Pan X, Xiong K (2015) PredcircRNA: computational classification of circular RNA from other long non-coding RNA using hybrid features. Mol Biosyst 11:2219. doi:10.1039/c5mb00214a
Pang KC, Frith MC, Mattick JS (2006) Rapid evolution of noncoding RNAs: lack of conservation does not mean lack of function. Trends Genet 22:1–5. doi:10.1016/j.tig.2005.10.003
Patro R, Mount SM, Kingsford C (2014) Sailfish enables alignment-free isoform quantification from RNA-seq reads using lightweight algorithms. Nat Biotechnol 32:462–464. doi:10.1038/nbt.2862
Picardi E, Horner DS, Chiara M et al (2010) Large-scale detection and analysis of RNA editing in grape mtDNA by RNA deep-sequencing. Nucleic Acids Res 38:4755–4767. doi:10.1093/nar/gkq202
Picardi E, D'Erchia AM, Gallo A et al (2014) Uncovering RNA editing sites in long non-coding RNAs. Front Bioeng Biotechnol 2:64. doi:10.3389/fbioe.2014.00064
Pollard KS, Hubisz MJ, Rosenbloom KR, Siepel A (2010) Detection of nonneutral substitution rates on mammalian phylogenies. Genome Res 20:110–121. doi:10.1101/gr.097857.109
Pruitt KD, Brown GR, Hiatt SM et al (2014) RefSeq: an update on mammalian reference sequences. Nucleic Acids Res 42:D756–D763. doi:10.1093/nar/gkt1114
Quail MA, Smith M, Coupland P et al (2012) A tale of three next generation sequencing platforms: comparison of Ion Torrent, Pacific Biosciences and Illumina MiSeq sequencers. BMC Genomics 13:341. doi:10.1186/1471-2164-13-341
Quek XC, Thomson DW, Maag JLV et al (2015) lncRNAdb v2.0: expanding the reference database for functional long noncoding RNAs. Nucleic Acids Res 43:D168–D173. doi:10.1093/nar/gku988
Ramaswami G, Li JB (2014) RADAR: a rigorously annotated database of A-to-I RNA editing. Nucleic Acids Res 42:D109–D113. doi:10.1093/nar/gkt996
Ramaswami G, Lin W, Piskol R et al (2012) Accurate identification of human Alu and non-Alu RNA editing sites. Nat Methods 9:579–581. doi:10.1038/nmeth.1982
Ramsköld D, Kavak E, Sandberg R (2012a) How to analyze gene expression using RNA-sequencing data. Methods Mol Biol 802:259–274. doi:10.1007/978-1-61779-400-1_17
Ramsköld D, Luo S, Wang Y-C et al (2012b) Full-Length mRNA-Seq from single cell levels of RNA and individual circulating tumor cells. Nat Biotechnol 30:777–782. doi:10.1038/nbt.2282
Rapaport F, Khanin R, Liang Y et al (2013) Comprehensive evaluation of differential gene expression analysis methods for RNA-seq data. Genome Biol 14:R95. doi:10.1186/gb-2013-14-9-r95
Rinn JL (2014) lncRNAs: linking RNA to chromatin. Cold Spring Harb Perspect Biol. doi:10.1101/cshperspect.a018614
Rinn JL, Chang HY (2012) Genome regulation by long noncoding RNAs. Annu Rev Biochem 81:145–166. doi:10.1146/annurev-biochem-051410-092902
Ritchie ME, Phipson B, Wu D et al (2015) limma powers differential expression analyses for RNA-sequencing and microarray studies. Nucleic Acids Res 43:e47. doi:10.1093/nar/gkv007
RNAcentral Consortium T (2015) RNAcentral: an international database of ncRNA sequences. Nucleic Acids Res 43:D123–D129. doi:10.1093/nar/gku991
Robertson G, Schein J, Chiu R et al (2010) *De novo* assembly and analysis of RNA-seq data. Nat Methods 7:909–912. doi:10.1038/nmeth.1517
Robinson MD, McCarthy DJ, Smyth GK (2010) edgeR: a Bioconductor package for differential expression analysis of digital gene expression data. Bioinformatics 26:139–140. doi:10.1093/bioinformatics/btp616
Rosenbloom KR, Armstrong J, Barber GP et al (2015) The UCSC genome browser database: 2015 update. Nucleic Acids Res 43:D670. doi:10.1093/nar/gku1177
Saliba A-E, Westermann AJ, Gorski SA, Vogel J (2014) Single-cell RNA-seq: advances and future challenges. Nucleic Acids Res 42:8845. doi:10.1093/nar/gku555

Schulz MH, Zerbino DR, Vingron M, Birney E (2012) Oases: robust *de novo* RNA-seq assembly across the dynamic range of expression levels. Bioinformatics 28:1086–1092. doi:10.1093/bioinformatics/bts094

Seyednasrollah F, Laiho A, Elo LL (2015) Comparison of software packages for detecting differential expression in RNA-seq studies. Brief Bioinform 16:59. doi:10.1093/bib/bbt086

Shalek AK, Satija R, Adiconis X et al (2013) Single-cell transcriptomics reveals bimodality in expression and splicing in immune cells. Nature 498:236–240. doi:10.1038/nature12172

Shen S, Park JW, Lu Z et al (2014) rMATS: robust and flexible detection of differential alternative splicing from replicate RNA-Seq data. Proc Natl Acad Sci U S A 111:E5593–E5601. doi:10.1073/pnas.1419161111

Siepel A, Bejerano G, Pedersen JS et al (2005) Evolutionarily conserved elements in vertebrate, insect, worm, and yeast genomes. Genome Res 15:1034–1050. doi:10.1101/gr.3715005

Stegle O, Teichmann SA, Marioni JC (2015) Computational and analytical challenges in single-cell transcriptomics. Nat Rev Genet 16:133–145. doi:10.1038/nrg3833

Sultan M, Amstislavskiy V, Risch T et al (2014) Influence of RNA extraction methods and library selection schemes on RNA-seq data. BMC Genomics 15:675. doi:10.1186/1471-2164-15-675

Sun K, Chen X, Jiang P et al (2013a) iSeeRNA: identification of long intergenic non-coding RNA transcripts from transcriptome sequencing data. BMC Genomics 14(Suppl 2):S7. doi:10.1186/1471-2164-14-S2-S7

Sun L, Luo H, Bu D et al (2013b) Utilizing sequence intrinsic composition to classify protein-coding and long non-coding transcripts. Nucleic Acids Res 41:e166. doi:10.1093/nar/gkt646

The UniProt Consortium (2015) UniProt: a hub for protein information. Nucleic Acids Res 43:D204–D212. doi:10.1093/nar/gku989

Tilgner H, Raha D, Habegger L et al (2013) Accurate identification and analysis of human mRNA isoforms using deep long read sequencing. G3 (Bethesda) 3:387–397. doi:10.1534/g3.112.004812

Trapnell C, Roberts A, Goff L et al (2012) Differential gene and transcript expression analysis of RNA-seq experiments with TopHat and Cufflinks. Nat Protoc 7:562–578. doi:10.1038/nprot.2012.016

Trapnell C, Cacchiarelli D, Grimsby J et al (2014) The dynamics and regulators of cell fate decisions are revealed by pseudotemporal ordering of single cells. Nat Biotechnol 32:381–386. doi:10.1038/nbt.2859

Trinklein ND, Aldred SF, Hartman SJ et al (2004) An abundance of bidirectional promoters in the human genome. Genome Res 14:62–66. doi:10.1101/gr.1982804

Uesaka M, Nishimura O, Go Y et al (2014) Bidirectional promoters are the major source of gene activation-associated non-coding RNAs in mammals. BMC Genomics 15:35. doi:10.1186/1471-2164-15-35

Ulitsky I, Shkumatava A, Jan CH et al (2011) Conserved function of lincRNAs in vertebrate embryonic development despite rapid sequence evolution. Cell 147:1537–1550. doi:10.1016/j.cell.2011.11.055

Vizcaíno JA, Côté RG, Csordas A et al (2013) The Proteomics Identifications (PRIDE) database and associated tools: status in 2013. Nucleic Acids Res 41:D1063–D1069. doi:10.1093/nar/gks1262

Volders P-J, Verheggen K, Menschaert G et al (2015) An update on LNCipedia: a database for annotated human lncRNA sequences. Nucleic Acids Res 43:D174–D180. doi:10.1093/nar/gku1060

Wagner GP, Kin K, Lynch VJ (2012) Measurement of mRNA abundance using RNA-seq data: RPKM measure is inconsistent among samples. Theory Biosci 131:281–285. doi:10.1007/s12064-012-0162-3

Wagner GP, Kin K, Lynch VJ (2013) A model based criterion for gene expression calls using RNA-seq data. Theory Biosci 132:159–164. doi:10.1007/s12064-013-0178-3

Wakano C, Byun JS, Di L-J, Gardner K (2012) The dual lives of bidirectional promoters. Biochim Biophys Acta 1819:688–693. doi:10.1016/j.bbagrm.2012.02.006

Wang K, Singh D, Zeng Z et al (2010) MapSplice: accurate mapping of RNA-seq reads for splice junction discovery. Nucleic Acids Res 38:e178. doi:10.1093/nar/gkq622

Wang L, Park HJ, Dasari S et al (2013) CPAT: Coding-Potential Assessment Tool using an alignment-free logistic regression model. Nucleic Acids Res 41:e74. doi:10.1093/nar/gkt006

Wei W, Pelechano V, Järvelin AI, Steinmetz LM (2011) Functional consequences of bidirectional promoters. Trends Genet 27:267–276. doi:10.1016/j.tig.2011.04.002

Wernersson R (2006) Virtual Ribosome—a comprehensive DNA translation tool with support for integration of sequence feature annotation. Nucleic Acids Res 34:W385–W388. doi:10.1093/nar/gkl252

Wright MW (2014) A short guide to long non-coding RNA gene nomenclature. Hum Genomics 8:7. doi:10.1186/1479-7364-8-7

Wu TD, Nacu S (2010) Fast and SNP-tolerant detection of complex variants and splicing in short reads. Bioinformatics 26:873–881. doi:10.1093/bioinformatics/btq057

Xiao Y, Lv Y, Zhao H et al (2015) Predicting the functions of long noncoding RNAs using RNA-Seq based on Bayesian network. Biomed Res Int 2015:839590. doi:10.1155/2015/839590

Xie C, Yuan J, Li H et al (2014) NONCODEv4: exploring the world of long non-coding RNA genes. Nucleic Acids Res 42:D98–D103. doi:10.1093/nar/gkt1222

Yan L, Yang M, Guo H et al (2013) Single-cell RNA-Seq profiling of human preimplantation embryos and embryonic stem cells. Nat Struct Mol Biol 20:1131–1139. doi:10.1038/nsmb.2660

Zhang Z, Qi S, Tang N et al (2014) Discovery of replicating circular RNAs by RNA-seq and computational algorithms. PLoS Pathog 10:e1004553. doi:10.1371/journal.ppat.1004553

Chapter 8
Ribosome Profiling

Anze Zupanic and Sushma Nagaraja Grellscheid

8.1 Introduction

In the last decade gene expression profiling by microarrays (Brown and Botstein 1999), and more recently by RNA-Seq (Mortazavi et al. 2008), has become one of the most important and widely used tools of molecular biology. However, recent studies have shown that mRNA levels only imperfectly correlate with protein levels (Vogel et al. 2010), and that regulation at the level of translation and the level of protein stability plays a very important role (Sonnenberg and Hinnebusch 2009) in influencing the final outcome of gene expression. Ribosome profiling (also called Ribo-Seq), i.e. next-generation sequencing of mRNA fragments protected by the translating ribosome, pioneered in the Weissman lab in 2009 (Ingolia et al. 2009), is a method that closes some of the gap between the mRNA molecule and the protein. Since 2009, ribosome profiling has been used to shed light on many open questions in several different species (Table 8.1), from the mechanisms behind miRNA regulation (Bazzini et al. 2012) to experimental determination of translation initiation sites (Ingolia et al. 2011). Perhaps surprisingly, and most probably due to a very demanding and labour intensive protocol behind the ribosome profiling, since the first publication only 56 published studies have presented new ribosome profiling datasets. This means that although already 6 years old, ribosome profiling is still very much in the development phase and although a detailed protocol has been

A. Zupanic, Ph.D. (✉)
Department of Environmental Toxicology, Eawag – Swiss Federal Institute for Aquatic Research and Technology, Überlandstrasse 133, Dübendorf 8600, Switzerland
e-mail: anze.zupanic@eawag.ch

S.N. Grellscheid, Ph.D.
School of Biological and Biomedical Sciences, Durham University, Mountjoy Science Site, Durham DH1 3LE, UK
e-mail: s.n.grellscheid@durham.ac.uk; sushma@cantab.net

A.M. Aransay, J.L. Lavín Trueba (eds.), *Field Guidelines for Genetic Experimental Designs in High-Throughput Sequencing*,
DOI 10.1007/978-3-319-31350-4_8

Table 8.1 Studies that provided new ribosome profiling datasets from 2009 to 2014

Application	Species	Sequencing platform	Reference
Methodology	*S. cerevisiae*	Illumina GAII	Ingolia et al. (2009)
Methodology	*M. musculus*	Illumina GAII, HiSeq	Ingolia et al. (2011)
Methodology	*E. coli*	Illumina HiSeq2000	Li et al. (2014a)
Oxidative stress	*S. cerevisiae*	Illumina HiSeq2000	Gerashchenko et al. (2012)
Chemotherapy	*H. sapiens*	Illumina HiSeq2000	Wiita et al. (2013)
Light exposure	*A. thaliana*	Illumina (model not given)	Liu et al. (2013a)
Methodology	*S. cerevisiae*	Illumina HiSeq2000	Gerashchenko and Gladyshev (2014)
Meiosis	*S. pombe*	Illumina GAII, HiSeq	Duncan and Mata (2014)
Meiosis	*S. cerevisiae*	Illumina GAII	Brar et al. (2012)
Cell cycle	*M. musculus, H sapiens*	Illumina HiSeq2000	Stumpf et al. (2013)
Development	*C. elegans*[a]	Illumina HiSeq2000	Stadler and Fire (2013)
Development	*D. melanogaster*	Illumina HiSeq2000/2500	Lee et al. (2013)
Development	*A. suum*	Illumina HiSeq	Wang et al. (2014)
Development	*P. falciparum*	Illumina HiSeq2000	Caro et al. (2014)
Development	*T. brucei*	Illumina GAII	Jensen et al. (2014)
Antibiotics	*E. coli*	Illumina GAII	Kannan et al. (2014)
Ethanol stress	*E. coli*	Illumina HiSeq2000	Haft et al. (2014)
Lifespan	*S. cerevisiae*	Illumina HiSeq2000	Labunsky et al. (2014)
Sarcoma	*Herpesvirus, H. sapiens,*	Illumina HiSeq2000	Arias et al. (2014)
Viral infection	*Bacteriophage lambda E. coli*	Illumina HiSeq2000	Liu et al. (2013b)
Elongation	*E. coli*	Illumina HiSeq2000	Li and Weissman (2012)
Elongation	*C. elegans*	Illumina HiSeq2000	Stadler and Fire (2011)
Elongation	*S. cerevisiae*	Illumina HiSeq2000	Gardin et al. (2014)
Elongation	*S. cerevisiae*	Illumina GAII	Lareau et al. (2014)
Elongation	*S. cerevisiae*	platform not given	Pop et al. (2014)
Elongation	*E. coli*	Illumina GAII	Nakahigashi et al. (2014)
Elongation	*D. melanogaster*	Illumina HiSeq	Dunn et al. (2013)
miRNA	*C. elegans*	Illumina GAII	Stadler et al. (2012)
miRNA	*D. rerio*	Illumina GAII	Bazzini et al. (2012)
miRNA	*H. sapiens, M. musculus*	Illumina GAII	Guo et al. (2010)

(continued)

Table 8.1 (continued)

Application	Species	Sequencing platform	Reference
Leaders	*S. cerevisiae, S. paradoxus*	Illumina HiSeq2000	McManus et al. (2014)
Evolution	*S. cerevisiae*	Illumina HiSeq2000	Artieri and Fraser (2014)
Selenoproteins	*M. musculus*	Illumina HiSeq2000	Howard et al. (2013)
eIF4A	*H. sapiens*	Illumina HiSeq2000	Rubio et al. (2014)
P53	*H. sapiens*	Illumina HiSeq2000	Loayza-Puch et al. (2013)
mTOR	*M. musculus*	Illumina GAII	Hsieh et al. (2012)
mTOR	*M. musculus*	Illumina GAII	Thoreen et al. (2012)
ORFs	*S. cerevisiae*	Illumina HiSeq	Smith et al. (2014)
ORFs	*C. albican*	Illumina GAII, HiSeq	Muzzey et al. (2014)
ORFs	*D. rerio*	Illumina HiSeq2000	Bazzini et al. (2014)
ORFs	*T. brucei*	platform not given	Vasquez et al. (2014)
ORFs	*M. musculus,*	Illumina HiSeq2000	Ingolia (2014)
ORFs	*S. cerevisiae*	Illumina HiSeq2000	Albert et al. (2014)
ORFs	*C. crescentus*	Illumina GAII, HiSeq	Schrader et al. (2014)
ORFs	*D. rerio*	Illumina HiSeq2000	Chew et al. (2013)
ORFs	*H sapiens*	Illumina HiSeq	Koch et al. (2013)
ORFs	*D. melanogaster*	Illumina HiSeq2000, Illumina MiSeq	Aspden et al. (2014)
Hypoxia	*A. thaliana*	Illumina HiSeq2000	Juntawong et al. (2013)
Proteotoxic stress	*H. sapiens*	Illumina HiSeq2000	Liu et al. (2013c)
Heat Shock	*M. musculus*	Illumina GAII	Shalgi et al. (2013)
Prion stress	*S. cerevisiae*	Illumina (model not given)	Baudin-Bailleau et al. (2014)
Mito-translation	*H. sapiens*	Illumina HiSeq2000	Rooijers et al. (2013)
Mito-translation	*S. cerevisiae*	platform not given	Williams et al. (2014)
ER-translation	*S. cerevisiae*	platform not given	Jan et al. (2014)
ER-translation	*H. sapiens*	SOLiD 4	Reid and Nicchitta (2012)
Chaperones	*E. coli*	Illumina GAII	Oh et al. (2011)

[a]Other Caenorhabditis species also used

published (Ingolia et al. 2012), individual procedures, such as the use of cycloheximide for translation inhibition, have recently come under intense scrutiny. In the following pages, we present different ways in which ribosome profiling has been put to use, different biological application of the methods and the current state-of-the-art experiment guidelines, with special attention put to alternative protocols and still open questions.

8.2 Applications

Different applications of ribosome profiling have been recently reviewed (Ingolia 2014). In this section, we report on all studies (identified by using the search term *ribosome profiling or Ribo-Seq* in Web of Knowledge) that have generated ribosome profiling data until December 2014, together with information of the species, main application and sequencing platform (Table 8.1). There have been several specific exciting discoveries made with ribosome profiling in the last 5 years and it is beyond this chapter to name all of them; however from all the studies some very general conclusions can be made. Perhaps most important is that the studies have demonstrated that global and specific regulation of gene expression at the translational level is ubiquitously present in all biological processes, from development to defence against oxidative stress. The mechanisms behind specific regulation are most likely sequence features on the 5′ and 3′-UTRs of individual transcripts that are subject to different translation initiation regimes, but more research is needed until firmer conclusions can be made. A second important conclusion is that translation often involves initiation from alternative initiation codons on single transcripts, and thirdly, apparently translated RNAs correspond to surprising regions of the genome, such as 5′UTRs or noncoding RNAs. It is reasonable to assume that ribosome profiling will in the future significantly increase the number of discovered peptide and proteins.

In the abovementioned studies, ribosome profiling has generally been used in three different ways: (1) identification of translated RNA regions, (2) calculation of single transcript and global translation efficiency as a measure of protein synthesis, and (3) comparing ribosome occupancy along single transcripts and along the transcriptome. Each of these takes advantage of different ribosome profile properties and is described in more detail below.

8.2.1 Identification of Translated Regions

Traditionally eukaryotic protein-coding regions were identified based on cDNA sequence data generated from known transcripts or from peptide sequences. Normally the longest possible ORF in a transcript is assumed to be the coding region (CDS). Today, despite the fact that a combination of ab initio transcriptomic, comparative genomic and machine learning approaches have increased the accuracy of gene prediction above 95 %, the prediction of coding regions still lags behind (Yip et al. 2013). Ribosome profiling provides a very promising alternative to the current state-of-the-art (Ingolia et al. 2011) by (1) assuming that ribosome-protected regions of the mRNA are also translated and (2) taking advantage of the near nucleotide precision of ribosome profiling—since ribosome-protected fragments are of quite uniform size it is possible to assign the position of ribosomal A site to a particular nucleotide or at least codon.

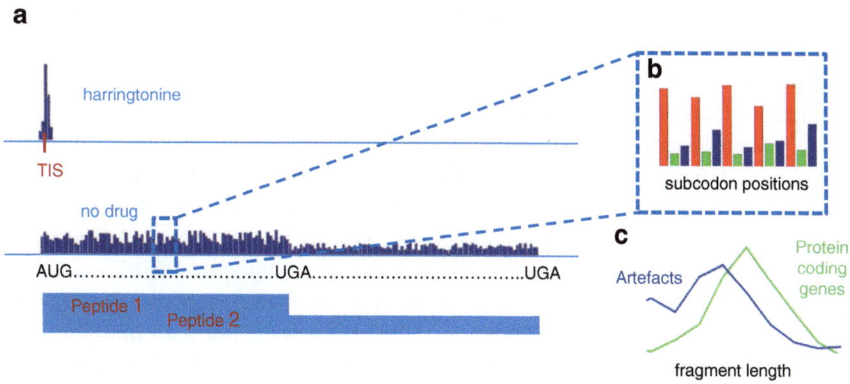

Fig. 8.1 Strategies for detecting translated mRNA regions based on ribosome profiling. (**a**) Typical ribosome profile obtained with or without translation inhibition with harringtonine. The harringtonine profile is high over the putative TIS, while the no drug profile is high throughout the translated region. Segmentation of the profile identified two separate translated regions, both starting at the same TIS: it seems a shorter and a longer peptide are produced from this genomic region, as would be the case of selenoprotein translation (Zupanic et al. 2014). (**b**) If the ribosome density over the subcodon positions does not follow a standard subcodon pattern (usually high low low), then translation is questionable. (**c**) Distribution of RNA fragment length in a protein-coding region and RNA not covered by ribosomes (but by, e.g. telomerase) (Ingolia et al. 2014)

One of the strategies (see Fig. 8.1 for a schematic of all strategies) used for identification of coding region by several groups was to detect all translation initiation sites (TIS), by using a translational initiation inhibitor, such as harringtonine (Ingolia et al. 2011) or lactimidomycin (Lee et al. 2012), before sequencing ribosome-protected fragments. The result is a very sparse ribosome coverage, which is assumed to coincide with translational initiation sites. To further reduce the number of false positives, machine learning methods are used to recognize patterns of ribosome coverage similar to pattern of known initiation sites. In all studies that have used this strategy so far, a surprisingly high number of translational initiation sites was discovered in 5′-UTRs and in noncoding RNAs, leading to the hypothesis that current annotation misses a large part of the translated transcriptome (Ingolia et al. 2011). This proved a very controversial hypothesis and many following studies have tried to confirm or repudiate it using alternative strategies.

One of the arguments against prevalent translation of UTR regions and noncoding RNAs was that although the discovered TISs do show translation initiation, this does not necessarily also lead to elongation. In one study, the predicted TISs were compared to regions predicted to be translated by a segmentation algorithm, which identified genetic regions with uniform ribosome coverage, indicating uninterrupted translation (Zupanic et al. 2014). The study showed that less than 1 % of the alternative identified TISs were found to initiate robust translation. The segmentation method was also able to detect alternative initiation in cases when more than one TIS is used for a given transcript.

Other strategies have also been developed. In one, the size distribution of the ribosome fragment aligned to the putative translated region is compared to a standard distribution of fragment sizes and significant deviation from the standard was deemed artefacts not connected with translation (Ingolia et al. 2014). To profile only actively translating ribosome complexes Poly-Ribo-Seq was developed, in which polysomes (actively translated RNA-ribosomes complexes) are biochemically purified prior to ribosome footprinting (Aspden et al. 2014). Another strategy was to analyse the nucleotide periodicity of ribosome profiling (the first nucleotide position in a codon has higher ribosome density than the second and third)—a broken periodicity points to artefacts or a possible frameshift during translation (Michel et al. 2012). Other strategies for defining coding regions include searching for a stop codon after the putative TIS (Zupanic et al. 2014; Albert et al. 2014; Howard et al. 2013; Guttman et al. 2013), which enables detection of premature termination during translation, and confirmation of the putative translated peptide sequences by mass spectrometry (Schrader et al. 2014; Smith et al. 2014; Menschaert et al. 2013). There has so far been no standardized comparison of the different strategies using common or comparable datasets, so it is currently not clear whether any single of them is superior or a combination would provide the best result.

8.2.2 Translational Efficiency

Biological systems react to perturbation by employing appropriate regulatory pathways. In most cases, the regulation consists of changes in gene expression; however these changes occur at both the transcriptional and translational level. To differentiate between regulation that occurs at the translational level from that at the transcriptional level, a measure called translational efficiency (TE) was developed (Ingolia et al. 2009; Guo et al. 2010):

$$TE = \frac{C'/N'L'}{C/NL}$$

where C' is the number of ribosome profiling reads aligned to an individual coding region of a gene, N' is the total number of ribosome profiling reads aligned to all coding regions, L' is the length of the coding region of the gene, C is the number of RNA-Seq reads aligned to a transcript, N is the total number of RNA-Seq reads aligned to all the transcripts and L is the length of the transcript. TE can only be calculated if RNA-Seq and ribosome profiling were both performed by taking samples from the same source. For lower counts, the TE metric is associated with a large error; therefore all genes with a low number of aligned reads (usually, below an average of at least 1 read per nucleotide—(Guo et al. 2010)) are disregarded in the analysis. As defined above, TE does not account for error due to alternative splicing of individual genes or alternative protein-coding regions on individual transcripts;

however if RNA-Seq and ribosome profiling are also used to estimate these two events (Zupanic et al. 2014), it can easily be adjusted.

Although most studies performed so far use the above definition of translational efficiency, it lacks statistical robustness. This can be improved by using a linear modelling approach that also leverages both RNA-Seq and ribosome profiles—the method has been provided as an R package (Larsson et al. 2011). Another, more recently published approach called Babel relies on error-in-variables regression model for estimation of unexpected patterns in ribosome occupancy, and the Fisher's exact test to calculate significance levels (Olshen et al. 2013).

The outcome of a translational efficiency study is a list of genes that are differentially regulated at the translational level, and this list can be used analogously to RNA-Seq to determine differentially expressed pathways and processes regulated at the translational level or resolve sequence features of groups of genes to establish mechanisms behind their differential translation (Thoreen et al. 2012; Hsieh et al. 2012). Another option is to use ribosome profiling datasets as estimates of protein production rates and perform downstream analysis on these alone (Li et al. 2014a).

8.2.3 Ribosome Speed

A number of studies thus far have focused, not on detecting translated regions or translational efficiency, but on using the nucleotide precision of ribosome profiling to try to understand what controls ribosomal speed along a transcript (Ingolia et al. 2011; Gardin et al. 2014; Stadler and Fire 2011; Pop et al. 2014; Li and Weissman 2012; Charneski and Hurst 2013; Artieri and Fraser 2014; Dana and Tuller 2012, 2014; Shah et al. 2013). The assumption behind this is that ribosomes spend more time on slower codons; therefore there is a higher probability that a ribosome will be found on these codons and the ribosome density on these codons will be bigger than on their faster counterparts.

So far, studies have come to very different conclusions, and it is not clear whether these results depend on the species studied or are due to different analysis methods. Heterogeneity in tRNA availability across tissues and cell types used in different experiments is also likely to contribute to the biases observed (Dittmar et al. 2006). Although all studies have used a similar methodology, there is as yet no consensus on how to account for the biases (see later sections) inherent to ribosome profiling: some studies have excluded regions at the beginning and end of coding regions from the analysis, others have used these regions (but used normalization) and again others have not accounted for bias at all.

In short, some studies have found a strong effect of codon bias on elongation speed (Pop et al. 2014), others of tRNA availability (Dana and Tuller 2014), again other effects of positive amino acids (Charneski and Hurst 2013), strong control asserted by proline alone (Artieri and Fraser 2014), specific stalling sequences (Li and Weissman 2012) or even none of the above (Ingolia et al. 2011). A systematic evaluation of a large number of ribosome profiling datasets with the whole set of methodologies is needed to evaluate different contributions to elongation speed.

8.3 Experimental Design Guidelines

With regard to sequencing, ribosome profiling is little different from the more traditional RNA-Seq; therefore the guidelines established in the last decade for RNA-Seq (and described in other sections in this book) should also be valid for ribosome profiling (SEQC/MAQC-III Consortium 2014; Li et al. 2014b). In any case, no systematic comparison of different sequencing platform for ribosome profiling is available, and those few studies that made any sort of comparison between RNA-Seq and ribosome profiling properties have found clear correlations between different properties of both types of datasets (Zupanic et al. 2014; Artieri and Fraser 2014).

There are, however, important differences between ribosome profiling and RNA-Seq with respect to the preparation of samples for sequencing and in bioinformatic analysis after sequencing. In the following pages we, therefore, focus particularly on those parts of the ribosome profiling protocols that are different from RNA-Seq counterparts. In the description, we mostly follow the ribosome profiling protocol published by Ingolia et al. in 2012, and its modifications as proposed by various studies.

8.3.1 Technical and Biological Replicates, Sequencing Depth

In the recent large-scale assessment of RNA-Seq accuracy, it has been found that technical variation due to sequencing artefacts is low, while biological variation is high (SEQC/MAQC-III Consortium 2014). The study thus emphasized the value of biological replicates to increase the quality of RNA-Seq studies. Although the **minimum number of biological replicates** required in some studies has been 2, the study suggests big improvements can be made with each additional biological replicates, with the biggest influence of the first 4–5. There is currently no reason to expect that ribosome profiling would have different requirements.

The same study also evaluated the importance of sequencing depths and concluded that increasing the depth up to 500 million aligned reads still contributes significantly to the number of detected genes, but that the improvements with further increase are smaller (SEQC/MAQC-III Consortium 2014). While the first ribosome sequencing studies feature lower total read counts, some of the later studies have already taken the **number of aligned reads towards 100 million** and this has significantly increased the number of detected genes (McManus et al. 2014). As for the number of biological replicates, there is currently no reason to provide any recommendation that would differ from RNA-Seq guidelines.

8.3.2 Wet Lab Protocol

A detailed ribosome profiling protocol for mammalian cells, together with a list of necessary reagents, reagent setup, equipment and equipment setup, has recently been published (Ingolia et al. 2012). In the following sections, we follow the published protocol, but also describe alternatives and point out those parts that have received criticism from the community.

8.3.2.1 Cell Lysis

Following cell culture according to conditions relevant to the study, cells must undergo lysis. The most contentious issue during this first phase of the protocol is the timing and use of translation elongation inhibitors. In the original ribosome profiling study, cycloheximide was used to stabilize the polysomes before performing lysis (Ingolia et al. 2009). The study found an increase in ribosome density immediately after the TISs and postulated that an elevated 5′ ribosome density (ramp) is a general feature of translation. It was later discovered that different translation inhibitors (i.e., emetine vs cycloheximide vs anisomycin vs chloramphenicol vs tetracycline) lead to different distribution of sizes of ribosome-protected fragment and also different shapes of the ramp, while the ramp even disappears when using no drugs (Ingolia et al. 2011; Lareau et al. 2014; Nakahigashi et al. 2014).

Recently, a critical study has cast some doubt on some of the previous discoveries and put them down to a bias caused by inappropriate cycloheximide use (Gerashchenko and Gladyshev 2014). They discovered that the nature of the ramp also depends on the used concentrations of the translation inhibitors: the ramp effect gets smaller with higher concentration and disappears completely when the concentration used is high enough. This concentration dependence was explained by slow passive diffusion of the drug into the cells—at low concentrations cycloheximide is only partly effective and allows for some extra movement of the ribosomes. For this reason many of the newer studies avoid the use of translation inhibitors and rather opt for flash freezing (Oh et al. 2011) of the samples to stabilize the ribosome positions.

8.3.2.2 Translation Initiation Site Profiling

While the use of translation elongation inhibitors, such as cycloheximide and emetine, can bias the position of ribosomal fragment and should be used with caution, nothing similar has been reported for translation initiation inhibitors, such as harringtonine (Ingolia et al. 2011) or lactimidomycin (Lee et al. 2012). These inhibitors, which need to be used immediately before adding cycloheximide and lysis of the cells, are used to enrich ribosomes on TISs and thus enable discovery of new coding regions. While their use might still bias the distribution of ribosome around the TIS, this was shown not to be critical for TIS identification.

8.3.2.3 Nuclease Footprinting

After lysis, the next step is ribosome footprinting, using endonucleases to digest the unprotected RNA. While most studies use bacterial RNAse I for digestion (Ingolia et al. 2009), some recent studies also use micrococcal nuclease (MNAse)

(Dunn et al. 2013; Nakahigashi et al. 2014). The choice of nuclease depends on the studied species, with most higher eukaryote studies so far using RNAse I, but in those studies that used both nucleases no significant differences were found (Nakahigashi et al. 2014). Recently, a ribosome profiling kit has become available for both yeast and mammalian cells (ARTseq/TruSeq Ribo Profile Kit) and it has been successfully used in a few studies (Bazzini et al. 2014).

While the use of different endonucleases does not seem to affect the results, it has been shown that the lysis buffer can have an important effect. Buffers with lower salt and magnesium content result in narrower ribosome fragment size distributions, and fragments whose termini show more specific positioning relative to the reading frame being decoded (Ingolia et al. 2012). These can then be aligned to the genome with a higher positional resolution, making inference of the coding regions easier. Recent studies have shown that ribosome complexes are not maintained in all buffer compositions, the result being loss of a part of the ribosome footprint population (Aspden et al. 2014).

8.3.2.4 Ribosome and RNA Fragment Recovery

After nuclease digestion, ribosome-RNA complexes need to be isolated from cell lysates. In earlier studies this was performed by sucrose density gradient purification (Ingolia et al. 2009); however due to the need of special equipment and methodological difficulties this was then replaced by sucrose cushion sedimentation (Ingolia et al. 2012). This includes laying the lysate on top of a 1 M sucrose cushion in an ultracentrifuge tube, followed by centrifugation to pellet ribosomes.

Alternative methods for ribosome recovery include translating ribosome affinity purification (TRAP) (Heiman et al. 2008; Oh et al. 2011; Becker et al. 2013) and size exclusion chromatography (Bazzini et al. 2014). TRAP takes advantage of genetically modified, epitope tagged ribosomal proteins, and chromatography using strongly specific antibodies. Size-exclusion spin column chromatography, on the other hand, separates the ribosome-RNA complexes from other lysate content purely based on size. The speed and convenience of size exclusion chromatography could very well make it the preferred method for ribosome recovery in the future; however so far, it has not been used in many studies.

After recovery of ribosome-RNA complexes, the ribosomes need to be removed from the RNA fragments, which is usually done using one of the widely available RNA purification kits, such as miRNeasy kit (Ingolia et al. 2012). Care must be taken to avoid any ribonuclease contamination from this point on, as this will lead to RNA fragment digestion. Finally, the remaining RNA fragments of sizes ranging from 26 to 34 nt for mammalian cells (Ingolia et al. 2012) or shorter for prokaryotes (Li et al. 2014a) are separated from the rest using RNA gels and electrophoresis followed by gel extraction. Recently, at least in *E. coli* it has been shown that a larger range of mRNA foot print sizes can also be used without significantly affecting the final results. Indeed, another recent study in yeast showed that in the absence of cycloheximide, there exist two different populations of ribosome-protected fragments, one of size 28–30 nt and a shorter of size 20–22 nt (Lareau et al. 2014). Contrary to cycloheximide, the 20–22 nt fragments were seen in case of using anisomycin as translation inhibitors, indicating that the ribosome-RNA complex can exist in two different configurations. It therefore seems best that the size inclusion of RNA fragments is defined according to the translation inhibitor used and that if no inhibitor is used, a wider fragment size distribution is taken for further analysis.

8.3.2.5 Library Preparation

Linker Ligation

Since most of the studies performed so far used Illumina platforms for the sequencing, linker ligation is mostly performed according to the Illumina prescribed protocols, which include the addition of a polyA tail to each sequence. Alternatively, optimized RNA ligation of a preadenylated linker can be used to achieve similar results (Ingolia et al. 2012). Ligation is followed by reverse transcription, polyacrylamide gel electrophoresis and circularization of the reverse transcription products to get the cDNA molecules used in the following procedures.

Barcoding

Following the circularization it is optional to add barcode sequences for each sample (multiplexing) (Ingolia et al. 2012; Duncan and Mata 2014), followed by several cycles of PCR amplification. The amplification reactions can either be purified by magnetic bead-based methods or are loaded on to polyacrylamide nondenaturating gels, separated by electrophoresis and the amplified PCR product excised. The latter step is now widely available as an automated process via pippin prep, E-gels and other similar products. The libraries thus generated are finally characterized using one or more of the following methods such as qPCR, Bioanalyzer, and Tape-station to ensure library quality and concentration, before using for sequencing.

rRNA Depletion

At this point, cDNA molecules derived from rRNA still represent a significant amount of the sample. In most studies, it turned out that a few (species specific) rRNA molecules are responsible for the bulk of the contamination and it was thus possible to remove most of the contamination by focusing on a few specific molecules. This was mostly done using hybridization to biotinylated sense-strand oligonucleotide followed by removal of duplexes through streptavidin affinity (Ingolia et al. 2012). Alternatively, more general removal of rRNA via rRNA removal kits before the library preparation step was also used with good results.

8.3.2.6 Sequencing

All ribosome profiling studies conducted so far, with the exception of one (Reid and Nicchitta 2012), have used the Illumina Platforms (GAII or HiSeq2000) for the sequencing, with the same basic protocol that is no different from the one used in RNA-Seq (Ingolia et al. 2012). The output of a Illumina sequencing run is a FASTQ format file, which includes both the sequence and the quality all the sequenced read and is the basis for computational analysis which follows the sequencing.

8.3.3 Computational Analysis

Although it takes quite some time and effort to get from the initial samples to the sequences, without proper interpretation the sequences are not worth much. Computational analysis enables one to first align the sequenced reads to a genome and then to evaluate whether the number of aligned reads to particular genetic regions has an important biological function.

8.3.3.1 Alignment

The alignment of the reads to the genome is also no different than for RNA-Seq. First, the sequencing data are pre-processed by discarding low quality reads, removing the 3′ linker sequence and removing the first nucleotide from the 5′ end of each read. This can be done, e.g. using the FastX Toolkit. Note that although the outputs of different sequencing platforms are not all the same, FastX Toolkit and most similar tools can read most of the formats if these are correctly specified. The trimmed sequences are then first aligned to an rRNA reference, using any of the available aligners (Bowtie, Subread, Burrows-Wheeler). The non-rRNA reads are then aligned to the genome using a splicing-aware aligner (e.g., Tophat2).

Because the ribosome-protected fragments are quite short, the alignment is not always perfect, i.e. many reads align to more than one genomic segment. Different studies have applied different strategies to remove the bias potentially arising from such multiple alignments: (Guo et al. 2010) simply discarded all reads with multiple alignments, (Ingolia et al. 2011) kept all alignments, thereby counting a single read multiple times, (Dana and Tuller 2012) suggested an iterative approach, in which first only uniquely aligned reads are kept, then in the second round each multiple aligned read is assessed for the presence of neighbouring reads from the first round, and keeping only those with neighbours, while discarding the rest. Since reads with no neighbours are excluded from the analysis in any case at later stages, the iterative procedure should lead to the least bias and is recommended. An iterative approach is usually implemented by running the alignment algorithms several times, with different input files. The output of the alignment is either a BAM or a SAM (human readable version of BAM) file, which is the basis for all further computational analysis.

Another general occurring problem with alignment shared with RNA-Seq is assignment of a read to the correct transcript for alternatively spliced genes. Although none of the ribosome profiling studies used alternative splicing detection, several studies have shown that this can bias the final analysis (Zupanic et al. 2014). We therefore recommend that an algorithm for detection of alternative splicing, such as rMATs (Shen et al. 2014), is used during analysis.

8.3.3.2 Biases

Since both ribosome profiling and RNA-Seq are based on the same sequencing procedures, it is reasonable to assume they would also suffer from the same biases. This was demonstrated by a recent study that used RNA-Seq profiles to normalize ribosome profiles. The study showed that the obtained normalized average profiles are a better representation of our current understanding of translation than the non-normalized profiles: ribosome density was quite smooth and slowly decreasing from the 5′ to the 3′ region, which was to be expected if occasional ribosome drop-offs occur (Zupanic et al. 2014). Another study took a similar approach and discovered that normalization with RNA-Seq significantly changes the previous analyses of ribosome speed, implicating proline as an important ribosome pausing factor (Artieri and Fraser 2014). In none of these studies did normalization completely remove the increased ribosome density observed in the first couple of codons in coding regions, when using translational inhibitors. This bias can be eliminated either by using correction factors for the biased region (Li et al. 2014a) or by simply ignoring the biased regions in the analysis. In any case, bias removal by RNA-Seq normalization and accounting for translation inhibition artefacts is necessary before any further analysis.

8.3.3.3 Functional Analysis

Once the sequences have been aligned and bias has been taken care of, the visualization and the functional interpretation of data can begin. For easy visualization, we recommend the riboseqR Bioconductor package, which produces a genome browser type of a visualization which can be useful for analysis of open reading frames (Hardcastle 2014). The most common application of Ribo-Seq is to find translated regions, changes in translational efficiency after a perturbation or follow ribosomal speed across the genome to study codon bias. Regardless of the application, there are currently no standard methods that the community would use nor specifically developed and widely available computational packages. Currently, the optimal strategy for a researcher is to carefully study the work performed by others and then test the proposed methods. Most papers have made the algorithms they developed available as supplementary material, but even when this is not the case the community gladly shares their computational resources.

In case of using Ribo-Seq to determine differentially translated transcripts after a perturbation, it is possible to use the differential expression packages developed for RNA-Seq, such as edgeR and DESeq (Robinson et al. 2010; Anders and Huber 2010). Upon obtaining a list of differentially expressed genes, further functional analysis is possible, but this is beyond the scope of this chapter.

8.4 Databases

Currently, most ribosome profiling datasets are being deposited in the GEO database in the SRA format (Barrett et al. 2013); however GWIPS-viz a ribosome profiling specific database and genome browser is under development (Michel et al. 2013). Currently, the database features some preloaded datasets available from the GEO, but in the future the developers plan to include options to upload own datasets. In its latest update, they have made available a range of tools to help the researcher develop own workflows of the sequenced data.

Although there are no alternatives for publishing raw ribosome profiling data, except for a special section for ribosome profiling data in the *E. coli* PortEco database (Hu et al. 2014), the results of ribosome profiling analysis have been included in a few other databases. One such option is the TISdb, a database of mRNA alternative translation that followed studies that searched for TISs (Lee et al. 2012; Wan and Qian 2014). Another is HAltORF, a database of alternative out-of-frame open reading frames for human (Vanderperre et al. 2012, 2013).

8.5 Conclusion

Ribosome profiling is emerging as a powerful technique to gain a genome-wide snapshot of gene expression and translation control under a given cellular condition. The availability of positional information of ribosome occupancy facilitates the discovery of novel translational control elements such as alternative initiation at non-canonical start sites, upstream and multiple ORFs, stop codon readthrough such as in the case of selenoprotein translation and pause/regulation of elongation. In addition, coupled with RNA-seq it is a powerful tool to discover alternative splicing variants undergoing differential translation as well as measuring productive alternative splicing at the translational level. Thus, this technique is expected to have a far reaching impact on multiple biological investigations.

Acknowledgements The authors would like to thank Dr. Julie Aspden for critically reviewing the manuscripts and suggesting several improvements.

Annex: Quick Reference Guide

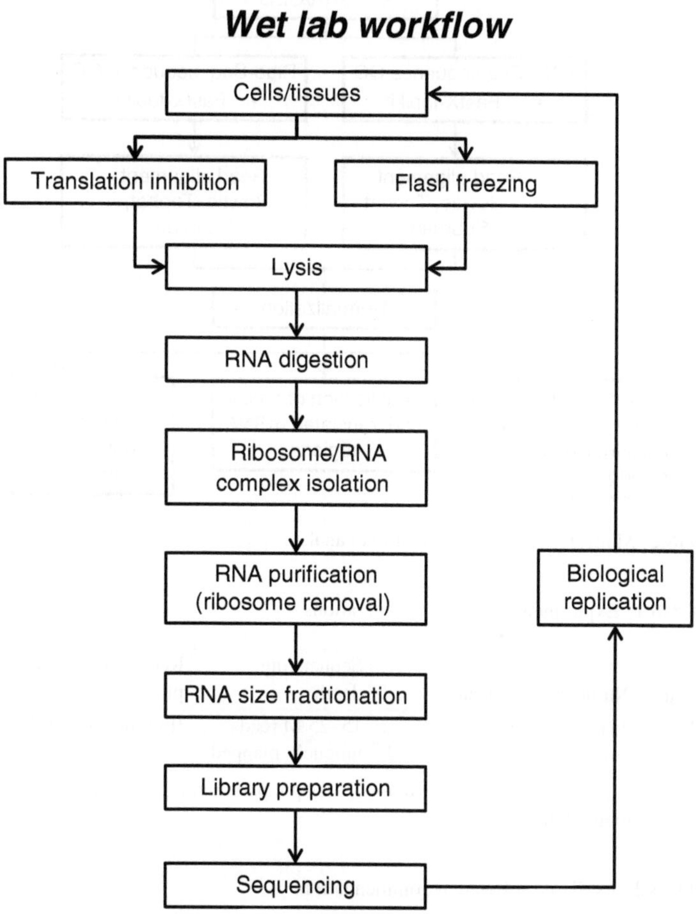

Fig. QG8.1 Representation of the wet lab procedure workflow

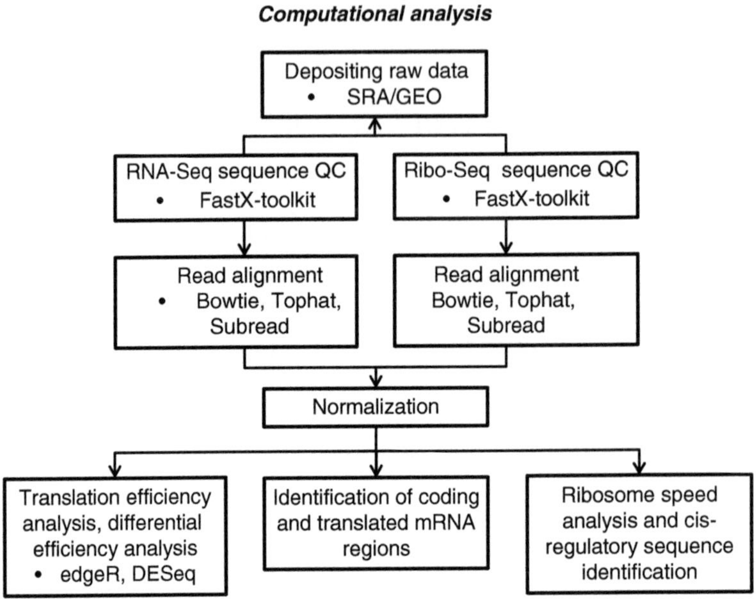

Fig. QG8.2 Main steps of the computational analysis pipeline

Table QG8.1 Experimental design considerations

Technique	Number of replicates	Sequencing depth	Recommended sequencing platforms
Ribo-Seq	3 (minimum per condition), 5 recommended	15–25 M reads uniquely mapped	Illumina HiSeq, Solid 5500

Table that comprises relevant experimental design parameters, to carefully consider before applying this methodology

Table QG8.2 Available software recommendations

Results reporting					
Application	Method	Software	Reference	Output	Format
Translated regions	– ribosome reads map counting (threshold)	riboseqR (R package)	Hardcastle (2014)	– bar plots of number of reads along the genome	–png, pdf
Translational efficiency	– Ribo-Seq/ RNA-Seq	edgeR, DESeq (R packages)	Ingolia et al. (2009)	– gene lists	–txt, xls
	– differential expression analysis			– graphics	
Ribosome speed	– ribosome density	riboseqR (R package)	Dana and Tuller (2012)	– bar plots of number of reads along the genome	–png, pdf
			Hardcastle (2014)		

Table displaying a selection of the recommended software available for the computational analysis of data yielded by this technique

References

Albert FW, Muzzey D, Weissman JS, Kruglyak L (2014) Genetic influences on translation in yeast. PLoS Genet 10:e1004692

Anders S, Huber W (2010) Differential expression analysis for sequence count data. Genome Biol 11:R106

Arias C, Weisburd B, Stern-Ginossar N, Mercier A, Madrid AS, Bellare P, Holdorf M, Weissman JS, Ganem D (2014) KSHV 2.0: a comprehensive annotation of the Kaposi's sarcoma-associated herpesvirus genome using next-generation sequencing reveals novel genomic and functional features. PLoS Pathog 10:e1003847

Artieri CG, Fraser HB (2014) Accounting for biases in riboprofiling data indicates a major role for proline in stalling translation. Genome Res 24:2011–2021

Aspden JL, Eyre-Walker YC, Phillips RJ, Amin U, Mumtaz MA, Brocard M, Couso JP (2014) Extensive translation of small Open Reading Frames revealed by Poly-Ribo-Seq. eLife 3:e03528

Barrett T, Wilhite SE, Ledoux P, Evangelista C, Kim IF, Tomashevsky M, Marshall KA, Phillippy KH, Sherman PM, Holko M, Yefanov A, Lee H, Zhang N, Robertson CL, Serova N, Davis S, Soboleva A (2013) NCBI GEO: archive for functional genomics data sets - update. Nucleic Acids Res 41:D991–D995

Baudin-Bailleau A, Legendre R, Kuchly C, Hatin I, Demais S, Mestdagh C, Gautheret D, Namy O (2014) Genome-wide translational changes induced by the prion [PSI+]. Cell Rep 8:439–448

Bazzini AA, Lee M, Giraldez A (2012) Ribosome profiling shows that miR-430 reduces translation before causing mRNA decay in zebrafish. Science 336:233–237

Bazzini AA, Johnstone TG, Christiano R, Mackwiak SD, Obermayer B, Fleming ES, Vejnar CE, Lee MT, Rajewski N, Walther TC, Giraldez AJ (2014) Identification of small ORFs in vertebrates using ribosome footprinting and evolutionary conservation. EMBO J 33:981–993

Becker AH, Oh E, Weissman JS, Kramer G, Bukau B (2013) Selective ribosome profiling as a tool for studying the interaction of chaperones and targeting factors with nascent polypeptide chains and ribosomes. Nat Protoc 8:2212–2239

Brar GA, Yassour M, Friedman N, Regev A, Ingolia NT, Weissman JS (2012) High-resolution view of the yeast meiotic program revealed by ribosome profiling. Science 335:552–557

Brown PO, Botstein D (1999) Exploring the new world of the genome with DNA microarrays. Nat Genet 21:33–37

Caro F, Ahyong V, Betegon M, DeRisi JL (2014) Genome-wide regulatory dynamics of translation in the Plasmodium Falciparum asexual blood stages. eLife 3:e04106

Charneski CA, Hurst LD (2013) Positively charged residues are the major determinants of ribosome velocity. PLoS Biol 11:e1001508

Chew GL, Pauli A, Rinn JL, Regev A, Schier AF, Valen E (2013) Ribosome profiling reveals resemblance between long non-coding RNAs and 5′ leaders of coding RNAs. Development 140:2828–2834

Dana A, Tuller T (2012) Determinants of translational elongation speed and ribosomal profiling biases in mouse embryonic stem cells. PLoS Comput Biol 8:e10022755

Dana A, Tuller T (2014) The effect of tRNA levels on decoding times of mRNA codons. Nucleic Acids Res 42:9171–9181

Dittmar KA, Goodenbour JM, Pan T (2006) Tissue-specific differences in human transfer RNA expression. PLoS Genet 2:e221

Duncan CDS, Mata J (2014) The translational landscape of fission-yeast meiosis and sporulation. Nat Struct Mol Biol 21:641–647

Dunn JG, Foo CK, Belletier NG, Gavis ER, Weissman JS (2013) Ribosome profiling reveals pervasive and regulated stop codon readthrough in Drosophila melanogaster. eLife 2:e01179

Gardin J, Yeasmin R, Yurovsky A, Cai Y, Skiena S, Futcher B (2014) Measurement of average decoding rates of the 61 sense codons in vivo. eLife 3:e03735

Gerashchenko MV, Gladyshev VN (2014) Translation inhibitors cause abnormalities in ribosome profiling experiments. Nucleic Acids Res 42:e134

Gerashchenko MV, Lobanov AV, Gladyshev VN (2012) Genome-wide ribosome profiling reveals complex translational regulation in response to oxidative stress. Proc Natl Acad Sci U S A 109:17394–17399

Guo H, Ingolia NT, Weissman JS, Bartel DP (2010) Mammalian microRNAs predominantly act to decrease target mRNA levels. Nature 466:835–840

Guttman M, Russell P, Ingolia NT, Weissman JS, Lander ES (2013) Ribosome profiling provides evidence that large noncoding RNAs do not encode proteins. Cell 154:240–251

Haft RJF, Keating DH, Schwaegler T, Schwalbach MS, Vinokur J, Tremaine M, Peters JM, Kotlajich MV, Pohlmann EL, Ong IM, Grass JA, Kiley PJ, Landick R (2014) Correcting direct effects of ethanol translation and transcription machinery confers ethanol tolerance in bacteria. Proc Natl Acad Sci U S A 111:E2576–E2585

Hardcastle TJ (2014). riboSeqR: analysis of sequencing data from ribosome profiling experiments. R package version 1.2.0

Heiman M, Schaefer A, Gong S, Peterson JD, Day M, Ramsey KE, Suarez-Farinas M, Schwarz C, Stephan DA, Surmeier DJ, Greengard P, Heintz N (2008) A translational profiling approach for the molecular characterization of CNS cell types. Cell 1135:738–748

Howard MT, Carlson BA, Anderson CB, Hatfield DL (2013) Translational redefinition of UGA codons is regulated by selenium availability. J Biol Chem 288:19401–19413

Hsieh AC, Liu Y, Edlind MP, Ingolia NT, Janes MR, Sher A, Shi EY, Stumpf CR, Christensen C, Bonham MJ, Wang S, Ren P, Martin M, Jessen K, Feldman ME, Weissman JS, Shokat KM, Rommel C, Ruggero D (2012) The translational landscape of mTOR signalling steers cancer initiation and metastasis. Nature 485:55–61

Hu JC, Sherlock G, Siegele DA, Aleksander SA, Ball CA, Demeter J, Gouni S, Holland TA, Karp PD, Lewis JE, Liles NM, McIntosh BK, Mi H, Muruganujan A, Wymore F, Thomas PD (2014) PortEco: a resource for exploring bacterial biology through high-throughput data and analysis tools. Nucleic Acids Res 42:D677–D684

Ingolia NT (2014) Ribosome profiling: new views of translation, from single codons to genome scale. Nat Rev Genet 15:205–213

Ingolia NT, Ghaemmaghami S, Newman JR, Weissman JS (2009) Genome-wide analysis in vivo of translation with nucleotide resolution using ribosome profiling. Science 324:218–223

Ingolia N, Lareau L, Weissman J (2011) Ribosome profiling of mouse embryonic stem cells reveals the complexity and dynamics of mammalian proteomes. Cell 147:789–802

Ingolia NT, Brar GA, Rouskin S, McGeachy AM, Weissman JS (2012) The ribosome profiling strategy for monitoring translation in vivo by deep sequencing of ribosome-protected mRNA fragments. Nat Protoc 7:1534–1550

Ingolia NT, Brar G, Stern-Ginossar N, Harris NS, Talhouarne GJS, Jackson SE, Wills MR, Weissman JS (2014) Ribosome profiling reveals pervasive translation outside of annotated protein-coding genes. Cell Rep 8:1365–1379

Jan CH, Williams CC, Weissman JS (2014) Principles of ER cotranslational translocation revealed by proximity-specific ribosome profiling. Science 346:1257521

Jensen BC, Ramasamy G, Vasconcelos EJR, Ingolia NT, Myler PJ, Parsons M (2014) Extensive stage-regulation of translation revealed by ribosome profiling of Trypanosoma brucei. BMC Genomics 15:911

Juntawong P, Girke T, Bazin J, Bailey-Serres J (2013) Translational dynamics revealed by genome-wide profiling of ribosome footprints in Arabidopsis. Proc Natl Acad Sci U S A 111:E203–E212

Kannan K, Kanabar P, Schryer D, Florin T, Oh E, Bahroos N, Tenson T, Weissman JS, Mankin AS (2014) The general mode of translation inhibition by macrolide antibiotics. Proc Natl Acad Sci U S A 111:15958–15963

Koch A, Gawron D, Steyaert S, Ndah E, Crappe J, De Keunlenaer S, De Meester E, Ma M, Shen B, Gevaert K, Van Criekinge W, Van Damme P, Menschaert G (2013) A proteogenomics approach integrating proteomics and ribosome profiling increases the efficiency of protein identification and enables the discovery of alternative translational start sites. Proteomics 14:2688–2698

Labunsky VM, Gerashchenko MV, Delaney JR, Kaya A, Kennedy BK, Kaeberlein M, Gladyshev VN (2014) Lifespan extension conferred by endoplasmic reticulum secretory pathway deficiency requires induction of the unfolded protein response. PLoS Genet 10:e1004019

Lareau LF, Hite DH, Hogan GJ, Brown PO (2014) Distinct stages of the translation elongation cycle revealed by sequencing ribosome-protected mRNA fragments. eLife 3:e01257

Larsson O, Sonenberg N, Nadon R (2011) Anota: analysis of differential translation in genome-wide studies. Bioinformatics 27:1440–1441

Lee S, Liu B, Lee S, Huang S-X, Shen B, Qian S-B (2012) Global mapping of translation initiation sites in mammalian cells at single-nucleotide resolution. Proc Natl Acad Sci U S A 109:E2424–E2432

Lee MT, Bonneau AR, Takacs CM, Bazzini AA, DiVito KR, Fleming ES, Giraldez AJ (2013) Nanog, Pou51 and SoxB+ activate zygotic gene expression during the maternal-to-zygotic transition. Nature 503:360–364

Li GW, Weissman JS (2012) The anti-Shine-Dalgarno sequence drives translational pausing and codon choice in bacteria. Nature 484:538–541

Li GW, Burkhardt D, Gross C, Weissman JS (2014a) Quantifying absolute protein synthesis rates reveals principles underlying allocation of cellular resources. Cell 157:624–635

Li S, Labaj PP, Zumbo P, Sykacek P, Shi W, Shi L, Phan J, Wu PY, Wang M, Wang C, Thierry-Mieg D, Thierry-Mieg J, Kreil DP, Mason CE (2014b) Detecting and correcting systematic variation in large-scale RNA sequencing data. Nat Biotechnol 32:888–895

Liu MJ, Wu SH, Wu JF, Lin WD, Wu YC, Tsai TY, Tsai HL, Wu SH (2013a) Translational landscape of photomorphogenic Arabidopsis. Plant Cell 25:3699–3710

Liu X, Jiang H, Gu Z, Roberts JW (2013b) High-resolution view of bacteriophage lambda gene expression by ribosome profiling. Proc Natl Acad Sci U S A 110:11928–11933

Liu B, Han Y, Qian SB (2013c) Cotranslational response to proteotoxic stress by elongation pausing of ribosomes. Mol Cell 49:1–11

Loayza-Puch F, Drost J, Rooijers K, Lopes R, Elkon R, Agami R (2013) P53 induces transcriptional and translational programs to suppress cell proliferation and growth. Genome Biol 14:R32

McManus CJ, May GE, Spealman P, Shteyman A (2014) Ribosome profiling reveals post-transcriptional buffering of divergent gene expression in yeast. Genome Res 24:422–430

Menschaert G, Van Criekinge W, Notelaers T, Koch A, Crappe J, Gevaert K, Van Damme P (2013) Deep proteome coverage based on ribosome profiling aids mass spectrometry-based protein and peptide discovery and provides evidence of alternative translation products and near-cognate translation initiation events. Mol Cell Proteomics 12:1780–1790

Michel AM, Choudhury KR, Firth AE, Ingolia NT, Atkins JF, Baranov PV (2012) Observation of dually decoded regions of the human genome using ribosome profiling data. Genome Res 22:2219–2229

Michel AM, Fox G, Kiran AM, De Bo C, O'Connor PBF, Heaphy SM, Mullan JPA, Donohue CA, Higgins DG, Baranov PV (2013) GWIPS-viz: development of a ribo-seq genome browser. Nucleic Acids Res 42:D859–D864

Mortazavi A, Williams BA, McCue K, Schaeffer L, Wold B (2008) Mapping and quantifying mammalian transcriptomes by RNA-Seq. Nat Methods 5:621–628

Muzzey D, Sherlock G, Weissman JS (2014) Extensive and coordinated control of allele-specific expression by both transcription and translation in Candida albicans. Genome Res 24:963–973

Nakahigashi K, Takai Y, Shiwa Y, Wada M, Honma M, Yoshikawa H, Tomita M, Kanai A, Mori H (2014) Effect of codon adaptation on codon-level and gene-level translation efficiency in vivo. BMC Genomics 15:1115

Oh E, Becker AH, Sandikci A, Huber D, Chaba R, Gloge F, Nichols RJ, Typas A, Gross CA, Kramer G, Weissman JS, Bukaun B (2011) Selective ribosome profiling reveals the cotranslational chaperone action of trigger factor in vivo. Cell 147:1295–1308

Olshen AB, Hsieh AC, Stumpf CR, Olshen RA, Ruggero D, Taylor BS (2013) Assessing gene-level translational control from ribosome profiling. Bioinformatics 29:2995–3002

Pop C, Rouskin S, Ingolia NT, Han L, Phizicky EM, Weissman JS, Koller D (2014) Causal signals between codon bias, mRNA structure, and the efficiency of translation and elongation. Mol Syst Biol 10:770

Reid DW, Nicchitta CV (2012) Primary role for endoplasmic reticulum-bound ribosomes in cellular translation identified by ribosome profiling. J Biol Chem 287:5518–5527

Robinson MD, McCarthy DJ, Smyth GK (2010) edgeR: a Bioconductor package for differential expression analysis of digital gene expression data. Bioinformatics 26:139–140

Rooijers K, Loayza-Puch F, Nijtmans LG, Agami R (2013) Ribosome profiling reveals features of normal and disease-associated mitochondrial translation. Nat Commun 4:2886

Rubio CA, Weisburd B, Holderfield M, Arias C, Fang E, DeRisi JL, Fanidi A (2014) Transcriptome-wide characterization of the eIF4A signature highlights plasticity in translation regulation. Genome Biol 15:476

Schrader JM, Zhou B, Li GW, Lasker K, Childers WS, Williams B, Long T, Crosson S, McAdams HH, Weissman JS, Shapiro L (2014) The coding and noncoding architecture of the Caulobacter crescentus genome. PLoS Genet 10:e1004463

SEQC/MAQC-III Consortium (2014) A comprehensive assessment of RNA-Seq accuracy, reproducibility and information content by the Sequencing Quality Control Consortium. Nat Biotechnol 32:903–914

Shah P, Ding Y, Niemczyk M, Kudla G, Plotkin JB (2013) Rate-limiting steps in yeast protein translation. Cell 153:1589–1601

Shalgi R, Hurt JA, Krykbaeva I, Taipale M, Lindquist S, Burge CB (2013) Widespread regulation of translation by elongation pausing in heat shock. Mol Cell 49:439–452

Shen S, Park JW, Lu ZX, Lin L, Henry MD, Wu YN, Zhou Q, Xing Y (2014) rMATS: robust and flexible detection of differential alternative splicing from replicate RNA-Seq data. Proc Natl Acad Sci U S A 111:e5593–e5601

Smith JE, Alvarez-Dominguez JR, Kline N, Huynh NJ, Geisler S, Hu W, Coller J, Baker KE (2014) Translation of small open reading frames within unannotated RNA transcripts in Saccharomyces cerevisiae. Cell Rep 7:1858–1866

Sonenberg N, Hinnebusch AG (2009) Regulation of translation initiation in eukaryotes: mechanisms and biological targets. Cell 136:731–745

Stadler M, Fire A (2011) Wobble base-pairing slows in vivo translation elongation in metazoans. RNA 17:2063–2073

Stadler M, Fire A (2013) Conserved translatome remodelling in nematode species executing a shared developmental transition. PLoS Genet 9:e1003739

Stadler M, Artiles K, Pak J, Fire A (2012) Contributions of mRNA abundance, ribosome loading, and post- or peri-translational effects to temporal repression of C. Elegans heterochronic miRNA targets. Genome Res 22:2418–2426

Stumpf CR, Moreno MV, Olshen AB, Taylor BS, Ruggero D (2013) The translational landscape of the mammalian cell cycle. Mol Cell 52:1–9

Thoreen CC, Chantranupong L, Keys HR, Wang T, Gray NS, Sabatini DM (2012) A unifying model for mTORC1-mediated regulation of mRNA translation. Nature 485:109–113

Vanderperre B, Lucier JF, Roucou X (2012). HAltORF: a database of predicted out-of-frame alternative open reading frames in human. Database 2012:bas025.

Vanderperre B, Lucier JF, Bissonnette C, Motard J, Tremblay G, Vanderperre S, Wisztorski M, Salzet M, Boisvert FM, Roucou X (2013) Direct detection of alternative open reading frames translation products in human significantly expands the proteome. PLoS One 8:e70698

Vasquez JJ, Hon CC, Vanselow JT, Schlosser A, Siegel TN (2014) Nucleic Acids Res 42:3623–3637

Vogel C, Abreu R, Ko D, Le S-Y, Shapiro B, Burns S, Sandhu D, Boutz D, Marcotte E, Penalva L (2010) Sequence signatures and mRNA concentration can explain two-thirds of protein abundance variation in a human cell line. Mol Syst Biol 6:400

Wan J, Qian SB (2014) TISdb: a database for alternative translation initiation in mammalian cells. Nucleic Acids Res 42:845–850

Wang J, Garrey J, Davis RE (2014) Transcription in Pronuclei and One- to Four-Cell Embryos Drives Early Development in a Nematode. Curr Biol 24:124–133

Wiita AP, Ziv E, Wiita PJ, Urisman A, Julien O, Burlingame AL, Weissman JS, Wells JA (2013) Global cellular response to chemotherapy-induced apoptosis. eLife 2:e01236

Williams CC, Jan CH, Weissman JS (2014) Targeting and plasticity of mitochondrial proteins revealed by proximity-specific ribosome profiling. Science 346:748–751

Yip KY, Cheng C, Gerstein M (2013) Machine learning and genome annotation: a match meant to be? Genome Biol 14:205

Zupanic A, Meplan C, Grellscheid SN, Mathers JC, Krikwood TB, Hesketh JE, Shanley DP (2014) Detecting translational regulation by change point analysis of ribosome profiling data sets. RNA 20:1507–1518

Chapter 9
Genome-Wide Analysis of DNA Methylation Patterns by High-Throughput Sequencing

Tuncay Baubec and Altuna Akalin

9.1 Principles of Genome Regulation by DNA Methylation

DNA methylation is a highly relevant epigenetic mark associated with transcriptional repression in mammals, plants, and various other organisms (Suzuki and Bird 2008). In mammals, this epigenetic modification occurs predominantly at cytosines in the CpG dinucleotide context, although non-CpG methylation has been observed in human embryonic stem and neuronal cells (Lister et al. 2009, 2013). Once established by the de novo methyltransferases DNMT3A and DNMT3B, the maintenance methyltransferase DNMT1 secures stable inheritance of CpG methylation during cell division (Goll and Bestor 2005), until removed by passive or active processes including DNA repair or TET-mediated conversion to 5-hydroxymethylcytosine (5-hmC) (Kohli and Zhang 2013). Functional evidence for the involvement of DNA methylation in gene regulation and genome function comes from knock-out studies of DNMTs (Okano et al. 1999) or chemical interference with these enzymes (Jones and Taylor 1980), resulting in embryonic lethality, chromosomal aberrations, or transcriptional derepression of repetitive elements.

Recent advances in genome-wide analysis were instrumental to broaden our knowledge on the genomic distribution and the developmental dynamics of DNA methylation in various species, tissues, and cancer cells (Lister et al. 2009; Zemach et al. 2010; Stadler et al. 2011; Ziller et al. 2013). These important studies identified

T. Baubec, Ph.D.
Epigenomics and Chromatin Biology Lab, Institute of Veterinary Biochemistry and Molecular Biology, University of Zurich, Winterthurerstrasse 190, 8057 Zurich, Switzerland
e-mail: tuncay.baubec@uzh.ch

A. Akalin, Ph.D. (✉)
Bioinformatics Platform, Berlin Institute for Medical Systems Biology, Max Delbrück Centre, Robert Rössle Strasse 10, 13125 Berlin, Germany
e-mail: altuna.akalin@mdc-berlin.de

A.M. Aransay, J.L. Lavín Trueba (eds.), *Field Guidelines for Genetic Experimental Designs in High-Throughput Sequencing*,
DOI 10.1007/978-3-319-31350-4_9

various features of *DNA methylomes*: While the majority of CpG dinucleotides in mammalian genomes are methylated, CpGs within regulatory sites such as active promoters or enhancers are depleted of methylation. Importantly, changes in DNA methylation at such regulatory sites correlate with transcriptional activity and dynamic binding of transcription factors during biological processes. This suggests a tightly regulated interplay between DNA methylation and transcriptional regulation with relevance for developmental regulatory processes and disease.

Taken together, these relevant findings highlight the emerging requirement for quantitative and high-resolution measurements of cytosine methylation at a genome-wide scale followed by downstream analysis using computational and statistical tools. Here we will discuss current wet-lab and computational approaches for quantitative DNA methylation analysis.

9.2 Methods for Quantitative DNA Methylation Analysis

Prior to the advent of microarray and high-throughput sequencing technologies, DNA methylation analysis could only be performed for single genomic sites individually. These measurements utilized either restriction enzymes sensitive to DNA methylation or methylcytosine-specific antibodies and methyl-CpG-binding domains to enrich for methylated DNA at sites of interest. These methods however did not address the methylation status of single CpGs, but rather indicated the methylation over the measured region (depending on the assay ranging from 100 to 1000 bp). A significant contribution was made by the discovery of sodium bisulfite treatment, which converts unmethylated cytosines to thymine (via uracil), whereby methylated cytosines remain protected (Wang et al. 1980) (Fig. 9.1). In combination

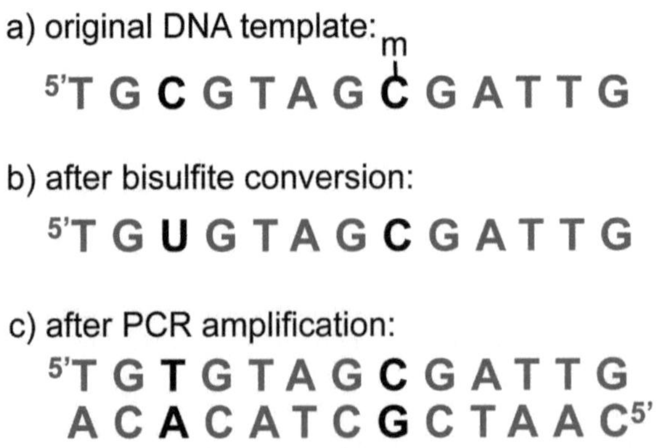

Fig. 9.1 Schematic representation of bisulfite conversion of methylated DNA molecules. (**a**) The original DNA template (shown as single strand) is converted by bisulfite treatment. (**b**) Cytosines (C) are converted to uracil (U), whereas the methyl group (m) inhibits this reaction. (**c**) PCR amplification results in a double-strand molecule where uracil is replaced by thymine (T)

with Sanger sequencing of cloned PCR products, this allowed readout of DNA methylation at nucleotide resolution (Clark et al. 1994) (Fig. 9.1).

The development of microarray and sequencing technologies allowed to profile DNA methylation at high resolution in a genome-wide manner, partially by building up on the approaches previously used to study single sites. Here we will discuss the most commonly utilized methods (Table 9.1 and Fig. 9.2). For a comprehensive overview of all methods designed to measure 5mC, including derivates from TET-mediated oxidation (5-hmC), we would like to draw attention to a recent review describing these approaches (Plongthongkum et al. 2014).

9.2.1 The Infinium 450K BeadChip Array

The Infinium 450K BeadChip from Illumina is a microarray-based readout that is widely applied for rapid profiling of DNA methylation at 450,000 CpG sites in the human genome. These CpG sites are located within nearly all promoters, CpG islands, genes, and numerous enhancers. Using bisulfite-converted DNA as template, this assay detects the methylation status of cytosines via hybridization of probes specific to the methylated or unmethylated locus. Extension of these probes by fluorescently labeled nucleotides allows quantification of DNA methylation at single CpGs, which is reported as beta values ranging from 0 to 1. The benefit of this approach is easy accessibility, low input requirements (500 ng–1 μg of DNA), parallel analysis of up to 12 samples, and a great amount of available datasets from various tissues, ages, and diseases generated using the same standardized protocol (>10,000 samples). The 450K BeadChip is routinely used in clinical settings or large-scale mapping initiatives (ENCODE and Cancer Genome Atlas), which makes this platform a great resource for comparison between samples of interest, or for data mining. The downside of the 450K BeadChip is the restriction to a predefined set of CpGs and availability for human genomes only.

9.2.2 MeDIP/MBD-Affinity Enrichments and Sequencing

MeDIP/MBD-seq relies on affinity enrichment of methylated DNA followed by microarray hybridization or high-throughput sequencing. In both approaches, genomic DNA is randomly sheared to 100–500 base pairs by sonication or restriction enzymes. For MeDIP, the DNA is denatured and captured using monoclonal antibodies specific to 5-methylcytosine (Weber et al. 2005) (or to 5-hydroxymethylcytosine for hydroxyl-MeDIP (Ficz et al. 2011)). Alternatively, methylated DNA can be precipitated using domains cloned from methyl-CpG-binding domain (MBD) proteins, without the need of denaturation (Cross et al. 1994). MBD proteins specifically bind methylated CpGs in vitro and in vivo (Hendrich and Bird 1998; Baubec et al. 2013) and several enrichment techniques have been developed based

Table 9.1 Comparison of most commonly used DNA methylation analysis techniques

Technique	Protocol	Readout	Replicates	Seq. depth	Advantage	Disadvantage
450K array	Sodium bisulfite conv.	Array	3	n.a.	Large database	Preselected
MeDIP/MBD-seq	Antibody/MBD domain IP	Array or sequencing	3	>30 mio. aligned reads	Protocol handling	Strong CG bias
RRBS	Restriction digest and bisulfite conv.	Sequencing	2	>30× coverage	Reduced complexity	Reduced representation
WGBS	Sodium bisulfite conv.	Sequencing	2	>15× coverage	Whole genome	Cost intensive
Target-BS	Sequence enrichment and bisulfite conv.	Sequencing	3	>30× coverage	Targeted to regions of interest	Requires target design

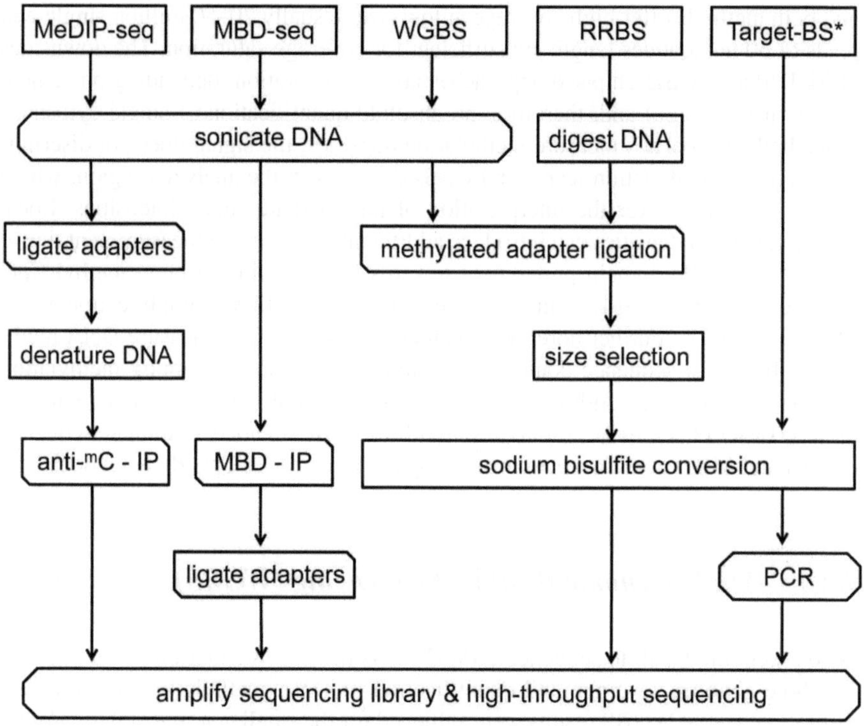

Fig. 9.2 Workflow for high-throughput sequencing methods for DNA methylation analysis. Target-BS: only PCR-based amplification of target regions is shown

on MBD domains (Rauch and Pfeifer 2005; Gebhard et al. 2006; Jørgensen et al. 2006). MBD-based enrichment using salt gradients further allows stratification of CpG densities (MethylCap, Brinkman et al. 2010).

The DNA enriched by both methods can be used for genomics approaches using microarray hybridization or high-throughput sequencing. For microarray hybridization, low amounts of DNA require amplification by whole-genome amplification (WGA) protocols. However, since WGA could introduce biases, pooling of multiple MeDIP samples is recommended (Mohn et al. 2009). The amplified material is subsequently labeled with fluorophores for microarray hybridization.

Alternatively, the precipitated material can be used to generate libraries for high-throughput sequencing. For MeDIP, library adapters have to be ligated to the sonicated DNA prior to denaturation and immunoprecipitation. Library amplification is performed after the MeDIP step. Since MBD-enrichment does not require denaturation, all MBD-seq library preparation steps can be performed on the precipitated material. Standard protocols using Illumina reagents require 1–5 μg of DNA, but MeDIP-seq protocols have been optimized to use 100-fold less (10–50 ng) material (Taiwo et al. 2012). MeDIP and MBD-seq are both quick and simple approaches to measure DNA methylation genome-wide. Numerous kits and protocols are available and both methods offer high coverage at low cost. Usually 20–50 million single-end reads of 50 nucleotides length are sufficient for coverage saturation. The downsides of MeDIP and MBD-chip/seq approaches are low resolution (depending on sonication or array design) and, therefore, no absolute quantification of single cytosines. Since both approaches measure methylation densities, the signal does not discriminate between methylation levels and CpGs densities in the analyzed region, which can be problematic for the interpretation of intermediate signal intensities. Local variation in CpG densities can introduce additional biases in the measurement due to antibody or MBD domain preferences towards regions with different methyl-CpG densities (Nair et al. 2014) or, in case of sequencing, from library amplification biases (Aird et al. 2011). Furthermore, both methods do not allow to measure DNA methylation based on sequence context, as required for tissues containing methylation outside of CpG dinucleotides, such as plants (Cokus et al. 2008) or human neuronal tissues (Lister et al. 2013). Antibodies used for MeDIP do not discriminate sequence context and MBD domains recognize methylation mainly at CpG dinucleotides.

9.2.3 Whole-Genome Bisulfite Sequencing (WGBS)

Whole-genome bisulfite sequencing (WGBS) is the gold standard method in terms of whole-genome coverage and quantification of DNA methylation at nucleotide resolution. Initial WGBS results were achieved for the small genome of *Arabidopsis thaliana* (Cokus et al. 2008). Increasing sequencing depth and read length facilitated generation of high-resolution DNA methylation maps from larger genomes such as human (Lister et al. 2009), mouse (Stadler et al. 2011; Hon et al. 2013), maize (Gent et al. 2013), and numerous other organisms (Feng et al. 2010; Zemach et al. 2010). The benefits of WGBS are clearly the unbiased representation of the entire genome. This for instance allowed the genome-wide identification of megabase-sized, partially methylated domains (PMDs, Lister et al. 2009) or low methylated regions at distal regulatory elements (LMRs, Stadler et al. 2011), features otherwise undetected by previous approaches. In addition, the single nucleotide resolution readout gives insight into previously unnoticed sequence-specific deposition of methylcytosines (Cokus et al. 2008; Lister et al. 2009, 2013). Individual reads mirror the methylation state of single DNA molecules. For instance, WGBS can be used to compare the methylation state of neighboring CpGs on the same DNA molecule (Baubec et al. 2015).

WGBS requires extensive coverage (Ziller et al. 2015). Depending on the genome size, several sequencing reactions are required in order to reliably call methylation frequencies for the majority of CpGs. A precise recall of cytosine methylation does not only require sufficient sequencing depth but also strongly depends on the quality of bisulfite conversion and library amplification. In brief, 1–5 μg of fragmented genomic DNA (100–500 bp, by sonication) is sufficient to generate WGBS libraries. However, lower amounts can also be used, as recently shown for single cell methylomes (Smallwood et al. 2014). In order to prevent artifacts during bisulfite conversion, RNAs and proteins should be completely removed prior to treatment (Warnecke et al. 2002). We recommend to spike-in low amounts of sonicated, unmethylated Lambda phage DNA and in vitro methylated T7 dsDNA (by the CpG-methyltransferase *SssI*) before library preparation. These controls allow measuring non-conversion and over-conversion rates respectively, and therefore are useful for estimating bisulfite conversion quality. Combined genomic DNA and spike-ins are end-repaired and ligated to methylated paired-end library adaptors. Libraries are size selected on agarose gels (300–400 bp) and bisulfite converted using commercially available kits (e.g., Epitect from Qiagen, EZ DNA methylation from Zymo Research, and Imprint from Sigma). These kits are also designed for lower starting material. Alternatively, newer protocols and bisulfite sequencing library kits allow adapter ligation post-conversion (EpiGnome from Epicentre). In both cases, prepared libraries are amplified via PCR using polymerases that tolerate uracil templates (e.g., Pfu Cx Turbo from Agilent Technologies). Number of PCR cycles have to be determined according to input material. Usually 6–10 cycles should be sufficient. These can be increased for low starting material; however this can introduce library amplification biases (see Sect. 9.3.2). Libraries are subsequently sequenced using paired-end reads of 100–150 nucleotides length in order to obtain sufficient coverage of the entire template.

9.2.4 Reduced Representation Bisulfite Sequencing (RRBS)

Enzymatic restriction of genomic DNA methylation-insensitive enzymes that cut in a CG context allows enrichment of genomic regions with mid to high CpG densities (Meissner 2005; Gu et al. 2011). Since CpG distribution in mammalian genomes is nonuniform and with higher CpG densities at promoters and CpG islands, this digest allows for preferential enrichment of such genomic elements. The digested DNA is end-repaired and ligated to methylated sequencing adapters followed by gel-based size selection (40–220 base pairs). Sequencing libraries are treated with sodium bisulfite for conversion and finally amplified by PCR, cleaned up and sequenced similar to WGBS (for a detailed protocol see (Gu et al. 2011)). Due to the reduced representation of the genome (~1 %), RRBS requires only modest sequencing read numbers for sufficient coverage, which makes it more affordable than whole-genome sequencing. However, this benefit is also the disadvantage of RRBS. Biologically relevant changes in DNA methylation can occur beyond CpG islands (Doi et al. 2009; Lister et al. 2009; Stadler et al. 2011). Thus, bias towards CpG-rich regions results in limited coverage for relevant genomic regions such as distal enhancers and transcription binding sites, repetitive elements, or large regions in the genome located within PMDs. A more extended CpG representation can be achieved by increasing the range of DNA fragments (up to 400 bp) selected after gel purification (Akalin et al. 2012a).

9.2.5 Targeted Bisulfite Sequencing (Target-BS)

Targeted bisulfite sequencing (Target-BS) entails selection or amplification of predefined genomic regions in combination with bisulfite conversion and high-throughput sequencing. Frequently used protocols employ either PCR amplification of regions of interest (Taylor et al. 2007; Landan et al. 2012), padlock probes (Ball et al. 2009), or hybridization-based target enrichment (SureSelect, Ivanov et al. 2013).

PCR amplification from targeted genomic regions requires specialized primer design protocols that take in account DNA sequence conversion by bisulfite treatment. First, the primers need to be designed based on the converted DNA template, meaning that the first primer should anneal to one strand of the converted template, while the second primer should base pair to the DNA sequence synthesized in the first PCR reaction. In addition, numerous other parameters have to be taken into consideration that grant reliable PCR amplification. Primers need to have reasonably high melting temperatures (above 50 °C), should not exceed a product size of 500 bp, and should not be designed over regions with CpGs. The latter parameter can complicate the design over CpG-dense regions such as CpG islands. To overcome these limitations, some design parameters can be relaxed, such as allowing 1–2 CpGs within the primer site, or including mixed bases for pyrimidines (T and C) or purines (A and G) at CpG sites. Numerous useful tools are available online for bisulfite primer design (e.g., MethPrimer, Li and Dahiya 2002); however, given the complications that arise from the reduced nucleotide complexity after bisulfite conversion and the before-mentioned optimizations, none of them allows batch design of multiple target sites. Development of automated, iterative primer design algorithms (Komori et al. 2011, Schmidt et al. in prep) should aid in the design of primer pairs targeting multiple regions of interest in the genome, resulting in a set with similar properties (melting temperature, product length). PCR reactions are performed as individual reactions (e.g., 96-well or 384-well setup) from bisulfite-converted genomic DNA (1–2 μg DNA is sufficient for 96 reactions). PCR conditions have to be optimized in order to obtain homogenous representation of all targets. We recommend touchdown PCR using proofreading enzymes and testing various annealing temperatures and cycle numbers. Once established, the same PCR protocol can be used for all subsequent samples analyzed with the designed primer set. Amplified PCR products are pooled and purified on agarose gels based on the expected size distribution of the entire PCR library.

Target enrichment based on hybridization has been previously applied for exome capture followed by sequencing. Similar approaches can be used for methylated DNA. First the DNA is fragmented, ligated to methylated adapters, and enriched by hybridization using biotinylated oligonucleotides. Bisulfite conversion can be performed prior or after enrichment. Conversion after enrichment suffers from low DNA input and depending on the amount of material, this can be problematic for downstream library preparation steps (Lee et al. 2011). Conversion before enrichment requires design of oligonucleotides complementary to the converted DNA. Since genomic regions can be partially and low methylated (Stadler et al. 2011; Gaidatzis et al. 2014), bisulfite conversion of the same genomic location can generate DNA molecules with heterogeneous DNA sequences. In order to avoid mismatches, the probes need to be designed in such a manner that they take all possible combinations of C to T conversions in consideration (Ivanov et al. 2013). Commercial vendors provide predesigned kits for methylation capture and allow customers to design their own sets online (SureSelect by Agilent Technologies).

Depending on the number of target regions, libraries can be multiplexed for parallel sequencing. For example, 6–8 libraries of 96-well PCR-based targeted bisulfite sequencing can be sequenced on an Illumina MiSeq machine, yielding sufficient coverage. The benefits of complexity reduction by targeting methods are specific interrogation of a predesigned set of genomic regions, low genome complexity of sequencing libraries, and nucleotide resolution readout. Another benefit of both approaches is that the fragment length can be optimized to fit the sequencing platform read length, resulting in complete sequencing of the entire DNA molecule.

9.3 Computational Analysis of High-Throughput Bisulfite Sequencing

Since high-throughput bisulfite sequencing is the gold standard method and rapidly gaining popularity over other methods, we will describe the computational analysis of bisulfite sequencing over the next couple of sections.

9.3.1 Alignment and Methylation Calling for Bisulfite Sequencing Experiments

During bisulfite sequencing the unmethylated cytosines (C) are changed to thymines (T). This helps to pinpoint unmethylated Cs on the reads but this complicates the alignment process by introducing these artificial mutations. If one were able to align the reads reliably, percent methylation for a cytosine would be number of Cs divided by number of Cs+Ts (Fig. 9.3). The alignment methodologies mostly revolve around modifications of known short read alignment strategies accounting for possible $C \rightarrow T$ conversions. One of the most popular methods Bismark (Krueger and Andrews 2011) utilizes the popular Bowtie aligner (Langmead et al. 2009). The essential idea is that the aligner transforms the reads and the genome to bisulfite-treated versions in silico, then it aligns the converted reads to the converted genomes and resolves multi-mapping reads based on alignment quality calculated from mismatch rates. Other methods, such as MethylCoder (Pedersen et al. 2011), BS-Seeker2 (Guo et al. 2013), and BRAT-BW (Harris et al. 2012), also use similar approaches. However, there are methods that deviate from the strategy described above. Notably, BSMAP (Xi and Li 2009) masks thymines in the reads and regards them as potential

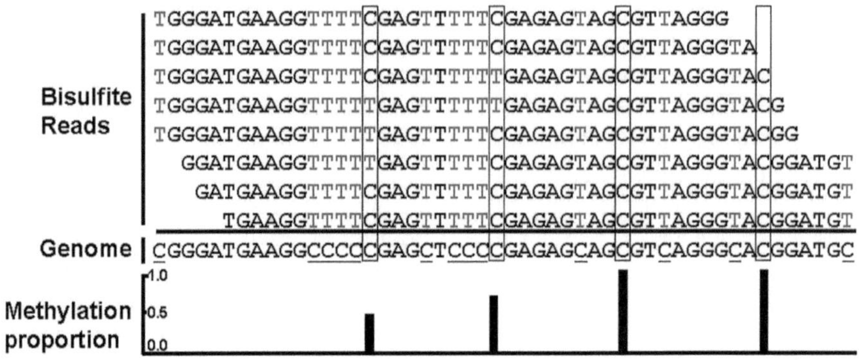

Fig. 9.3 Calling methylation from aligned reads. If bisulfite-converted reads can be aligned to the genome. Calling methylation proportion/percentage is simply counting number of Cs and dividing it with number of Cs + Ts for a given base. The illustration exemplifies that procedure. Note that non-CpG Cs have no methylation in the example. Ts that stem from bisulfite conversion are in gray color. Cytosines of CpGs are bound by rectangles

match to cytosines in the genome. In addition, Last (Frith et al. 2012) uses a more traditional approach where it makes use of a score matrix for alignment scoring. The matrix adjusts for possibility of C-T mismatches. Finally, BiSS (Dinh et al. 2012) uses a Smith-Waterman local alignment implementation for bisulfite-converted reads, allowing for increased mapping of sequencing reads.

Proceeding the alignment and methylation calling, the results can be exported as BigWig files (https://genome.ucsc.edu/goldenPath/help/bigWig.html) to allow visual inspection of results in a genome browser. At this point, the samples can also be merged to inspect correlation and clusters of the samples. This should further confirm the sample quality, similarity of the replicates, and existence of a biological effect that is of interest.

9.3.2 Issues with Methylation Calling

There are multiple caveats with methylation calling that stem from the nature of the experiment and specific variations of the bisulfite sequencing protocol. Below, we described some of the caveats and suggest mitigating strategies.

9.3.2.1 Base Qualities

First issue is ubiquitous in most sequencing experiments. Each base on a read coming from a high-throughput sequencing experiments is associated with a quality value that indicates the confidence in the called base. Low quality bases might harbor sequencing errors and, therefore, should be removed from the analysis. This can either be done by sequence trimming since quality tends to decrease towards the end of the reads using tools such as FASTX toolkit (http://hannonlab.cshl.edu/fastx_toolkit/) or by discarding bases during methylation calling based on their quality score (Akalin et al. 2012a).

9.3.2.2 Coverage

In addition, coverage per base becomes an important issue for methylation calling and subsequent differential methylation analysis. The greater the number of reads covering a certain position, the greater the precision for estimating differential methylation will be. The recommended coverage optimized for sensitivity and cost-effectiveness is 10× (Ziller et al. 2015). However, it should be noted that high coverage in some regions may be due to PCR bias; therefore for WGBS reads, we recommend discarding reads with overlapping coordinates. For RRBS, due to use of a restriction enzyme, most reads will have the same start and end coordinates and such a removal of duplicated reads is not feasible. However, one can remove the very high coverage regions from the analysis (above 1000× or top 5 %) in order to minimize PCR bias in RRBS experiments (Akalin et al. 2012a).

9.3.2.3 Adapter Sequencing

In certain cases parts of the adapter are sequenced. This could be due to decay mediated by bisulfite conversion or problems with size selection, where shorter fragments slip into the sample. In these cases, removing sequencing adapters prior to alignment increases the number of mapped reads. Removal of adapters can be achieved with tools like FLEXBAR (Dodt et al. 2012) or cutadapt (Martin 2011) where adapters can be partially aligned to reads and aligning parts can be excised from the read. Removing adapters will most likely increase the mapping rates and therefore the overall data quality.

9.3.2.4 SNPs

The SNPs that are cytosine in the genome but thymine in the sample will be regarded as authentic conversion events during methylation calling. To avoid such inaccuracies, one can remove such SNPs from the dataset if there is available genomic polymorphism data. If there is no SNP information available, one can try to use SNP callers that are designed for bisulfite sequencing experiments (Bis-SNP, Liu et al. 2012). However, it will not be possible to recover all C → T SNPs.

9.3.2.5 Conversion Rate

Another important issue is the conversion rate, which denotes the efficiency of unmethylated cytosines being converted to thymine. This can be calculated from the number of non-CpG C → T conversions divided by the total coverage of non-CpG Cs. This relies on the fact that non-CpG methylation is rare or of low prevalence in many mammalian cell types except in embryonic stem cells, oocytes, and the brain (Lister et al. 2009, 2013). For a more reliable conversion rate, spike-in sequences with fully unmethylated DNA could be introduced before bisulfite conversion and subsequent sequencing. Measuring C → T conversions in those spike-in samples would give a better understanding of conversion rate (Stadler et al. 2011). The samples with low conversion rate should be discarded, as methylation measurement will not be reliable. Although there are no systematic studies on the effect of conversion rate to differential methylation calculations, it is better to maximize conversion rate close to 100 % and certainly not below 95 %.

9.3.2.6 Assay-Specific Issues

In addition to the issues described above, there are a couple of other issues that stem from variations in the experimental protocol. First, it should be noted that bisulfite sequencing can not discriminate between hydroxy-methylation and methylation (Huang et al. 2010). Therefore, methylation measurements for tissues having high 5-hydroxy-methylation will not be reliable at least in certain genomic regions. Here, specific measurement protocols are required that discriminate hydroxy-methylation from methylation (Plongthongkum et al. 2014). Other issues may arise due to bisulfite sequencing protocol variations. For example, RRBS introduces biased methylation at C in a 5′-CCGG-3′ motif, and this should be removed before calling methylation (Gu et al. 2011).

9.3.3 Segmentation-Based Methods for Discovering Genome-Wide DNA Methylation Patterns

Distinct patterns in methylation profiles are associated with other epigenomic marks and consequently gene regulation (Smith and Meissner 2013). For example, regions with low methylation are usually associated with H3K4me3 on active promoters. In contrast, high methylation for a promoter or regulatory region is associated with repression. However, many variations of this exist. Recently, low methylated regions in mouse embryonic stem cell methylomes were shown to mark enhancers genome-wide (Stadler et al. 2011).

As regulatory regions can be cell type specific and their locations can be discovered by methylome analysis, it is of interest to explore all possible patterns in an unsupervised way. One of the popular methods is to segment methylomes into distinct categories such as regions with low methylation or high methylation. The most popular method for categorical methylome segmentation is based on hidden Markov models (HMM). The method is widely used in bioinformatics for applications in sequence analysis such as finding patterns in genomic DNA or protein sequences (Durbin 1998). However, it can also be used for analyzing quantitative and continuous signals from genome-wide experiments, including whole-genome bisulfite sequencing. In very basic terms, when applied to the methylomes, the method labels each CpG based on its methylation and the methylation status of neighboring CpGs. The labels could be as simple as high or low methylated regions (sometimes referred to as hyper- and hypo-methylated, respectively).

In essence, the method tries to find the optimal statistical model (HMM) that could have generated the observed data. The statistical model is defined over a sequence of methylation proportion values and consists of methylation states and the transition probabilities between states. In addition, each state generates a distribution of methylation proportion values. Figure 9.4 shows the observed sequence of methylation values and HMM states that fit the observed methylation values. The HMM model learned from the data is depicted on the left side. The hypo-methylation or low methylation state generates mostly low methylation values (Gray distribution Fig. 9.4b) and the hyper-methylation state gives rise to higher methylation values (Black distribution Fig. 9.4). During the optimization process, the model

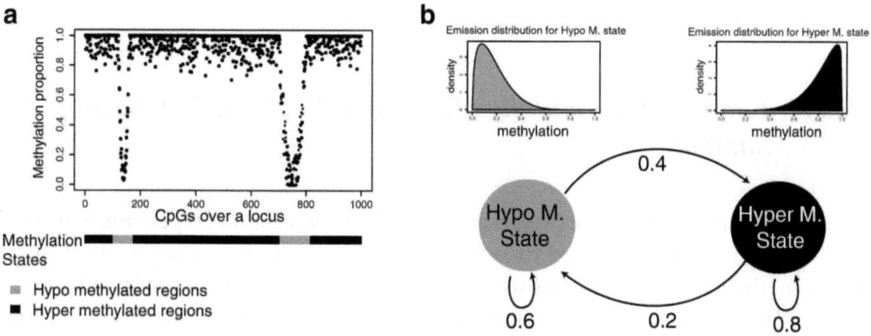

Fig. 9.4 (**a**) Hidden Markov model-based segmentation of the depicted methylation profile into hypo-methylated (*gray segments*) and hyper-methylated segments (*black segments*). Each CpG methylation proportion is shown as a *black dot*. (**b**) The segmentation is achieved by learning the parameters of the HMM that best fit the observed methylation profile. The procedure learned the transition probabilities between states (probabilities depicted on the arcs) and emission probability distributions associated with each state. In this case, the method learned that when a CpG is in hypo-methylated state its methylation proportion is distributed as shown in the *gray density plot* and when it is associated with a hyper-methylation state, the methylation proportion is distributed as the *black density plot*

learns the parameters of gray and black distributions (Fig. 9.4), the most likely state labels for each CpG and the transition probabilities between states. Those states correspond to hyper- or hypo-methylated CpGs in the simplest case (Molaro et al. 2011), and methPipe tool (http://smithlabresearch.org/software/methpipe/) implements this approach.

There are also variations along this theme where one can allow more methylation states than just two. Stadler et al. (Stadler et al. 2011) fitted a three state HMM model upon observation of lowly methylated regions along with fully methylated and unmethylated regions. This allowed them to discover all low methylated regions in the genome, which turned out to be enhancer regions.

Other segmentation strategies include variants of changepoint analysis where change points in a signal across the genome is recorded and the genome is segmented to regions between two change points. These methods are typically used in CNV (copy number variation) detection but have applications in this context as well (Klambauer et al. 2012). In the context of methylation, segments separated by change points can be found and those segments can be clustered into groups based on the methylation similarity of the segment (implemented in methylKit (Akalin et al. 2012b)) In addition, hybrid approaches between HMMs and simple data modeling such as MethylSeekeR (Burger et al. 2013) are also useful for segmentation. MethylSeekeR first identifies partially methylated domains using an HMM and removes them from the rest of the analysis. Following that low, fully methylated and unmethylated regions are identified by dataset-specific cutoffs.

All in all, the users can use multiple publicly available tools to do the segmentation. The expected output of all these programs are at the very least a tabular output and BED files that can be used to visualize the segments on a genome browser.

9.3.4 Finding Differentially Methylated Regions: Comparing Samples

Differences in methylation between samples indicate changes in epigenomic structure and therefore could be related to gene regulation. In addition, many loci are aberrantly methylated in cancer cells when compared to normal cells (Laird and Jaenisch 1994). Therefore, it is usually of interest to compare different samples and find differentially methylated regions or bases in the genome. The comparisons employ various statistical tests to assess the statistical confidence associated with the difference seen between samples for a given region or base. There are multiple ways to model the methylation from samples in a comparative manner when there are replicates. However, when there are no replicates one of the most appropriate method is Fisher's Exact test where methylation ratios between two samples for a given loci can be compared (Akalin et al. 2012b). Even in the presence of replicates, the replicates can be pooled and Fisher's exact test can be applied. However, this process will lead to loss of variation between replicates that could be leveraged by other tests.

When there are replicates, regression-based frameworks are mostly used to model the variation and methylation levels in relation to the sample groups. The differences in methods usually come from the choice of underlying distribution to model the data and the variation associated with it. Using regression-based frameworks have the added benefit of being able to model other covariates into the tests. For example, it has been shown that age is a contributing factor for methylation values at some CpGs. By adding covariates into the model, their contribution can be accounted when deciding if the observed difference between methylation levels of two sample groups is indeed due to biological differences.

In the simplest case, linear regression can be used to model methylation per given CpG or loci across sample groups. The model fits b_0 and b_1 values, which model the expected methylation proportion values (denoted as P in Eq. (9.1)) for each x.

$$\begin{aligned} P &= b_0 + b_1 X + \varepsilon \\ \varepsilon &\sim N(0,\sigma^2) \end{aligned} \tag{9.1}$$

In this case, x is a factor variable that is either 0 or 1 (this variable controls if the samples are from the test group or control group in a simple case–control experimental design). When $x=0$ (assume this is the control group), methylation is b_0. When $x=1$ methylation equals to b_0+b_1. The error term models the variation along the fitted values. It is expected to have a normal distribution with 0 mean and variation σ^2 to be estimated from the data. If there is a good fit to the data, where the methylation values are considerably different in test and control groups, the fitted model will be better than a null model where only the intercept term b_0 is fitted. These two models can be tested using F-tests within the analysis of variance framework. The idea depends on the accurate estimation and comparison of variances between and within samples. If the model fits well, variance substantially drops in fitted model versus the null model. Estimating accurate variances may not be always possible when sample sizes are small. Empirical Bayes methods then can be used to estimate variances while borrowing information across all loci (Smyth 2004).

However, linear regression-based methods to model methylation levels for a given sample group might produce predictions out-of-bounds, meaning methylation levels that are beyond the 0 and 1 bracket can be fitted. Furthermore, the variance estimated by these models will also not be bound by 0 and 1 and it is assumed to be constant. An alternative for linear regression-based models is logistic regression (a generalized linear model with binomial errors). Logistic regression for modeling methylation per given CpG or loci across samples is a more appropriate choice since it can deal with data that is strictly bounded between 0 and 1, with nonconstant variance, and it also is a go-to modeling method for proportion data such as methylation proportion values (denoted as P in Eq. (9.2)).

$$\log\left(\frac{P}{1-P}\right) = b_0 + b_1 X \tag{9.2}$$

Table 9.2 Differential methylation software for bisulfite sequencing experiments

Software	Method	Language/platform	Reference
methylKit	Logistic regression with/without overdispersion correction/Fisher's Exact test	R package	Akalin et al. (2012b)
BSseq	Smoothing + Linear regression + Empirical Bayes	R package	Hansen et al. (2012)
BiSeq	Beta regression	R package	Hebestreit et al. (2013)
DSS	Beta-binomial with Empirical Bayes	R package	Feng et al. (2014)
MOABS	Beta-binomial with Empirical Bayes	C/C++/Command line	Sun et al. (2014)
RADMeth	Beta-binomial regression	C/C++/Command line	Dolzhenko and Smith (2014)

In the logistic regression case, the data is fitted to a model by optimizing the b_1 and b_0 parameters, but this time the dependent variable is not the methylation proportion but the logarithm of $P/(1-P)$ where P is the methylation proportion (see Eq. (9.2)). The procedure essentially maximizes the likelihood of observing the data coming from a specific binomial distribution with parameters related to b_1 and b_0 values in the regression. Similar to the linear regression case, again a statistical test can be used to compare the fitted model versus the null model to see if the fitted model explains the data better.

Further enhancements can be made by tinkering with the variance assumptions of the logistic regression model. In logistic regression, fitted values assumed to have variation of $n(p)(1-p)$ where p is the fitted value for methylation proportion for a given sample and n is the read coverage. This assumption sometimes tends to underestimate the variance. This can be amended by calculating a scaling factor and using that factor to adjust the variance and/or other estimates that could be used in the statistical tests.

More complex models are also available for methylation data. They are particularly useful for better modeling of the variance. One natural choice is to use beta-binomial models. It is similar to logistic regression where the data (number of methylated and unmethylated Cs) is binomial distributed but methylation proportion is distributed according to a beta distribution. Practically, this amends the $n(p)(1-p)$ variance assumption, thus performing better when there is more variance than expected by the simple logistic model. Further enhancements to this include using Empirical Bayes methods to better estimate variance-related parameters by borrowing information from other bases or regions in the genome. Although detailed statistical explanations are beyond the scope of this text, some of the available tools using different methods described in this section are summarized in Table 9.2.

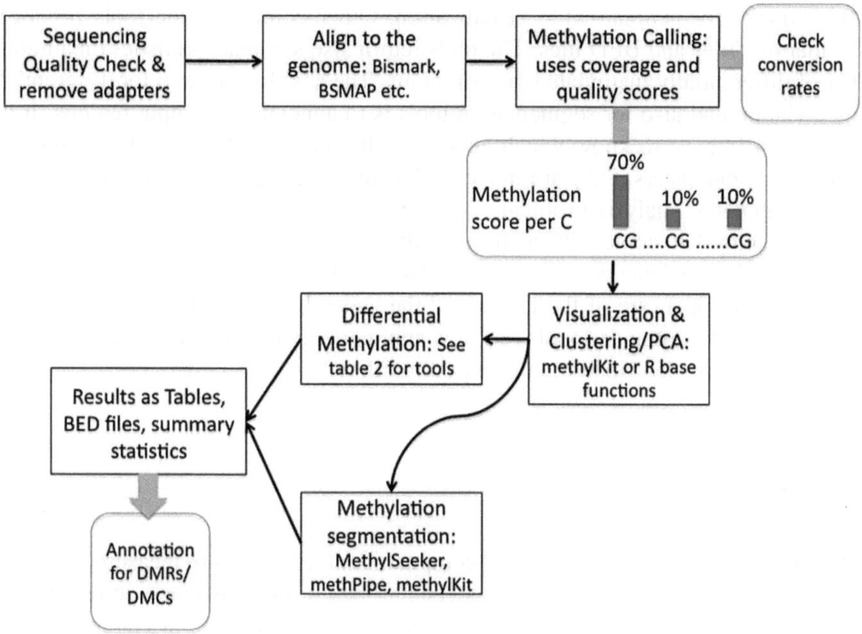

Fig. 9.5 Workflow for analysis of DNA methylation using data from bisulfite sequencing experiments

9.4 Regional vs. Base-Pair Resolution Differential Methylation

Most of the methods mentioned here can operate in both base-pair level and regional level. When the input for differential methylation functions are regions, the data should be summarized adequately per region. Normally, this is done by counting methylated and unmethylated bases per C of a CpG in a given region. The regions may be chosen arbitrarily such as promoters of interest or tilling windows that cover the whole genome. Another way of getting differentially methylated regions is to first get differentially methylated bases and combine the differentially methylated bases to differentially methylated regions. Several methods use this strategy. RADmeth (Dolzhenko and Smith 2014) and eDMR (Li et al. 2013) groups *P*-values of adjacent CpGs and produce differentially methylated regions based on distance between differential CpGs and combination of their *P*-values by weighted Z-test. MOABS (Sun et al. 2014) proposes to use an HMM to segment the differential CpGs into hypo- and hyper-methylated regions; however this is not implemented in the software.

All methods described in Table 9.2 uses methylation profiles from multiple sample groups to detect differentially methylated regions or bases. The output of R-based tools are R objects that are in tabular format and can easily be exported as

BED tracks for genome-wide visualization. Other command line tools produce tabular text files and BED files. The tools return also summary statistics on a number of differentially methylated bases/regions. The next logical step for differential methylation and also for segmentation tools is to annotate the output regions. It is usually of interest to know which genes or other genomic features (CpG islands, promoters, enhancers, etc.) are associated with those regions. Although this is a general genome analysis problem not specific to methylation data, certain tools, such as methylKit, come with such capability.

The analysis of methylation data involves multiple steps and checkpoints. The users have to be aware of the issues and the general analysis flow. Therefore, we have prepared a computational workflow summarizing major steps of analysis and quality checking described in this section in Fig. 9.5.

9.5 Conclusion

We have discussed experimental and computational techniques to measure genome-wide methylation levels. Bisulfite sequencing-based methods come across as the state of the art for detecting methylation patterns genome-wide or in a targeted manner. We further described computational methods to deal with downstream analysis of bisulfite sequencing experiments. Coincidently, most of the tools described here use the R framework for downstream analysis. We believe through this guideline experimental biologists not only will have an idea about experimental protocols and best practices in the wet-lab but also they will be able to get into hands-on analysis if other priorities allow.

Annex: Quick Reference Guide

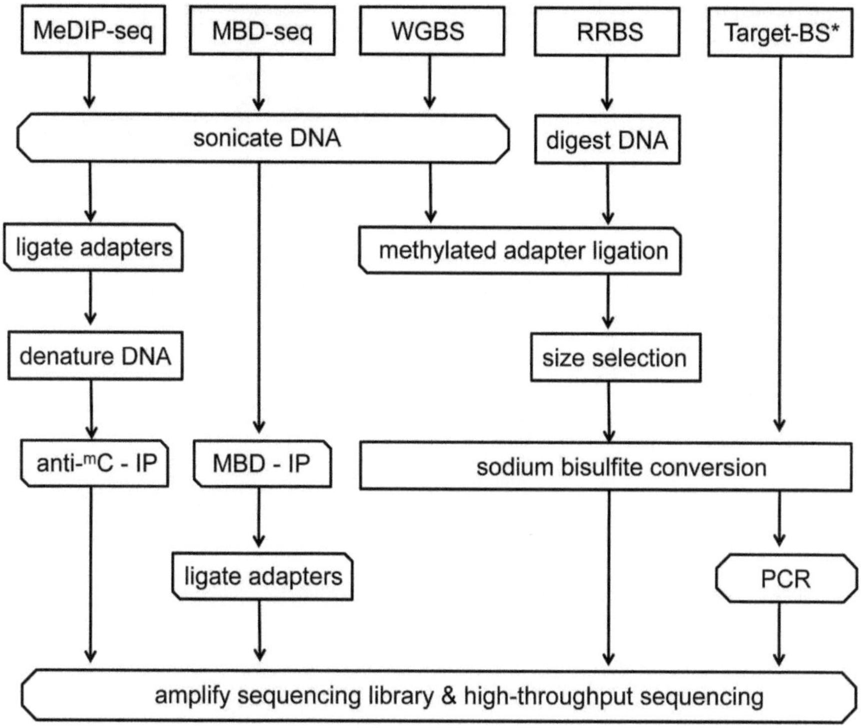

Fig. QG9.1 Representation of the wet-lab procedure workflow

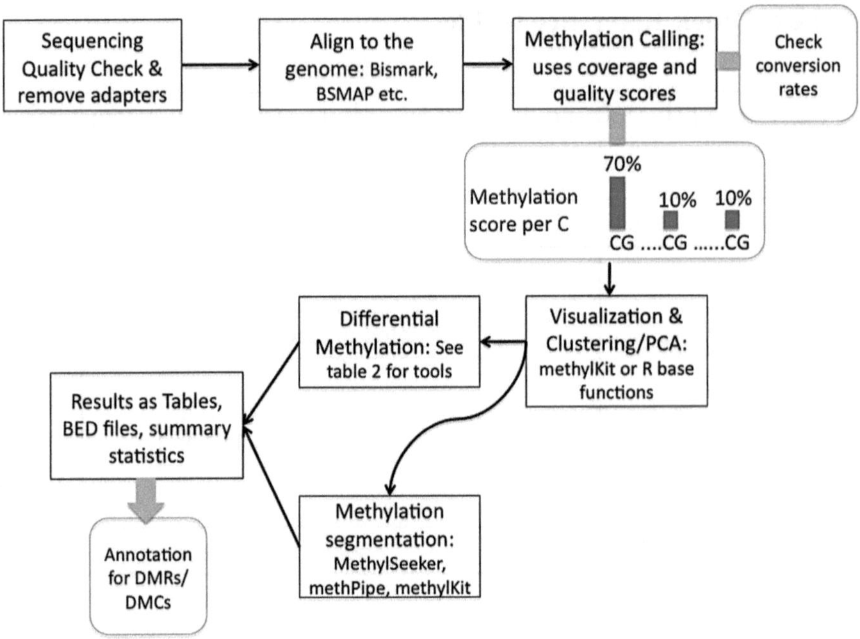

Fig. QG9.2 Main steps of the computational analysis pipeline

Table QG9.1 Experimental design considerations

Technique	Protocol	Readout	Replicates	Seq. depth	Advantage	Disadvantage
450K array	Sodium bisulfite conv.	Array	3	n.a.	Large database	Preselected
MeDIP/MBD-seq	Antibody/MBD domain IP	Array or sequencing	3	>30 mio. aligned reads	Protocol handling	Strong CG bias
RRBS	Restriction digest and bisulfite conv.	Sequencing	2	>30× coverage	Reduced complexity	Reduced representation
WGBS	Sodium bisulfite conv.	Sequencing	2	>15× coverage	Whole genome	Cost intensive
Target-BS	Sequence enrichment and bisulfite conv.	Sequencing	3	>30× coverage	Targeted to regions of interest	Requires target design

Table that comprises relevant experimental design parameters, to carefully consider before applying this methodology

Table QG9.2 Available software recommendations

Software	Method	Language/platform	Reference
methylKit	Logistic regression with/without overdispersion correction/ Fisher's Exact test	R package	Akalin et al. (2012b)
BSseq	Smoothing + Linear regression + Empirical Bayes	R package	Hansen et al. (2012)
BiSeq	Beta regression	R package	Hebestreit et al. (2013)
DSS	Beta-binomial with Empirical Bayes	R package	Feng et al. (2014)
MOABS	Beta-binomial with Empirical Bayes	C/C++/Command line	Sun et al. (2014)
RADMeth	Beta-binomial regression	C/C++/Command line	Dolzhenko and Smith (2014)

Table displaying a selection of the recommended software available for the computational analysis of data yielded by this technique

References

Aird D, Ross MG, Chen W-S et al (2011) Analyzing and minimizing PCR amplification bias in Illumina sequencing libraries. Genome Biol 12:R18. doi:10.1186/gb-2011-12-2-r18

Akalin A, Garrett-Bakelman FE, Kormaksson M et al (2012a) Base-pair resolution DNA methylation sequencing reveals profoundly divergent epigenetic landscapes in acute myeloid leukemia. PLoS Genet 8:e1002781. doi:10.1371/journal.pgen.1002781.s011

Akalin A, Kormaksson M, Li S et al (2012b) methylKit: a comprehensive R package for the analysis of genome-wide DNA methylation profiles. Genome Biol 13:R87. doi:10.1186/gb-2012-13-10-r87

Ball MP, Li JB, Gao Y et al (2009) Targeted and genome-scale strategies reveal gene-body methylation signatures in human cells. Nat Biotechnol 27:361–368. doi:10.1038/nbt.1533

Baubec T, Ivanek R, Lienert F, Schübeler D (2013) Methylation-dependent and -independent genomic targeting principles of the MBD protein family. Cell 153:480–492. doi:10.1016/j.cell.2013.03.011

Baubec T, Colombo DF, Wirbelauer C et al (2015) Genomic profiling of DNA methyltransferases reveals a role for DNMT3B in genic methylation. Nature 520:243. doi:10.1038/nature14176

Brinkman AB, Simmer F, Ma K et al (2010) Whole-genome DNA methylation profiling using MethylCap-seq. Methods 52:232–236. doi:10.1016/j.ymeth.2010.06.012

Burger L, Gaidatzis D, Schübeler D, Stadler MB (2013) Identification of active regulatory regions from DNA methylation data. Nucleic Acids Res 41:e155. doi:10.1093/nar/gkt599

Clark SJ, Harrison J, Paul CL, Frommer M (1994) High sensitivity mapping of methylated cytosines. Nucleic Acids Res 22:2990–2997

Cokus SJ, Feng S, Zhang X et al (2008) Shotgun bisulphite sequencing of the Arabidopsis genome reveals DNA methylation patterning. Nature 452:215–219. doi:10.1038/nature06745

Cross SH, Charlton JA, Nan X, Bird AP (1994) Purification of CpG islands using a methylated DNA binding column. Nat Genet 6:236–244. doi:10.1038/ng0394-236

Dinh HQ, Dubin M, Sedlazeck FJ et al (2012) Advanced methylome analysis after bisulfite deep sequencing: an example in Arabidopsis. PLoS One 7:e41528. doi:10.1371/journal.pone.0041528.s026

Dodt M, Roehr JT, Ahmed R, Dieterich C (2012) FLEXBAR-flexible barcode and adapter processing for next-generation sequencing platforms. Biology (Basel) 1:895–905. doi:10.3390/biology1030895

Doi A, Park I-H, Wen B et al (2009) Differential methylation of tissue- and cancer-specific CpG island shores distinguishes human induced pluripotent stem cells, embryonic stem cells and fibroblasts. Nat Genet 41:1350–1353. doi:10.1038/ng.471

Dolzhenko E, Smith AD (2014) Using beta-binomial regression for high-precision differential methylation analysis in multifactor whole-genome bisulfite sequencing experiments. BMC Bioinformatics 15:215. doi:10.1186/1471-2105-15-215

Durbin R (1998) Biological sequence analysis. Cambridge University Press, Cambridge

Feng S, Cokus SJ, Zhang X et al (2010) Conservation and divergence of methylation patterning in plants and animals. Proc Natl Acad Sci U S A 107:8689–8694. doi:10.1073/pnas.1002720107

Feng H, Conneely KN, Wu H (2014) A Bayesian hierarchical model to detect differentially methylated loci from single nucleotide resolution sequencing data. Nucleic Acids Res 42:e69. doi:10.1093/nar/gku154

Ficz G, Branco MR, Seisenberger S et al (2011) Dynamic regulation of 5-hydroxymethylcytosine in mouse ES cells and during differentiation. Nature 473:398–402. doi:10.1038/nature10008

Frith MC, Mori R, Asai K (2012) A mostly traditional approach improves alignment of bisulfite-converted DNA. Nucleic Acids Res 40:e100. doi:10.1093/nar/gks275

Gaidatzis D, Burger L, Murr R et al (2014) DNA sequence explains seemingly disordered methylation levels in partially methylated domains of mammalian genomes. PLoS Genet 10:e1004143. doi:10.1371/journal.pgen.1004143.g005

Gebhard C, Schwarzfischer L, Pham T-H et al (2006) Rapid and sensitive detection of CpG-methylation using methyl-binding (MB)-PCR. Nucleic Acids Res 34:e82. doi:10.1093/nar/gkl437

Gent JI, Ellis NA, Guo L et al (2013) CHH islands: de novo DNA methylation in near-gene chromatin regulation in maize. Genome Res 23:628–637. doi:10.1101/gr.146985.112

Goll MG, Bestor TH (2005) Eukaryotic cytosine methyltransferases. Annu Rev Biochem 74:481–514. doi:10.1146/annurev.biochem.74.010904.153721

Gu H, Smith ZD, Bock C et al (2011) Preparation of reduced representation bisulfite sequencing libraries for genome-scale DNA methylation profiling. Nat Protoc 6:468–481. doi:10.1038/nprot.2010.190

Guo W, Fiziev P, Yan W et al (2013) BS-Seeker2: a versatile aligning pipeline for bisulfite sequencing data. BMC Genomics 14:774. doi:10.1186/1471-2164-14-774

Hansen KD, Langmead B, Irizarry RA (2012) BSmooth: from whole genome bisulfite sequencing reads to differentially methylated regions. Genome Biol 13:R83. doi:10.1186/gb-2012-13-10-r83

Harris EY, Ponts N, Le Roch KG, Lonardi S (2012) BRAT-BW: efficient and accurate mapping of bisulfite-treated reads. Bioinformatics 28:1795–1796. doi:10.1093/bioinformatics/bts264

Hebestreit K, Dugas M, Klein H-U (2013) Detection of significantly differentially methylated regions in targeted bisulfite sequencing data. Bioinformatics 29:1647–1653. doi:10.1093/bioinformatics/btt263

Hendrich B, Bird A (1998) Identification and characterization of a family of mammalian methyl-CpG binding proteins. Mol Cell Biol 18:6538–6547

Hon GC, Rajagopal N, Shen Y et al (2013) Epigenetic memory at embryonic enhancers identified in DNA methylation maps from adult mouse tissues. Nat Genet 45:1198–1206. doi:10.1038/ng.2746

Huang Y, Pastor WA, Shen Y et al (2010) The behaviour of 5-hydroxymethylcytosine in bisulfite sequencing. PLoS One 5:e8888. doi:10.1371/journal.pone.0008888

Ivanov M, Kals M, Kacevska M et al (2013) In-solution hybrid capture of bisulfite-converted DNA for targeted bisulfite sequencing of 174 ADME genes. Nucleic Acids Res 41:e72. doi:10.1093/nar/gks1467

Jones PA, Taylor SM (1980) Cellular differentiation, cytidine analogs and DNA methylation. Cell 20:85–93

Jørgensen HF, Adie K, Chaubert P, Bird AP (2006) Engineering a high-affinity methyl-CpG-binding protein. Nucleic Acids Res 34:e96. doi:10.1093/nar/gkl527

Klambauer G, Schwarzbauer K, Mayr A et al (2012) cn.MOPS: mixture of Poissons for discovering copy number variations in next-generation sequencing data with a low false discovery rate. Nucleic Acids Res 40:e69. doi:10.1093/nar/gks003

Kohli RM, Zhang Y (2013) TET enzymes, TDG and the dynamics of DNA demethylation. Nature 502:472–479. doi:10.1038/nature12750

Komori HK, LaMere SA, Torkamani A et al (2011) Application of microdroplet PCR for large-scale targeted bisulfite sequencing. Genome Res 21:1738–1745. doi:10.1101/gr.116863.110

Krueger F, Andrews SR (2011) Bismark: a flexible aligner and methylation caller for Bisulfite-Seq applications. Bioinformatics 27:1571–1572. doi:10.1093/bioinformatics/btr167

Laird PW, Jaenisch R (1994) DNA methylation and cancer. Hum Mol Genet 3 Spec No: 1487–1495

Landan G, Cohen NM, Mukamel Z et al (2012) Epigenetic polymorphism and the stochastic formation of differentially methylated regions in normal and cancerous tissues. Nat Genet 44:1207–1214. doi:10.1038/ng.2442

Langmead B, Trapnell C, Pop M, Salzberg SL (2009) Ultrafast and memory-efficient alignment of short DNA sequences to the human genome. Genome Biol 10:R25. doi:10.1186/gb-2009-10-3-r25

Lee EJ, Pei L, Srivastava G et al (2011) Targeted bisulfite sequencing by solution hybrid selection and massively parallel sequencing. Nucleic Acids Res 39:e127. doi:10.1093/nar/gkr598

Li L-C, Dahiya R (2002) MethPrimer: designing primers for methylation PCRs. Bioinformatics 18:1427–1431

Li S, Garrett-Bakelman FE, Akalin A et al (2013) An optimized algorithm for detecting and annotating regional differential methylation. BMC Bioinformatics 14(Suppl 5):S10. doi:10.1186/1471-2105-14-S5-S10

Lister R, Pelizzola M, Dowen RH et al (2009) Human DNA methylomes at base resolution show widespread epigenomic differences. Nature 462:315–322. doi:10.1038/nature08514

Lister R, Mukamel EA, Nery JR et al (2013) Global epigenomic reconfiguration during mammalian brain development. Science 341:1237905. doi:10.1126/science.1237905

Liu Y, Siegmund KD, Laird PW, Berman BP (2012) Bis-SNP: combined DNA methylation and SNP calling for Bisulfite-seq data. Genome Biol 13:R61. doi:10.1186/gb-2012-13-7-r61

Martin M (2011) Cutadapt removes adapter sequences from high-throughput sequencing reads. EMBnet J 17(1)

Meissner A (2005) Reduced representation bisulfite sequencing for comparative high-resolution DNA methylation analysis. Nucleic Acids Res 33:5868–5877. doi:10.1093/nar/gki901

Mohn F, Weber M, Schübeler D, Roloff T-C (2009) Methylated DNA immunoprecipitation (MeDIP). Methods Mol Biol 507:55–64. doi:10.1007/978-1-59745-522-0_5

Molaro A, Hodges E, Fang F et al (2011) Sperm methylation profiles reveal features of epigenetic inheritance and evolution in primates. Cell 146:1029–1041. doi:10.1016/j.cell.2011.08.016

Nair SS, Coolen MW, Stirzaker C et al (2014) Comparison of methyl-DNA immunoprecipitation (MeDIP) and methyl-CpG binding domain (MBD) protein capture for genome-wide DNA methylation analysis reveal CpG sequence coverage bias. Epigenetics 6:34–44. doi:10.4161/epi.6.1.13313

Okano M, Bell DW, Haber DA, Li E (1999) DNA methyltransferases Dnmt3a and Dnmt3b are essential for de novo methylation and mammalian development. Cell 99:247–257

Pedersen B, Hsieh T-F, Ibarra C, Fischer RL (2011) MethylCoder: software pipeline for bisulfite-treated sequences. Bioinformatics 27:2435–2436. doi:10.1093/bioinformatics/btr394

Plongthongkum N, Diep DH, Zhang K (2014) Advances in the profiling of DNA modifications: cytosine methylation and beyond. Nat Rev Genet 15:647–661. doi:10.1038/nrg3772

Rauch T, Pfeifer GP (2005) Methylated-CpG island recovery assay: a new technique for the rapid detection of methylated-CpG islands in cancer. Lab Invest 85:1172–1180. doi:10.1038/labinvest.3700311

Smallwood SEBA, Lee HJ, Angermueller C et al (2014) Single-cell genome-wide bisulfite sequencing for assessing epigenetic heterogeneity. Nat Methods 11:817. doi:10.1038/nmeth.3035

Smith ZD, Meissner A (2013) DNA methylation: roles in mammalian development. Nat Rev Genet 14:204–220. doi:10.1038/nrg3354

Smyth GK (2004) Linear models and empirical Bayes methods for assessing differential expression in microarray experiments. Stat Appl Genet Mol Biol 3: Article 3. doi:10.2202/1544-6115.1027

Stadler MB, Murr R, Burger L et al (2011) DNA-binding factors shape the mouse methylome at distal regulatory regions. Nature 480:490–495. doi:10.1038/nature10716

Sun D, Xi Y, Rodriguez B et al (2014) MOABS: model based analysis of bisulfite sequencing data. Genome Biol 15:R38. doi:10.1186/gb-2014-15-2-r38

Suzuki MM, Bird A (2008) DNA methylation landscapes: provocative insights from epigenomics. Nat Rev Genet 9:465–476. doi:10.1038/nrg2341

Taiwo O, Wilson GA, Morris T et al (2012) Methylome analysis using MeDIP-seq with low DNA concentrations. Nat Protoc 7:617–636. doi:10.1038/nprot.2012.012

Taylor KH, Kramer RS, Davis JW et al (2007) Ultradeep bisulfite sequencing analysis of DNA methylation patterns in multiple gene promoters by 454 sequencing. Cancer Res 67:8511–8518. doi:10.1158/0008-5472.CAN-07-1016

Wang RY, Gehrke CW, Ehrlich M (1980) Comparison of bisulfite modification of 5-methyldeoxycytidine and deoxycytidine residues. Nucleic Acids Res 8:4777–4790

Warnecke PM, Stirzaker C, Song J et al (2002) Identification and resolution of artifacts in bisulfite sequencing. Methods 27:101–107

Weber M, Davies JJ, Wittig D et al (2005) Chromosome-wide and promoter-specific analyses identify sites of differential DNA methylation in normal and transformed human cells. Nat Genet 37:853–862. doi:10.1038/ng1598

Xi Y, Li W (2009) BSMAP: whole genome bisulfite sequence MAPping program. BMC Bioinformatics 10:232. doi:10.1186/1471-2105-10-232

Zemach A, McDaniel IE, Silva P, Zilberman D (2010) Genome-wide evolutionary analysis of eukaryotic DNA methylation. Science 328:916–919. doi:10.1126/science.1186366

Ziller MJ, Gu H, Müller F et al (2013) Charting a dynamic DNA methylation landscape of the human genome. Nature 500:477–481. doi:10.1038/nature12433

Ziller MJ, Hansen KD, Meissner A, Aryee MJ (2015) Coverage recommendations for methylation analysis by whole-genome bisulfite sequencing. Nat Methods 12:230. doi:10.1038/nmeth.3152

Chapter 10
Characterization of DNA-Protein Interactions: Design and Analysis of ChIP-Seq Experiments

Rory Stark and James Hadfield

10.1 Introduction to Genome-Wide Analysis of DNA-Protein Interactions Using ChIP-seq

Within the last decade, advances in high-throughput sequencing have enabled extensive research into protein-DNA interactions on a genomic scale. These interactions include the binding of transcription factor proteins to localized positions on DNA, as well as proteins involved in other aspects of transcriptional regulation (e.g., methylases, acetylases) and in transcription itself (polymerases, etc.). The same methods can further be used to ascertain relevant aspects of chromatin state involved in transcriptional regulation, most notably key histone "marks" (including methylation and acetylation).

The primary experimental method used is chromatin immunoprecipitation followed by sequencing, or ChIP-seq. While ChIP assays have been utilized for some time, modern high-throughput sequencing has enabled the entire genome (rather than just a small number of genes or genomic loci) to be interrogated in a single experiment. Figure 10.1, generated by the ENCODE project (ENCODE Project Consortium 2011), shows the high-level picture of regulatory elements in the genome, including the aspects that may be examined using ChIP-seq. This chapter describes how to design, implement, and analyze ChIP-seq experiments to successfully address a range of biological questions involving DNA-protein interactions and transcriptional regulation.

R. Stark, B.A., M.Sc., M.Phil., D.Phil. (✉) • J. Hadfield, B.Sc., Ph.D.
Cancer Research UK Cambridge Institute, University of Cambridge,
Robinson Way, Cambridge CB2 0RE, UK
e-mail: rory.stark@cruk.cam.ac.uk; James.Hadfield@cruk.cam.ac.uk

A.M. Aransay, J.L. Lavín Trueba (eds.), *Field Guidelines for Genetic Experimental Designs in High-Throughput Sequencing*,
DOI 10.1007/978-3-319-31350-4_10

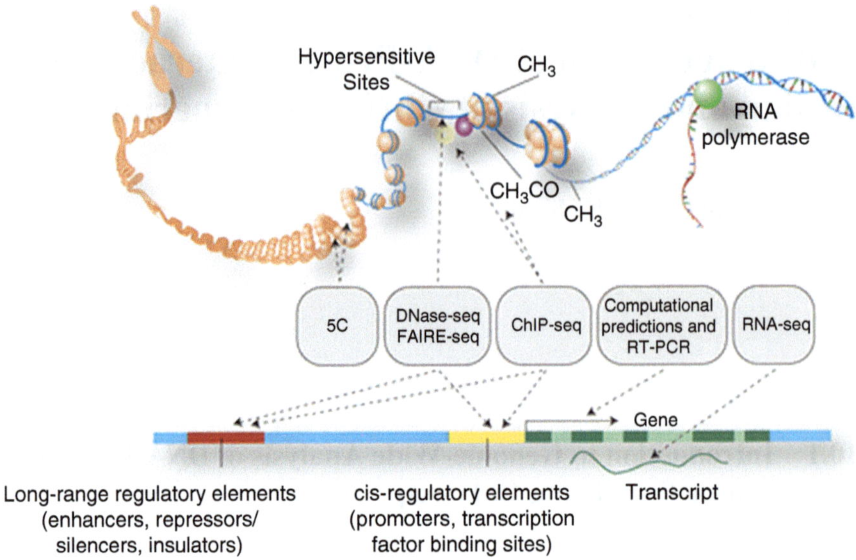

Fig. 10.1 Regulatory elements in the genome. Regulatory aspects of the genome include chromatin structure (open/closed chromatin), enhancers, repressors, silencers, insulators, promoters, transcription factor biding sites, histone modification, methylation of DNA nucleotides, presence of transcriptional proteins such as RNA Polymerase, etc. ChIP-seq and related techniques can be used to assay many of these. This figure is generated by the ENCODE project (ENCODE Project Consortium 2011)

10.1.1 What Is ChIP-seq?

ChIP-seq can be understood in terms of four definitional components contained in its name: **Ch**romatin **I**mmuno-**P**recipitation followed by **Seq**uencing:

- **Chromatin**, indicating that the assay requires not just purified DNA but also all the associated proteins;
- **Immuno**, indicating the use of antibodies that target specific proteins of interest;
- **Precipitation**, indicating that this is an enrichment assay, where a total pool of chromatin will be enriched for those parts that involve the protein of interest, leaving as much of the nonassociated chromatin as possible behind;
- **Sequencing**, indicating that the result of the precipitation will be subjected to high-throughput sequencing (which in turn implies that the precipitate should be purified to obtain sequenceable DNA).

Currently, ChIP-seq is performed on chromatin extracted from populations of cells (this chapter will not address potential issues involved in single-cell ChIP-seq), typically numbering in the tens of thousands. This has implications for the analysis phase as described in later sections.

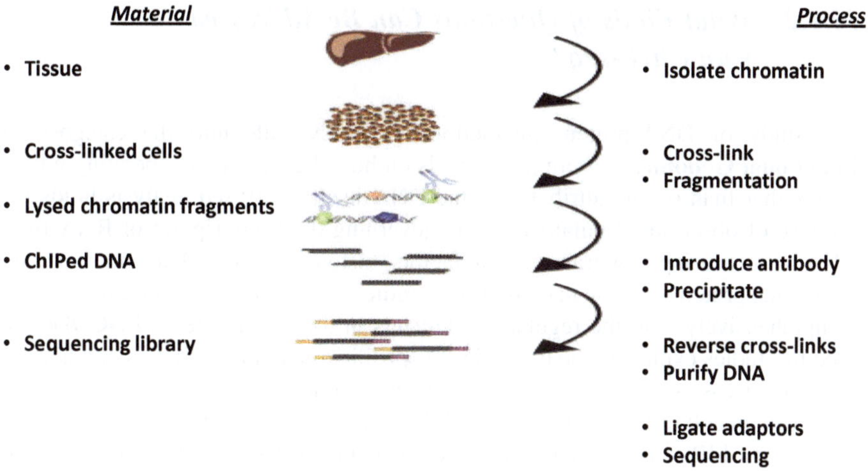

Fig. 10.2 Chromatin Immunoprecipitation (ChIP) procedure. Overview of ChIP processing steps, from whole cells to enriched DNA, ready for sequencing. Adapted from (Schmidt et al. 2009)

The main steps of a ChIP-seq assay are as follows (see Fig. 10.2):

- **Isolation of nuclear material** to extract chromatin from the cells.
- **Cross-linking** preserves the protein-DNA interactions.
- **Fragmentation** divides the sample into much smaller pieces of chromatin, some of which include the protein of interest (generally a very small proportion).
- **Introduce antibodies**, generally attached to beads, that will bind to the protein of interest, thus tagging the fragments associated with that protein.
- **Precipitate** the fragments by isolating the beads, and hence the chromatin fragments of associated with the protein of interest.
- **Reverse** the cross-linking of the precipitated fragments.
- **Purify** the fragments to obtain the associated DNA.
- **Sequence** the purified, precipitated DNA fragments. This includes preparing libraries by ligating sequencing adapters in a manner consistent with the sequencing technology platform to be used.
- **Process** the sequencing reads to determine quality and where they likely originated in the genome.
- **Analyze** the data to determine a result.

It is essential to recognize that while a well-performed ChIP-seq results in a library highly enriched for fragments associated with the protein of interest, this enrichment is far from perfect; the majority of resulting sequencing reads are generally from fragments that are not actually associated with the protein of interest.

10.1.2 What Kinds of Questions Can Be Addressed with ChIP-seq?

The study of DNA-protein interactions generally falls into the category of **Functional Genomics**, where the focus is on how the genome is operating within cells rather than on the attributes of the DNA itself. ChIP-seq analysis holds the promise of observing regulatory events governing the transcription of RNA from DNA by measuring how transcription factors, histone marks, and key nuclear transcriptional proteins behave in cells. These studies range from large-scale attempts to comprehensively map the regulatory elements in the genome [e.g., ENCODE] to specific studies exploring the dynamics of specific transcription factors and histone marks in disease states. Recent advances in epigenetics, such as work on the role of enhancers in cell differentiation and cellular function (Pennacchio et al. 2013), rely heavily on ChIP-seq experiments and are impacting nearly every aspect of molecular biology.

10.1.3 Overview of ChIP-seq Process

A successful ChIP-seq experiment goes well beyond the mechanics described above. The remainder of this chapter will discuss the key aspects of applying ChIP-seq to a specific biological question. The main three steps include:

- **Experimental design**: understanding exactly what biological question is being asked (optimally the testing of a specific hypothesis) is perhaps the most crucial step in a successful experiment. Once clearly explicated, refining an appropriate experimental design prior to preparing samples is the next most important aspect of obtaining a meaningful result;
- **Sample preparation and sequencing**: executing the steps of the assay at the bench and on the sequencing instrument, within the parameters of the experimental design, carefully and precisely;
- **Data Analysis**: performing an analysis of the data generated by the experiment involves many steps, each of which can impact the usefulness of the final result.

10.2 Design of ChIP-seq Experiments

Most interesting ChIP-seq experiments involve multiple samples that must be carefully coordinated to produce a usable result. This section looks at the key question of how to design multi-ChIP experiments to enhance the likelihood of a meaningful result. We consider different types of protein-DNA interactions and quantitative analyses, as well as technical considerations for determining the choice of antibody, numbers of replicates, experimental and technical controls, and sequencing parameters.

10.2.1 Types of DNA-Protein Interactions: Punctate vs. Broad Enrichment

There are two general types of interactions to consider, corresponding to how narrow or broad the enriched regions of DNA are expected to be. For classic transcription factors that bind directly to the DNA at specific locations (often marked by a sequence motif), the "binding sites" on the DNA are relatively narrow (generally between 4 and 24 bp, depending on the motif). However many assays explore broader regions of enrichment, such as histone marks that may be present on many contiguous nucleosomes covering longer stretches of DNA (anywhere from 100 bp to many thousands of base pairs long). The distinction is not as straightforward as DNA-binding proteins vs. marks on structural proteins such as histones, however; for example the histone mark H3K4me3 (an indicator of active expression when found in promoters) appears in very narrow ranges, forming peaks similar to that for transcription factors, while some DNA-binding proteins (e.g., polymerases) may bind over the full length of a gene. An additional complication is that even transcription factors may themselves not bind directly to the DNA, but bind to other co-factors that are DNA-associated.

One way of thinking about the punctate/broad distinction concerns the fragment size distribution obtained after the fragmentation step. These frequently have a mean between 200 and 300 bp in length. If the binding site is expected to be narrower than this, the enrichment can be considered punctate, while if the region of enrichment is wider, the enrichment can be considered broad.

In some experiments, multiple proteins will be ChIPed, some of which may be punctate and some broad. For example, when attempting to determine how a punctate transcription factor is impacting transcription, one may also assay active and repressive histone marks (such as H3K27me3 and H3K9me3), which have broad enrichment, as well as the polymerase PolII. Likewise, for transcription factors binding that occurs distal to gene promoters, it is often useful to ChIP histone marks and proteins that indicate active enhancers.

10.2.2 Occupancy Mapping vs. Quantitative Affinity

One of the most important aspects to consider in designing a ChIP-seq experiment is the type of analysis to be done. Specifically, a distinction can be made between mapping experiments looking primarily at identifying where in the genome a protein can bind and those looking at the relative strength and functional nature of that binding.

For example, when exploring the role of a specific transcription factor, it can be very helpful to know which genes it binds to. This can be used to narrow a set of genes that may be regulated by the transcription factor, help define its binding motif, identify potential co-factors, etc. We refer to this type of mapping experiment as being **occupancy** based, as the main question is what sites on the DNA are occupied by a protein of interest. The majority of ChIP-seq work done to date fall into this category, including the bulk of ChIP data generated by the ENCODE project, which is focused on identifying high-confidence binding sites for a range of factors across a range of cell types.

A related question is to determine how a factor's occupancy differs between two sample groups (such as different cell types, or treated vs. untreated cells, or diseased vs. normal cells). The most simplistic way of doing this is to generate occupancy maps for each sample group, then compare their overlap (often using Venn diagrams). In this way, three sets of binding sites can be identified: a set of sites bound in both sample groups, and two sets of "unique" sites bound in only one group.

However, this type of analysis, where binding sites are mapped independently in each sample and then overlapped, is inadequate to identify many of the most important differentially bound sites (see Sect. 10.4.3 for a detailed discussion). Consider how ChIP-seq data appears in a genome browser (Fig. 10.3). The binding sites are seen as "peaks" where there are pileups of sequencing reads in specific genomic locations. It is clear that the height of these pileups vary widely between binding sites. This is because binding sites are not fully characterized as either being occupied or unoccupied by a protein. As the assay is representing the binding over a population of cells, the height of the pileup is related to the proportion of the cells in the sample that have the protein bound at that location. This reflects the **affinity** of the protein to bind at that location in that cell population. Generally, a protein is

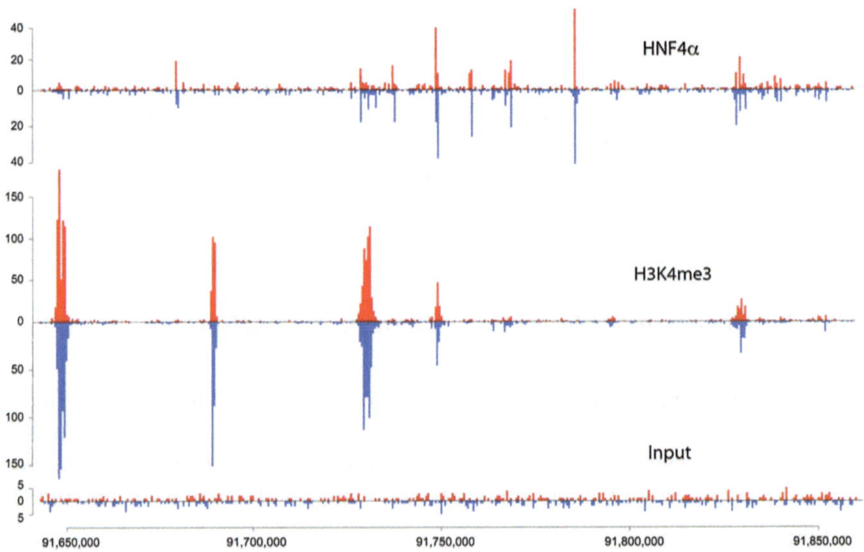

Fig. 10.3 ChIP-seq reads in genome browser. An example of sequencing reads for three ChIP-seq samples. The *top* shows a transcription factor, the *middle* shows a histone mark, and the *bottom* shows an Input control. Reads on the two strands are shown separately, with positive strand reads shown above the line in *red*, and negative strand reads shown below the line in *blue*

not bound in *every* cell in one condition and *none* of the cells in another condition. As occupancy mapping focuses on identifying sites where a protein can bind—even if it is in a small proportion of the cells—a differential analysis that relies on occupancy maps cannot distinguish between cases where the factor has a low-but-distinct binding affinity in one sample group and a very high binding affinity in another sample group (say, the difference between being bound in 10 % of the cells vs. 90 % of the cells).

If a robust affinity analysis is desired, this has an impact on experimental design, as more replicates are required (as discussed below) in order to capture within-group variance.

10.2.3 Antibody Specificity

ChIP-seq assays are deeply dependent on having an effective antibody that targets the protein of interest. While complete procedures for determining the efficacy of an antibody are out of the scope of this discussion, careful consideration should be given to ensure that the antibody being used effectively targets the protein of interest and has as high a degree of specificity as possible to minimize pull-down of fragments that are not associated with the protein of interest.

The ENCODE project maintains useful guidelines for characterizing antibodies (Landt et al. 2012), the most current of which can be found here: http://genome.ucsc.edu/ENCODE/experiment_guidelines.html

10.2.4 Replicates

For results to be meaningful, *all ChIP-seq experiments require some degree of replication*. Even the most straightforward experimental design, where a single antibody is used to map punctate binding sites in a single cell type, requires replication.

While there is general acceptance that RNA expression assays should never be done without replication, this is less widespread in the ChIP-seq arena. However, what we know of transcriptional regulation indicates that we should expect even higher variability in this area (Schmidt et al. 2010). In addition, the ChIP-seq assay itself is less reproducible than that for RNA-seq, as there are more steps to go from cells to ChIP-seq than to RNA-seq (each having its own associated variability and bias). Finally, the analysis component for ChIP-seq results in more variability as well, with peak calling in particular being highly sensitive to relatively small changes in the sequencing data (see Sect. 10.4.2).

As ChIP-seq experiments exhibit high variability, there is no way to know if the results of a single ChIP-seq are representative or an outlier; hence a nonreplicated ChIP-seq should never be considered definitive.

10.2.4.1 Types of Replicates: Biological, Experimental, and Technical

Depending on the types of samples being used, it is useful to think of replication in three categories:

- **Biological replication** is required to capture the inherent variance present in biological systems. In vivo experiments should involve multiple different source organisms. In vitro experiments should involve multiple cell lines, as the dominant signal in most ChIP-seq reflects the open chromatin, which is cell-type (or cell-line) dependent.
- **Experimental replication** is required to capture the differences in ChIP efficiency between different runs. For example, if a cell line is used, the ChIP should be repeated multiple times using the same antibody, preferably with cells grown separately (or a subsequent passage of the cells).
- **Technical replication**, referring mostly to the sequencing aspect of the assay, involves re-sequencing the same libraries to capture possible sequencing biases. While these biases are real, their variance is far less than that introduced by biological and experimental aspects. In the next section on sample preparation, large-scale multiplexing (pooled libraries) is advocated, which can help with technical replication as (a) any sequencing biases in a run will apply equally to all samples in the experiment and (b) it is more likely that multiple sequencing runs of the entire experiment will be required to obtain the necessary read depth.

10.2.4.2 Sample Groups

In order to determine the types and degree of replication appropriate for a ChIP-seq experiment, the sample groups used for the analysis must be identified up front.

In the simplest case, where there is no comparison being done (a pure occupancy mapping experiment), there may only be a single sample type.

If any sort of comparison is being made, there are at least two sample types representing the two conditions being compared. Examples include: two cell types; the same cell type in two states; untreated cells vs. ones treated with a drug; wild-type vs. knock-down cells; normal vs. diseased cells; two different disease subtypes, etc. For each of these sample groups, there may be more than one antibody being used to survey multiple factors or marks.

More complicated designs may include more elaborate comparisons (say, between multiple treatments or knock-downs, rather than just two), time series, or multi-level designs where multiple variables are being altered simultaneously (e.g., wild-type vs. knock-down, each untreated vs. treated with a drug).

In these designs, every sample group should have some degree of replication. If multiple antibodies are used to assay different proteins, each sample group-antibody pair should have some degree of replication.

10.2.4.3 Numbers of Replicates and Experimental Power

The first issue in determine how many replicates are required is whether the goal is purely occupancy mapping (identification of putative binding sites) or a quantitative affinity analysis (comparing differences in binding levels between groups).

For pure mapping exercises, the ENCODE project has established useful guideline for the identification of high-confidence peaks (Landt et al. 2012). These guidelines require two high-quality replicates for each cell type/antibody pair. For primary tissue these would be biological replicates, while for cultured cells these would be what we have referred to as experimental replicates. In order to get two high-quality replicates, it may be necessary to prepare more than two samples. Quality assessment and peak identification are discussed in Sect. 10.4 of this chapter.

For experiments where sample groups are to be compared by assessing quantitative differences in occupancy levels, more replicates are required. Determining the optimal number is potentially difficult, but echoes how sample sizes for RNA-seq are calculated. Optimally, power calculations can be carried out (Zuo and Keleş 2014) to determine how many samples are required to reliably detect differences reflecting a given effect size. However, these calculations rely on an accurate measure of variance in the data, which is rarely available when embarking on a new experiment. The analysis techniques outlined below rely on at least three high-quality replicates of each sample group (or of each sample group/antibody pair). For punctate transcription factors and well-studied histone marks with reliable antibodies, three or four replicates of each sample type for in vitro experiments have generated useful data (Ross-Innes et al. 2012; Mohammed et al. 2015). For certain in vivo experiments, for example using primary disease tissues from patients, many more biological samples may be required to obtain a useful result (although technical replicates are less necessary).

See Sect. 10.2.6 for an example of a published experimental design including replicate numbers.

10.2.5 Controls

Another key part of any experimental design is the use of controls to ensure the experiment is accomplishing its goals, detect technical biases, and calibrate confidence statistics. Like replicates, controls in ChIP-seq experiments may be thought of in three main categories:

- **Experimental controls** are used as part of the high-level experimental design to control for specific biological effects inherent to the experiment, such as introduction of siRNAs for knock-downs, or the effect of particular vectors utilized for introducing certain chemical treatments. These controls are not specific to ChIP-seq.

- **ChIP controls** are an essential part of the ChIP-seq assay. These are used to detect chromatin signatures specific to the cell types being studied and to establish background noise levels for separating signal from noise.
- **Technical controls** are generally specific to the sequencing portion of the assay and are used to detect sequencing biases, and possibly (via spike-ins) for quantitative normalization.

Detailed discussion of appropriate experimental and technical controls is out the scope of this chapter, so we will focus on the controls specific to ChIP-seq assays.

10.2.5.1 ChIP Controls

It is standard practice to generate control libraries alongside full ChIP preparations. The most common is an **Input control**, in which the ChIP protocol is followed except no antibody is introduced. The resulting pulled-down fragments should not be enriched for any specific binding protein.

There are two main purposes of generating these Input controls as part of a ChIP-seq experiment:

- **Input serves as a background model**. Most ChIP-seq analyses involve a peak identification step (see below). As previously stated, ChIP enrichment is far from perfect, and the majority of sequenced fragments are not actually associated with the protein of interest. For example, if a transcription factor binds on 0.01 % of the DNA, and the efficiency of the ChIP in enriching for these positions is 1000-fold, only 10 % of the fragments will actually be associated, and 90 % will be "background" or "noise." The ability to separate truly enriched regions requires a clean model of this background, which is provided by the Input control.
- **Input reveals chromatin signatures specific to a particular cell type**. While sequencing purified DNA results in relatively even coverage, sequenced Input controls exhibit coverage far from even (Park 2009). As the fragmentation step is conducted on integral chromatin, and not purified DNA, some DNA positions will be more likely to fragment than others, depending on how densely they are encased in proteins (particularly nucleosomes). Each cell type has an open-chromatin "signature" that is revealed by the Input control. Indeed, the open-chromatin signature tends to dominate the overall signal in ChIP-seq data; Input helps controlled for this.

While there is some debate about the necessity of ChIP controls, and whether use of a nonspecific antibody (such as IgG) is preferable to a pure Input, the vast majority of ChIP experiments utilize Input controls, which in our experience work best.

10.2.5.2 How Many ChIP Controls?

Determining how many Input controls to generate for a ChIP-seq experiment is a key step in the design. Optimally, every ChIP performed should be accompanied by a control. However in practice this may not be necessary; indeed, most published

ChIP-seq experiments with many samples do not have an equal number of Input controls.

It is however important to have distinct controls for each cell type. For example, when comparing binding of the same factor in different tissue types within a species (e.g., binding in liver vs. skin), separate Input controls for the two tissue types are required. However, if a single population of cultured cells is used to generate two ChIPs, with the only difference being the introduction of different antibodies (e.g., to compare two transcription factors), a single control serves for both ChIPs.

Now suppose this experiment were being done using three replicates, with each pair of ChIPs using a different passage. Optimally, there would be three controls generated, and we recommend using all three. In practice, often only one is used, as the cell type hasn't really changed, and the Inputs for each passage should be comparable. So, it may be acceptable to use a single Input control for all six ChIPs. If this experiment were to use three different cell lines, however, three Input controls should be generated, as these comprise true biological replicates.

Careful consideration is required when comparing different conditions and treatments. If a treatment is likely to change the chromatin signature of a cell type, an Input should be generated. For example, if one were to knock-down a chromatin remodeling gene in one condition, those replicates definitely require a separate Input control.

10.2.6 Example ChIP-seq Experimental Design

Here, we introduce an example ChIP-seq experiment, taken from Ross-Innes et al. (2012). This will be used not only to demonstrate an experimental design, but subsequently in the discussion of analyzing ChIP-seq data.

This experiment uses cultured estrogen-positive breast cancer cells to look at the role of the transcription factor ERα in resistance to the drug tamoxifen. The goal is to perform a differential analysis to isolate ERα sites that have significantly altered binding affinity in cells that are responsive to the drug vs. those that are resistant. Hence, there are two sample groups: **Responsive** and **Resistant**.

Two levels of replication are incorporated. Biological replication is achieved by using five different cell lines. Three of these are responsive to tamoxifen as evidenced by reduced cell growth. The other two are resistant to tamoxifen (cells grow in presence of tamoxifen), resulting in three biological replicates on the Responsive side and two on the Resistant side. There is an additional complication, discussed below, if that one of the Resistant cell lines is derived from one of the Responsive cell lines. There is also a level of experimental replication, in that the ChIP is repeated in two different passages of each cell line (with one of the Responsive cell lines having three experimental replicates). This gives 11 ChIPs in total: seven in the Responsive group (3 cell lines × 2 replicates + 1 additional replicate) and four in the Resistant group (2 cell lines × 2 replicates).

Table 10.1 Example data set (tamoxifen resistance in five breast cancer cell lines)

Sample	Tissue	Factor	Status	Rep#	Peaks
MCF71	MCF7	ERα	Responsive	1	74,029
MCF72	MCF7	ERα	Responsive	2	49,075
MCF73	MCF7	ERα	Responsive	3	67,130
T47D1	T47D	ERα	Responsive	1	28,713
T47D1	T47D	ERα	Responsive	2	23,575
ZR751	ZR75	ERα	Responsive	1	74,971
ZR752	ZR75	ERα	Responsive	2	70,560
MCF7r1	MCF7	ERα	Resistant	1	47,034
MCF7r2	MCF7	ERα	Resistant	2	52,517
BT4741	BT474	ERα	Resistant	1	41,924
BT4742	BT474	ERα	Resistant	2	40,783

Input controls are generated for each unique cell line (including the Responsive and Resistant versions of the shared cell line), for a total of five Input controls. In total, 16 libraries are required (11 ChIPs and 5 controls). Table 10.1 shows information for the samples.

10.3 Preparation and Sequencing of ChIP-seq Samples

10.3.1 Preparation of ChIP-seq Samples

This section is based on the methods used in the authors' previous work (Schmidt et al. 2009). The amount of starting material can be critical for any experiments and ChIP-seq is similarly affected. Unlike genome sequencing experiments, the amount of material available for library preparation is highly variable due to the variation in how much of the genome different DNA binding proteins will pull down in the immunoprecipitation step. We have successfully used around one million cells; others have reported using limited material (Acevedo et al. 2007; O'Neill et al. 2006) but the complexity of a library can be adversely affected during the "remove duplicates" step. We have had success with nonstandard library preparation technologies such as Thruplex [Rubicon Genomics, USA] in very low-input exome sequencing from cell-free tumor DNA in blood (Murtaza et al. 2013) and applied these to ChIP-seq with mixed results. The Thruplex technology does allow us to reduce DNA input significantly (10,000–100,000 cells) but only if chromatin fragmentation is carefully controlled.

Cells and/or tissue are cross-linked using formaldehyde before being homogenized and lysed to remove cytosolic proteins, leaving only the nucleus for ChIP. It is critical to treat all samples in an experiment in the same manner to avoid introducing confounding technical artifacts, cross-linking samples for different times, or shearing samples to different lengths could introduce different biases. All samples should be treated as a single batch where possible. Careful randomization of samples to each step can mitigate technical effects.

Chromatin from lysed cell nuclei is sheared, usually with sonication, although the Covaris system [Covaris Inc, USA] offers a less variable and more tunable alternative. Whichever system is used, settings for the shearing of chromatin must be predetermined to maximize the amount of DNA in the desired fragment size range most commonly 200 and 400 bp.

Sheared chromatin is incubated with a protein-specific antibody linked to protein-G magnetic beads; the specificity of this antibody is key to the success of ChIP-seq experiments. The antibodies need to be incubated with magnetic beads immediately prior to use; excess antibody is removed by washing to prevent competition of unbound antibodies for target proteins; the storage of antibody prepared beads is not recommended. ChIP with antibody bound beads is performed overnight at 4 °C. After washing to remove nonprecipitated chromatin, the ChIPed DNA is eluted and cross-links are reversed by incubation at 65 °C. It is critical to avoid overheating the DNA at this point to avoid denaturation, which will reduce the amount of material available for the double-stranded adapter ligation during library preparation.

10.3.2 Sequencing ChIP-seq Samples

Sequencing of ChIP samples proceeds based on the sequencing platform to be utilized. Generally there is a library preparation step first. For the Illumina Inc.'s HiSeq platforms, the sheared chromatin is used as the input to a standard end-repair adapter-ligation Illumina library preparation. It is more common to use kits from Illumina or other providers than perform a home-brew library preparation. Whichever method is used, individual libraries should be quality assessed and quantified. It is critical to carefully assess each library if these are to be pooled into a multiplexed sequencing run. Uneven pooling due to poor quantification or poor assessment of average library size can make the sequencing very inefficient as low yield libraries mean the pool needs significant oversequencing to achieve the specified minimum read depth.

The choice of sequencing parameters is based on the specifics of the experiment, but certain guidelines can be followed. The three main considerations are:

- **Single vs. Paired End**: The main goal of ChIP-seq is to maximize the number of uniquely mappable reads. In general this does not require paired-end sequencing. While paired-end sequencing can help in distinguishing PCR duplicates from "true" duplicate reads (see analysis section below), identifying more fragments in the sequencing pool has a greater impact on the quality of results. In addition, the most-used peak callers are designed to work with single-end reads. Unless there is a compelling reason to use paired-end sequencing, single end is the standard.

- **Read length**: the read length needs only be sufficient to optimally map fragments. As a result long reads do not meaningfully improve the results of a standard ChIP-seq experiment. 50 bp reads are more than sufficient for this purpose and generally represent a cost-efficient read length.
- **Read depth**: The ENCODE guidelines call for a minimum of 20 M reads with a goal of 30 M reads per sample. While this is sufficient for most ChIP-seq experiments, the optimal read depth can depend on the proteins being mapped. 30 M reads will provide plenty of depth for most punctate proteins like transcription factors, especially those that bind in very specific portions of the genome. Indeed with appropriate replication, 10–20 M reads can be sufficient to identify peaks and distinguish differences in enrichment. Some broader marks, such as H3K27me3, may be present to greater or lesser degrees across a much larger portion of the genome, and distinguishing between degrees of enrichment may rely on bigger depth.

For well-designed ChIP-seq experiments involving several samples, it is always better to multiplex the entire experiment in a single pool, and sequencing that pool as many times as necessary to obtain the desired depth, than is it to divide the samples and sequence them separately. This is in order to control for technical effects that may arise in the sequencing process itself. Sequencing in multiple lanes provides for technical replicates, while maintaining a single pool prevents batch effects that may be conflated with an experiment variable of interest.

10.4 Analysis of ChIP-seq Experiments

Depending on the purpose of the experiment, analysis of ChIP-seq data can follow a number of different paths. Here we discuss four main phases of analysis (not all of which apply to all analyses):

- **Read processing**, including alignment to a reference genome, application of filters, and quality assessment at the read level.
- **Enrichment analysis**, including peak calling and alternatives, derivation of consensus peaksets, and quality assessment at the peak level.
- **Differential analysis**, including binary (occupancy) based and quantitative (affinity) based analyses.
- **Downstream analysis**, including motif analysis and determination of target genes.

In the succeeding discussion, the examples are drawn from the previously described experiment looking at ERα binding in breast cancer cell lines that are responsive or resistant to treatment with the drug tamoxifen.

10.4.1 Read Processing: Alignment, Filtering, and Quality Assessment

The result of sequencing is in general a set of large files, most frequently in FASTQ format (Cock et al. 2010), containing the data for the many millions of sequencing reads. For each read, this includes the identified bases at each position with an associated confidence metric. For multiplexed experiments, there is a de-multiplexing step where a separate FASTQ is generated for each constituent sample, along with some metrics indicating the relative distribution of reads between the samples (as well as how many reads are unable to be assigned uniquely to a specific sample, and are discarded). If multiple sequencing runs (or multiple lanes on a run) are required to obtain necessary depth, there will be a set of such files for each sample for each lane. Coverage for each sample should be checked to ensure that targeted read levels have been reached. In multiplexed experiments, there will be a distribution of read numbers over the samples, so some samples will receive fewer reads (or drop out entirely). In some cases, libraries may need to be re-quantified and re-sequenced if read quantity targets are not reached.

It is a good idea to perform some quality assessment at this stage to determine that the sequencing phase was successful. A popular tool for this is FastQC (Andrews 2010), which can check for biases and other sequencing anomalies; a complementary tool, MGA (Hadfield and Eldridge 2014), can check for contamination, unalignable sequence, and presence of sequencing adapter dimers.

10.4.1.1 Alignment

The next step is to align the reads to a reference genome. The most popular current aligners are based on a Burrow-Wheeler transform (Li and Durbin 2009; Langmead et al. 2009). The alignment task for ChIP-seq is generally straightforward, without requiring local alignments to detect genomic anomalies.

The output of the alignment step are generally binary BAM files (Li et al. 2009), which includes the sequence information from the source FASTQ files in addition to alignment information. This information includes the genomic position of the best mapping (chromosome, start position, strand) and quality metrics for the confidence that the read is correctly and uniquely aligned. Separate BAM files for multiple sequencing runs can be combined for each individual sample at this stage.

10.4.1.2 Read Filtering

Once the reads are aligned, a number of filters may be applied to reduce them to a set appropriate for further analysis. Reads may be filtered based on mapping quality, duplicates, and overlap with backlists.

Mapping Quality

The most straightforward filter is on the mapping quality score. This score indicates the confidence that the correct origin location in the genome has been uniquely identified. Low values for this metric may arise from sequencing errors but are mostly expected to arise from the likelihood of an ambiguous mapping due to repeat regions in the genome. Reads that are unable to be uniquely mapped to a single location in the genome, sometimes called "multi-mapped" reads, can be problematic in ChIP-seq analysis. If the incorrect location is used, false positive enriched regions may be identified. Standard practice is to eliminate all multi-mapped reads from further analysis. However this can lead to other issues. The main one is that if a repeat occurs in the middle of a legitimate region of enrichment—for example at a motif identifying a binding site—the reads at that location will be eliminated, thus lessening the evidence for enrichment, or breaking a single large peak into two smaller ones. In some cases, the biology of the experiment is such that the most interesting reads are highly likely to occur in repeat regions. In such cases, multi-mapped reads may either be retained with one possible location chosen at random, or more sophisticated modeling may be used to distribute the reads to the most probable source locations (Kutter et al. 2011).

Duplication Rate and Handling of Duplicate Reads

The next issue to consider concerns remaining reads that align to identical genomic locations. In single-end sequencing, reads may appear to be duplicates if they were fragmented at the same point on only one end; paired-end sequencing can identify true duplicates covering an identical genomic interval. How these reads are treated can greatly influence final results.

It is common in ChIP-seq studies to consider all duplicate reads as artifacts of the PCR amplification stage of sequencing library preparation, and hence erroneous. The default step is to filter out all duplicated reads (leaving only a single exemplar of the read) before further analysis. This is meant to cut down on the false positive rate when identifying enriched regions, as a few highly overrepresented reads can give the false appearance of high enrichment.

However, this may be inappropriate, particularly if a differential analysis of quantitative affinity is desired (Carroll et al. 2014; Lun and Smyth 2014). While the likelihood of sampling multiple identical fragments is very low when conducting whole-genome sequencing, the nature of ChIP enrichment, particularly for a punctate factor, is such that true duplicate reads are expected at enriched areas, particularly when using short-read single-end sequencing. Removing duplicates "clips" the ChIP signal at relatively low levels and eliminates most of the relative quantitative information relating to binding affinity. If duplicates are not allowed, the highest number of unique fragments that include any one base position is twice the single-end read length (one fragment on each strand in each possible position of the read). So, for 50-bp single-end sequencing without duplicates, the greatest level of pileup at a peak summit cannot exceed a coverage rate of 100 reads. As cell populations used in the ChIP reactions involve many thousands of cells, each with more than one copy of the genome, much higher pileups should be expected (and indeed can easily be seen in a genome browser). While we cannot fully distinguish between duplicate reads attributable to technical effects and those associated with high binding affinity, proper use of replicate samples should enable sites with consistently high binding to be distinguishable, as amplification biases will be different in each replicate.

It is possible to remove duplicates for some portions of the analysis and retain them for others. For example, eliminating duplicates when peak calling can reduce the false positive rate, while true peaks will still have high enough read concentrations to be confidently identified. The full set of reads can then be used for the qualitative analysis to determine cases where a site is bound at detectable levels in multiple sample groups, but the affinity changes systematically between them.

From a quality assessment perspective, overall duplication rates should be checked after alignment. Input controls are useful here; lacking enrichment, they should exhibit low duplication rates (5 % or less). Higher duplication rates in their corresponding ChIPs support the hypothesis that the ChIP was successful in enriching specific regions. While there are no set values for the increase in enrichment, duplication rates 5× higher are not out of line (up to about 25 %). Libraries that exhibit very high duplication rates should be viewed with suspicion, particularly if they exceed 75 %.

Blacklists and Greylists

The third major criterion for filtering relates to known problem regions in specific genomes. Reads in these regions tend to have systematic anomalies that make analysis of enrichment unreliable across a variety of ChIP antibodies. These regions have been collected in genome-specific "blacklists" as part of the ENCODE project (Kundaje 2013). Unless there are specific experimental reasons for exploring these areas, reads overlapping these regions should be removed before processing further. Note that it is not advisable to keep the reads for peak calling and eliminate subsequently identified enriched regions as the presence of reads in these regions impacts all following processing negatively (Carroll et al. 2014).

Many samples also exhibit anomalous regions specific to their cell type. For example, immortalized cell lines have distinct karyotypes that result in unique issues, as do cells with high levels of genomic instability (such as cancer cells). In these cases, in addition to the blacklist for their genome, it may be advisable to compute a "greylist" unique to each cell type and filter reads from there as well. There is a Bioconductor package (Gentleman et al. 2004) called **GreyListChIP** (Brown 2015) that can aid in this. Note that if multiple cell types are to be used in an experiment, it is important to apply the same filters for all samples to be analyzed together.

10.4.1.3 Quality Assessment 1: Reads

Quality assessment should occur at each stage of data analysis. Once a set of aligned reads is identified, a number of quality assessment metrics may be computed and assessed. At this point in the analysis, the checks are primarily intended to verify appropriate enrichment of the ChIP samples.

10.4.1.3.1 ChIPQC Package

The example plots below were generated using the Bioconductor package **ChIPQC** (Carroll and Stark 2014). This package works on a sample sheet (similar to Table 10.1) and the BAM files (as well as, optionally, called peaks) to compute quality assessment metrics and generate plots. It is designed to work closely with the **DiffBind** package (Ross-Innes et al. 2012; Stark and Brown 2011), used below to illustrate differential binding analysis; together these are useful and flexible tools for processing and analyzing data from ChIP-seq experiments.

The following quality metrics may be of use in understanding data generated by ChIP-seq experiments:

Reads in Blacklists

For quality assessment, it can be valuable to compute a metric representing the proportion of total reads that are filtered out using blacklists. This metric can be useful in identifying outliers when comparing replicates or can be matched up to historical data for anomalous ChIPs.

Coverage Histogram and Computation of SSD

Evidence of enrichment can be seen by plotting a coverage histogram. Figure 10.4a shows example plots for the example dataset. In these plots, the *X*-axis represents the number of reads overlapping a single base, and the *Y*-axis shows the number (on a log scale) of base positions in the genome with exactly that level of coverage. Most of the genome shows low (or no) coverage, but successful ChIPs will have regions with higher coverage (representing enriched areas). By including the associated Input control on the same plot, there should be a clear area between the Input histogram (which should show lower levels of high coverage areas) and the ChIP histograms, representing enrichment of binding "peaks."

The Standardized Standard Deviation (SSD) metric can be calculated for each sample to indicate relative levels of enrichment (Planet et al. 2012). This is computed by taking the standard deviation of the coverage values for a sample and normalizing by dividing by the square root of the number of sequencing reads for the sample. Samples with high enrichment will have more variance in coverage

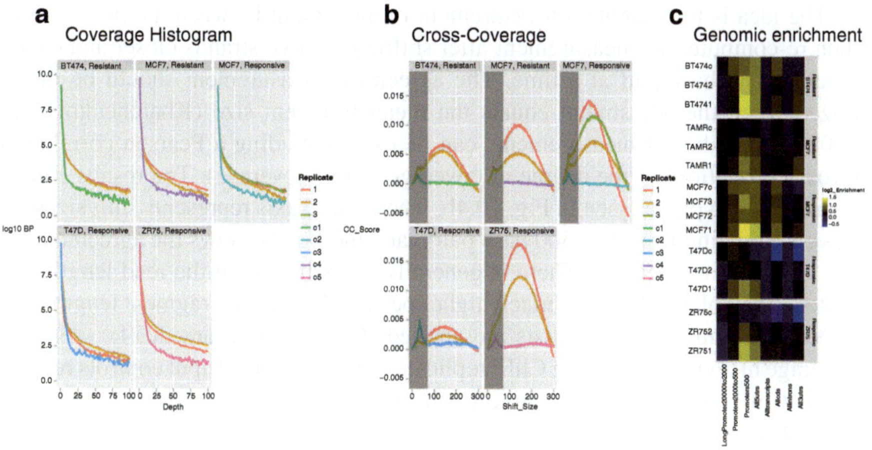

Fig. 10.4 Read-based quality control plots. This figure shows plots for the tamoxifen example dataset, generated using the **ChIPQC** package in **Bioconductor**. In each case, there is a separate part for each cell type with separate plots for Responsive and Resistant MCF7 cells. Within each plot, each replicate is plotted, along with the associated Input control. (**a**) Coverage histograms. The *X*-axis represents the range of pileup heights (coverage), while the *Y*-axis represents how many positions have this coverage (log 10 values). Most positions have low coverage, but successful ChIPs will have distinct subsets with high coverage (peaks), while Input controls should drop off more steeply. (**b**) Cross-coverage plots. The *X*-axis represents a range of shift sizes, while the *Y*-axis shows the genome coverage when reads on the two strands are shifted toward each other by that amount. ChIP peaks should converge at a sift-size equal to the mean fragment length, while Input controls (and failed ChIPs) have no coverage peak at the fragment length. (**c**) Genomic enrichment. Heatmaps showing the relative enrichment of reads in certain genomic features. *Bright yellow* indicates enrichment relative to background near the start of genes (promoters and 5′UTRs), while *blue* indicates some depletion in coding regions and the end of genes (3′UTRs)

between background and enriched regions and hence a higher standard deviation. By comparing SSD values, it can be seen if the ChIP samples have higher enrichment than their corresponding Input controls and if replicates of the same ChIP experiment have comparable enrichment.

Cross-Coverage and Computation of Fragment Length

Experiments performed using single-end sequencing are expected overall to have an equal proportion of reads mapping to each DNA strand. Likewise, around enriched areas, there should be on average an equal proportion of reads from each strand, although the proportion at individual sites will vary. Popular peak identification methods rely on this property in identifying high-quality peaks, as there should be distinct peaks on each strand offset by the mean insert size of the fragment (Zhang et al. 2008). This can be used to estimate the mean fragment size that is derived in the process as another metric that indicates enrichment in a sample (Kharchenko et al. 2008).

The idea is to compute a measurement of agreement between the two strands, then re-compute this measurement after shifting the two strands closer and closer together one base pair at a time. The agreement measurement should be maximized when the adjustment equals the mean fragment size (Kharchenko et al. 2008). A number of measurements can be used, including a Pearson correlation (cross-correlation) or the degree of coverage (cross-coverage). The results can be visualized in a plot as seen in Fig. 10.4b, where the Z-axis represents shift sizes (in this case ranging from 1 to 300 base pairs) and the Y-axis shows the agreement (in this case mean coverage). There is generally a small peak at the read length, but enriched samples will have a much higher peak at the correct fragment length (this effect is minimized if blacklists are applied (Carroll and Stark 2014)). As with coverage histograms, plotting ChIP replicates and associated Input controls reveals if the ChIPs are enriched relative to the Input and consistency between ChIP replicates.

A single metric for each sample can be computed by dividing the maximum agreement score by the score at the read length. ChIP samples should have values greater than 1.0, while Input controls should be close to that value. This metric works best for punctate enrichment (where the enriched regions are narrower than the fragment size).

Annotation and Genomic Distribution of Aligned Reads

At this point, it may be useful to annotate the reads, assigning them to categories of genomic features they may overlap. Feature types of interest may include promoters or other regions upstream of transcription start sites (TSSs), UTRs, exons, introns, known enhancers, and intergenic regions. Computing the enrichment of reads in certain type of features relative to an expected distribution based on unenriched genomic DNA can give insight into where the enrichment occurs. For example, when assaying a transcription factor that is expected to bind in promoters just upstream of TSSs, a higher proportion of reads would be expected to overlap these regions than would by chance. Figure 10.4c shows a heatmap representation of genomic enrichment for the example dataset. This plot comes from ChIPQC, but related, useful plots can be derived from other tools (Liu et al. 2011).

10.4.2 Peak Calling

After aligning, filtering, and quality assessment of the sequencing reads, the next step in many ChIP-seq analyses is to identify enriched regions for each sample. This step, often referred to as *peak calling*, attempts to separate the enrichment "signal" from the background "noise." As previously discussed, it is not unusual only a small minority of the reads overlap true enriched regions ("peaks"). Peak calling is feasible because this minority of "signal" reads is concentrated in very specific regions, while the larger set of "background" reads is distributed over the entire genome.

Methods for identification of peaks have been extensively researched and dozens of tools are available to aid in this endeavor (Rye et al. 2011; Massie and Mills 2012). While it is out of the scope of this chapter to describe their various statistical and computational approaches, there are salient points to keep in mind when deploying peak callers.

Crucial, given the profusion of competing methods, is the lack of agreement between the results from different peak callers. While there is generally a core set of peaks that will be consistently identified, corresponding to high pileups clearly visible in a genome browser, the majority of called peaks are more difficult to differentiate from background and less consistently identified. This is part of the nature of the peak-calling task, where genomic locations must conform to a binary classification of being either enriched (in a peak) or not enriched (not in a peak). Comparing different peak callers on different datasets shows significant differences in the numbers of peaks identified (Wilbanks and Facciotti 2010; Koohy et al. 2014). Besides differing in identification of enriched regions, the boundaries of such regions also vary from peak caller to peak caller, with some tending toward identifying wider or narrower regions. Indeed, most peak callers are oriented more toward identifying either punctate peaks where the enriched regions are narrower than the fragmented DNA size or broad peaks encompassing relatively long regions of enrichment.

As there is rarely a "gold standard" set of peaks by which to judge the performance of different methods, it is difficult to assess the accuracy of the peaks identified for a particular sample in a specific experiment. While some attempts have been made, using spike-ins, simulated data, and the presence of known binding motifs, no set of peak calls can be considered definitive. There are some steps that can be taken to increase the confidence in a set of peaks, generally by driving down the likelihood of false positive (but increasing false negatives). This includes using more than one peak caller and accepting only regions that are identified by all of them, or looking for overlaps in peaks called from different replicates (see discussion of IDR below).

Most popular peak callers use a control track in addition to the ChIP track to identify peaks. The MACS peak caller (Zhang et al. 2008) is the most popular of these for identifying punctate peaks; for broad peaks, popular choices include SICER (Zang et al. 2009). These are generally stand-alone pieces of software available for download, which must be installed and executed for each ChIP sample's BAM file and its corresponding control. The output of these programs is a set of peak intervals (chromosome, start and end location) along with some statistics, measures, and/or scores indicating the confidence of the peak call as well as some indication of its degree of enrichment.

For the tamoxifen resistance example, Table 10.1 shows the numbers of peaks identified by MACS for each sample.

10.4.2.1 Deriving Consensus Peaksets

In the types of ChIP-seq experiments we are discussing, with multiple samples groups and replicates, it is often useful to derive consensus sets of peaks to take forward in the analysis. Even in a simple case of an ENCODE-style exercise of mapping enrichment of a single protein in a single cell type, there should be at least two replicated samples. If only a single peak caller is used to identify peaks for each replicate, these intervals must be combined in some manner. Simple ways for doing this include taking the union of all identified peaks, or the more stringent method of taking their intersection, keeping only regions that are identified as enriched in both samples. The ENCODE project has outlined a more statistically robust procedure (Landt et al. 2012) using the Irreproducible Discovery Rate (IDR) (Li et al. 2011), which takes two sets of peaks with their confidence measures and computes a statistic corresponding to the confidence that the region is reproducibly identifiable.

This method has some limits, however, and is not easily generalizable to the case where there are many more than two replicates. In such a case, it may be useful to derive consensus peaksets separately by combing the peaks for the replicates in each sample group. Ultimately the choice of how to derive consensus peaksets depends on how they will be used for subsequent analysis, which is driven by the specific biological question being addressed. If the consensus peakset is itself a key deliverable of the analysis, it will be important to minimize false positive, and hence a conservative method (like intersection or IDR) should be used. If the subsequent analysis is robust with respect to noise, as is the quantitative differential analysis method described below, a more lenient approach may be used such as taking the union of all (or most) of the identified peaks.

10.4.2.2 Alternatives to Peak Calling

In certain cases, the use of peak callers (and their associated issues) can be avoided. Alternatives include annotation-based approaches and windowing schemes.

If the goal of the experiment involves cis-effects on transcription, the focus can be directed to enrichment in regions that encompass and/or are proximal to annotated genes. If interest lies in transcription factors (or certain properties of chromatin like the activating histone mark H3K4me3) that are known to bind in promoter regions, a set of potentially enriched intervals can be defined based on annotated transcription start sites. While other binding sites will be excluded, in many analyses, sites distal from genes are discarded anyway. As the annotation of regulatory elements becomes more widespread, more of the potentially functional sites can be identified in this way. Some (or even most) of these sites will not actually be enriched in specific experiments, but these can be filtered out or otherwise dealt with in subsequent steps, particularly in a quantitative analysis.

Another approach is to use windows across the genome and determine enrichment of each window (Lun and Smyth 2014). Different window sizes (and overlaps if using sliding windows) can be used based on the expected breadth of enriched regions. Adjacent enriched windows can be merged to define enriched intervals, and windows with low read counts across samples can be filtered out.

10.4.2.3 Quality Assessment 2: Peaks

Given a set of enriched regions defined by a peak caller, annotation, or windowing scheme, some further quality assessment metrics can be computed and checked. These can be done using either sets of peaks specific to each sample (i.e., the output of a peak caller) or using the same consensus peakset for all samples.

Reads in Peaks

A basic measure of ChIP efficiency is the proportion of sequencing reads that overlap peaks. These can be compared across replicates to identify technical outliers; across sample groups to identify difference in enrichment across treatments and conditions; and to determine baseline efficiency and consistency of different antibodies. Examining the distribution of reads across peaks can show whether the peaks have similar enrichment levels or vary considerably in coverage, and help identify outlier replicates. Figure 10.5a shows such a plot for the example experiment, with one of the cell line (ZR75) exhibiting higher variance than the others.

Peak Profiles

Generating a profile of the peaks can be useful in seeing the "shape" of the enrichment. Profiles are generated by either taking a window centered on certain point in each peak or dividing the peak into percentiles, and then computing the mean pileup at that position across all the peaks for a sample. Figure 10.5b shows peak profiles for the example data set (using 400 bp windows centered on the summit). ChIPs have more distinct peak shapes than their associated Input Samples, and some samples show greater peak heights, although at this point the data are not normalized (a simple normalization scheme, such as RPKM and derivative measures (Mortazavi et al. 2008), can be used for these plots). The profiles can be particularly helpful when using an annotation-based method for identifying enriched regions, for example to show the mean enrichment pattern of a histone mark before and after a transcription start site across all genes for differing sample groups.

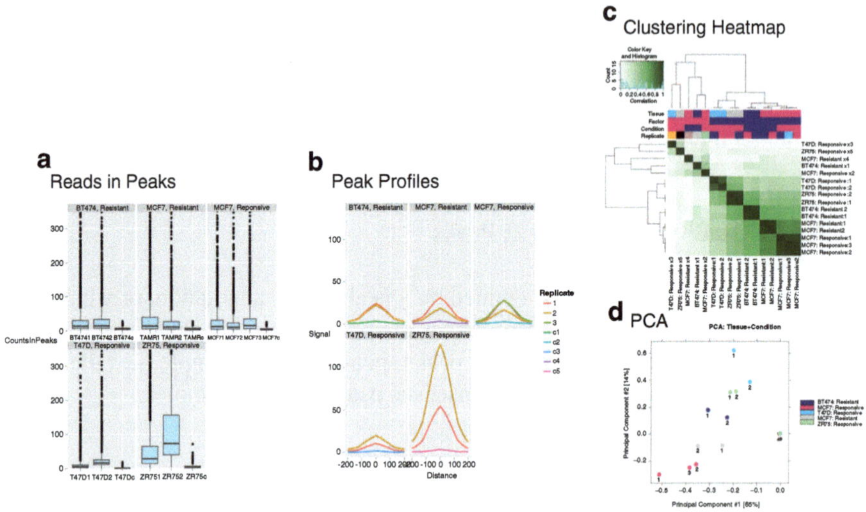

Fig. 10.5 Peak-based quality control plots. This figure shows plots for the tamoxifen example dataset, generated using the **ChIPQC** package in **Bioconductor**. In the first two plots, there is a separate part for each cell type with separate plots for Responsive and Resistant MCF7 cells. Within these plots, each replicate is plotted, along with the associated Input control. (**a**) Bar plots of distribution of the number of reads that overlap each peak. Successful ChIPs should show a range of enrichment values, while Input controls should be uniformly low. (**b**) Peak profiles. Mean number of reads across all peaks in a 400-bp window centered on the peak summit. ChIPs should show a distinct "peak" shape, while Input controls should be mostly flat. (**c**) Clustering correlation heatmap. Reads are counted for each consensus peak in every sample, and Pearson correlation coefficients are computed. The heatmap is plotted using the correlations cores, and hierarchical clustering is performed to determine the relationships between samples. Here, the Input controls form a distinct "outgroup" cluster, while the ChIP samples cluster by cell type, with replicates clustering most closely together. The two MCF7 cell types cluster together despite being in different response groups. (**d**) Principal Component Analysis plot using read count data. Input controls cluster very tightly together at one end of first component, while the replicates for the ChIP samples are close to each other, with the different cell types being distinguished in the second principal component

Sample Clustering

Identification of enriched regions can be used to explore clustering characteristics of the dataset. For this, a single consensus peakset must be used. A consensus peakset can be generated in a number of ways (see next section). The simplest way is to take all identified peaks (potentially merging overlapping ones) and create vectors for each sample with values of 1 if the peak was identified for that sample and 0 otherwise. These vectors can be used to cluster the samples. A more sophisticated method is to count all the reads for all enriched regions across all the samples in the experiment. A simple normalization method such as RPKM (reads per kilobase per million) should be applied, resulting in a vector for each sample, with length equal to the number of enriched regions in the consensus peakset, and with values equal to the normalized read count for each peak.

Unsupervised hierarchical clustering can be done by computing a correlation score between each pair of count vectors, giving a distance matrix for the experiment. Figure 10.5c shows a clustering heatmap for correlation scores of the example experiment. In this figure (generated using ChIPQC), all of the Input controls form a distinct cluster. The remainder of the samples cluster by cell type, with all of the replicates for each cell type clustering closely together. There is no inherent clustering dividing the samples group of primary interest (cells either Responsive or Resistant to treatment with the drug tamoxifen).

Another way of viewing clustering is to perform a principal component analysis (PCA) directly on the count vectors. Figure 10.5d shows a plot of the first two principal components for the example dataset. This shows all of the Input controls (which should have no enrichment) clustering extremely close to each other, separable in the first component. Replicates from the other cell lines are close to each other, and the cell line themselves are separable, particularly in the second component.

In ChIP-seq experiment involving multiple samples, it can be illuminating to study cluster pattern, particularly to identify outliers and possible batch effects. For comparative studies, it is useful to see if distinctions between elements of the contrast of interest are apparent even at this stage, before identifying differentially enriched peaks.

10.4.3 Differential Binding Analysis

Once a set of ChIP-seq samples have been processed, they can be used to address the original biological question of interest. Most functional genomics studies involve a comparison of some sort, where sample groups are contrasted to identify similarities and differences. For RNA transcription assays, this takes the form of differential gene expression analyses, which are well established. The ChIP-seq equivalent is a differential binding analysis. There are a number of ways to accomplish this, as discussed in this section.

10.4.3.1 Occupancy Analysis

The simplest method for isolating peaks unique to sample groups is to work directly with the peaks identified using a peak caller. The idea is to derive consensus peaksets for each sample group and overlap them to isolate common and unique peaks. Examples of this are common in the literature (e.g., (Ross-Innes et al. 2010)), but are becoming less common as more sophisticated techniques have been developed.

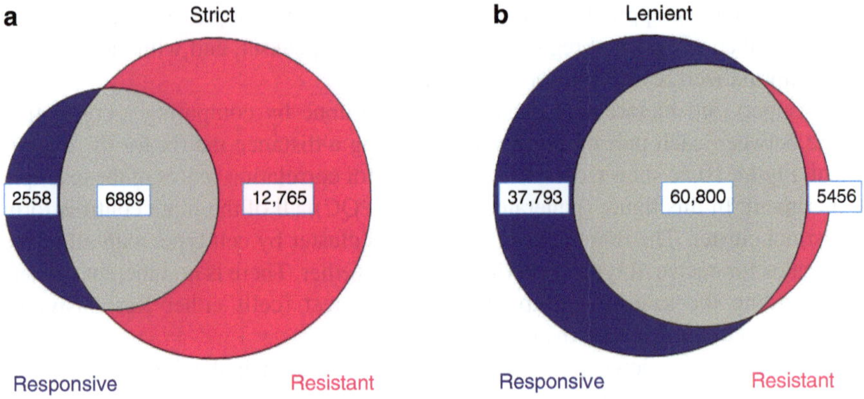

Fig. 10.6 Results from two overlap analyses. This figure shows the results from performing overlap analysis using strict and lenient criteria on the tamoxifen resistance dataset. (**a**) Using strict overlap criteria, where only peaks that are identified in either all the Responsive samples or all of the Resistant sample are included, results in 22,212 peaks to be considered, of which the majority show increased binding levels in the Resistant samples, suggesting a gain of ERα binding in tamoxifen resistance. 6889 peaks are identified in both the Responsive and Resistant samples and would be excluded from subsequent analysis as being nondifferentially bound. (**b**) Using more lenient criteria that includes all peaks identified in at least two samples, 104,049 peaks are considered, of which most (60,800) are identified in at least two samples in each group. Of the remaining peaks, the majority (37,793) are identified in at least two Responsive samples, with only 5456 peaks identified in at least two Resistant samples (but not Responsive samples). This suggests the opposite conclusion, a loss of ERα binding in tamoxifen resistance. Compare these results to Fig. 10.7b

For example, consider the tamoxifen resistance dataset described in Table 10.1. Using very strict criteria, a consensus peakset can be derived for each sample group (Responsive and Resistant) by including only peaks identified in **all** the samples in a group. For the samples in the Responsive group, there are 9456 peaks that are identified in all seven samples (where peaks that overlap by at least 1 bp are merged to form wider peaks). For the samples in the Resistant group, there are 19,941 peaks identified in all four samples. Figure 10.6a shows a Venn diagram of the overlap of these two peaksets. There are 6920 sites identified in both sample groups. If the goal is to identify binding sites that uniquely distinguish the Responsive or Resistant condition, these sites can be considered uninteresting. Of the remaining sites, there are 2558 regions uniquely enriched in the Responsive condition and a much higher number, 12,765, unique to the Resistant condition. This analysis suggests a substantial **gain** of ERα binding sites in tamoxifen resistant cells.

However, there are other ways to derive the consensus peaks that can change the conclusion. Consider a more lenient criterion, whereby all peaks that are identified in at least two samples are included. This yields a consensus peakset that include 104,051 ERα binding events. We can then consider these sites to be associated with the Resistant condition if they were identified in at least two tamoxifen resistant samples, and associated with the Responsive condition if they were identified in at least two responsive samples. This results in 98,593 of the sites being bound in the Responsive condition and 66,256 of the sites being bound in the Resistant condition. Figure 10.6b shows a Venn diagram of the overlap of these peaks. The greatest proportion of sites (60,800) is common to both conditions. Of the unique sites, the Resistant group has only 5456, while the Responsive group include 37,793 ERα binding sites. From this analysis, we would reach the opposite conclusion that we did previously: that tamoxifen resistance involves a large **loss** of ERα binding sites in tamoxifen resistant cells.

While there are a number of issues complicating this analysis, such as the noise inherent in peak calling and the imbalance in sample numbers between the groups, it is difficult to know what the "correct" answer is. What is needed is a more rigorous, statistically sound method for determining sites that change their binding profile between the sample groups.

10.4.3.2 Quantitative Analysis

While peak callers may be useful for identifying potentially interesting areas of enrichment, low agreement between peak callers suggests that they add a certain amount of noise to the experimental analysis. By comparing the identified peaks to the aligned reads in those regions, and particularly to the variance in enrichment between replicate samples, confidence statistics can be computed characterizing the likelihood of a difference in enrichment between sample groups at each binding site (Ross-Innes et al. 2012; Robinson and Oshlack 2010; Liang and Keleş 2012).

The steps to carrying out such an analysis are as follows:

- **Derive a consensus peakset for the experiment**. A variety of methods can be used to determine the consensus peakset. The "lenient" method above, where all or most of the identified peaks are included, can be utilized as the additional noise introduced by spurious peaks will be assigned very low confidence scores.
- **For each sample, count the reads that overlap each consensus peak**. A read count can be determined for every peak in every sample, *whether or not a peak was identified in that* sample. The result for each sample is a vector of read counts. These vectors form the columns of a **binding matrix**.
- **Utilize a negative binomial-based method for calculating differential expression**. Count-based differential expression tools, such as the Bioconductor (Gentleman et al. 2004) packages **edgeR** (Robinson et al. 2010) and **DESeq2** (Love et al. 2014), can be used directly on the binding matrix. There are four main steps to be followed in using these tools:
  - **Normalization**. As the different samples will be sequenced to different depths, may exhibit differences in antibody efficiency (Bao et al. 2013) and reflect varying degrees of enrichment, then, raw read counts must be normalized. Most of the read-based tools include normalization procedures, for

example the TMM method in edgeR (Robinson and Oshlack 2010). The choice of normalization method and parameters can have a significant effect on the ultimate conclusions reached. For example, the TMM method used in edgeR, developed primarily for RNA data, relies on an assumption that there is a core of sites that do not significantly change their affinity rates. If this assumption is not true for an experiment (e.g., if a transcription factor exhibits essentially no binding in one condition and high binding rate in the other), the normalization step can alter the data to the point of yielding invalid results.

- **Contrast modeling**. The simplest contrast is to compare one sample group against another. As the method uses a generalized linear model (GLM) (McCarthy et al. 2012), complex experimental designs can be modeled.
- **Dispersion estimation**. Each method has its own way of determining the dispersion of the negative binomial in fitting the GLM.
- **Computation of confidence statistics**. This includes applying an exact test to the GLM fit (Robinson and Smyth 2007), and performing a multiple testing correction on the resultant *p*-values (Benjamini and Hochberg 1995).

The Bioconductor package **DiffBind** (Ross-Innes et al. 2012; Stark and Brown 2011) encapsulates the entire process of working with ChIP-seq data, including deriving consensus peaksets, computing overlaps, counting and normalizing a binding matrix, establishing contrast, fitting linear models, generating reports of differentially bound sites, as well as including a variety of useful plotting tools.

Consider again the tamoxifen resistance example. Using DiffBind, we contrast the tamoxifen responsive and resistant samples using the 104,051 site consensus peakset described previously. Using the edgeR tool, 13,901 sites are identified as being differentially bound with $FDR < 0.1$. Figure 10.7a shows an MA plot of the result, with sites that show higher binding affinity in the Responsive case above the center line, and sites with greater binding affinity in the Resistant sample group below the line, and significantly differentially bound sites shown in magenta. Using the same association of sites to the Resistant and Responsive groups (sites that are identified in at least two samples in a group are associated with that group), we can categorize the 13,901 sites.

Figure 10.7b shows a Venn diagram of the results (compare with Fig. 10.7). Two observations are worth making. First, we see that neither of the original conclusions is valid as there is no dramatic gain in overall binding in one sample group over the other. Second, the largest single group contains sites that are common to both conditions, and hence would have been removed from further consideration in both of the previous overlap analyses. These are sites where there is some degree of ERα binding in both sample groups, *but the binding affinity changes significantly between the sample groups*.

This example is shown step-by-step in the DiffBind vignette, available online at http://bioconductor.org/packages/release/bioc/vignettes/DiffBind/inst/doc/DiffBind.pdf.

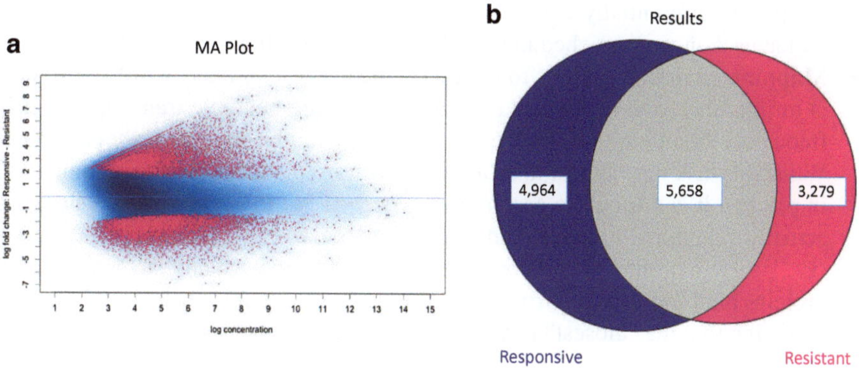

Fig. 10.7 Results from differential binding analysis. This figure shows the results of a differential binding analysis of the tamoxifen resistance dataset using the DiffBind package from Bioconductor. (**a**) MA plot of all peaks identified in at least two samples (all the sites in Fig. 10.6b). Peaks identified as being differentially bound (FDR < 0.1) are shown in *magenta*. Sites that gain binding affinity in the Resistant group have negative fold changes (below the center line). (**b**) Venn diagram of peaks identified as being differentially bound. Compare to Fig. 10.6 (especially Fig. 10.6b). Significant differences in binding affinity show no strong tendency toward ERα binding gain or loss in either the Responsive or Resistance groups. The largest subset of differentially bound sites includes peaks identified in both groups, and would be undetectable using an overlap analysis

10.4.4 Downstream Analysis

Identifying potentially interesting subsets of binding sites is generally the end of the first phase of analysis. Answering the underlying biological questions involves downstream analysis beyond the scope of this chapter. There is however a number of fairly standard steps that are useful in making sense of these sets, including the following:

- **Annotating peaks**: It is often useful to annotate peaksets with nearby genomic features (genes, promoters, etc.). There are a number of tools available for mapping enriched intervals to reference genome annotations, such as ChIPpeakAnno (Zhu et al. 2010), HOMER (Heinz et al. 2010), and Cistrome (Liu et al. 2011).
- **Motif analysis**: A useful analysis to run is a motif analysis to (a) discover sequence motifs associated with the binding sites and (b) identify known motifs enriched in a peakset. Tools that accomplish this include the HOMER (Heinz et al. 2010) and MEME (Machanick and Bailey 2011) suites. This can yield particularly interesting results in the case of a differential analysis, where peaks enriched in one condition may be associated with different motifs than in another conditions. In the tamoxifen resistance data, for example, it was discovered that

the peaks differentially enriched in the Responsive group included a different co-factor than was enriched in the other peaks (Ross-Innes et al. 2012).

- **Mapping enriched regions to genes**: As the purpose of many ChIP-seq studies is to illuminate the regulatory elements underlying genomic transcription, it is frequently desirable to associate binding sites with the genes that they regulate. While this task may appear to be a straightforward matter of genomic annotation, in practice it is difficult to accomplish. Many (if not most) binding sites are not actually functionally active, and while some peaks may be easier to associate (i.e., peaks in promoters of known protein-coding genes), many DNA-associated proteins bind some distant from known genes (for example in enhancers). Simply assuming that the "closest" gene is being regulated is generally incorrect (Wang et al. 2013).
- **Integration of ChIP-seq and RNA-seq data**: ChIP-seq assays have become more important in complementing RNA-based studies as the focus has been drawn toward understanding transcriptional regulation. Reliance primarily on mRNA levels requires regulatory components to be inferred, while, and how they may be associated with transcriptional output and observable phenotypes. If both ChIP-seq and RNA-seq data are available for an experiment, there is the possibility of integrating them in order to associate the regulatory events with transcription itself by finding correlations between changes in binding sites and changes in transcription. A number of tools are available for helping with this process, such as Binding and Expression target Analysis (BETA) (Wang et al. 2013).
- **Functional/Pathway enrichment**: While identified sites that have been mapped to genes can be tested for functional enrichment (i.e., GO analysis) or subjected to pathway analysis, tools also exist that can utilize binding peaksets directly to test for functional enrichment. Notable here is the GREAT tool (McLean et al. 2010).

10.5 Conclusions

Finally, it is important to keep in mind that ChIP-seq experiments are imperfect, and each step in the analysis process can result in noise and false positives. All interesting results need ultimately be validated using some other experimental method in order to have confidence in any conclusions.

Acknowledgements The authors wish to acknowledge the many people at the Cancer Research UK Cambridge Institute who have been involved in performing and analyzing ChIP-seq experiments since 2008. In particular we thank Suraj Menon and Gordon Brown for their feedback on this chapter. Venn diagram plots were generated using the Venn Diagram Plotting software from PNNL (omics.pnl.gov).

Annex: Quick Reference Guide

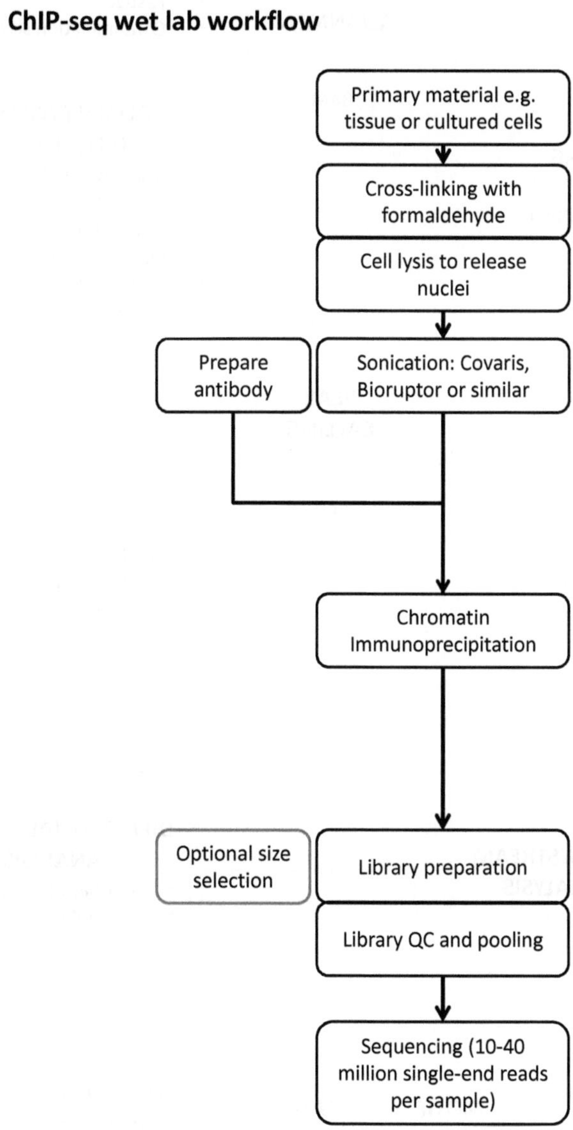

Fig. QG10.1 Representation of the wet-lab procedure workflow

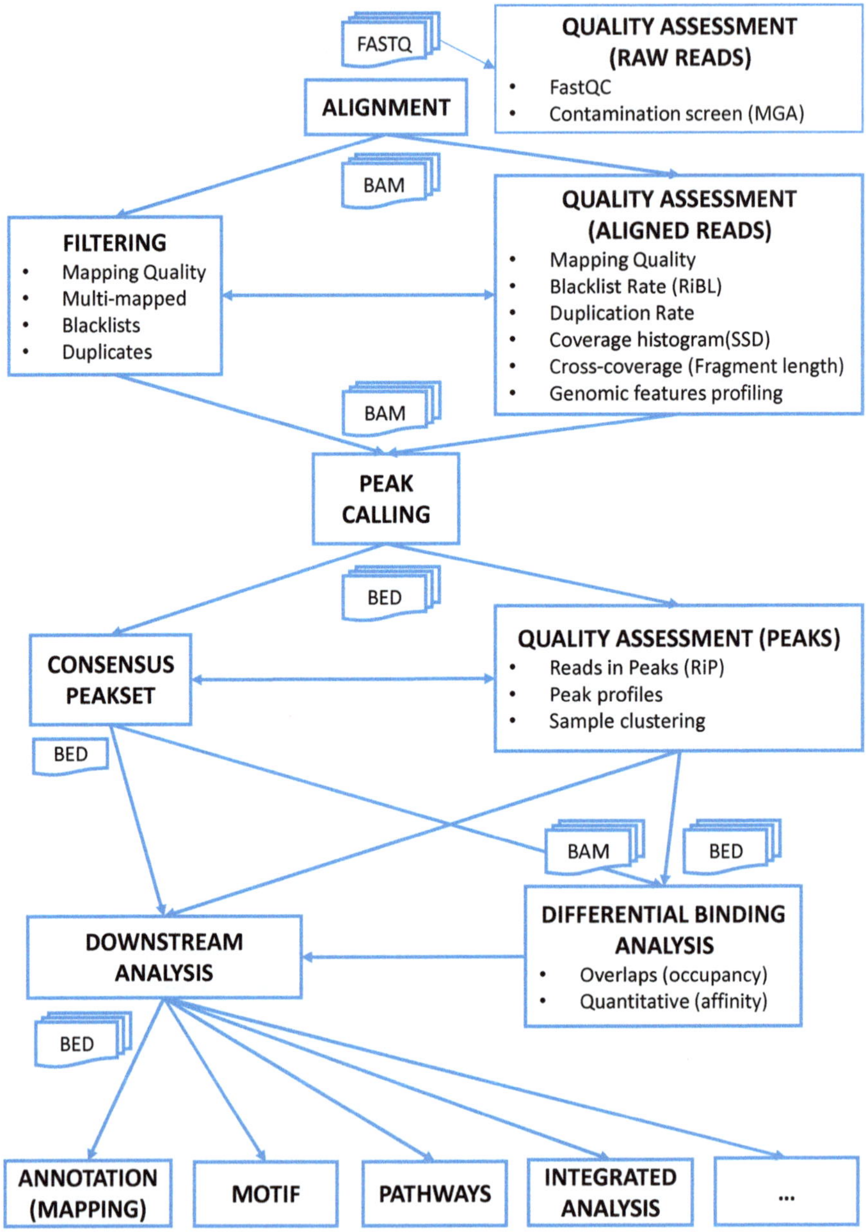

Fig. QG10.2 Main steps of the computational analysis pipeline

Table QG10.1 Experimental design considerations

Technique	Protocol	Control library (if applicable)	Recommended starting material (ng)	Number of replicates	Sequencing depth	Recommended sequencing platform and run (if applicable)	Reference
ChIP-seq (point source)	Antibody	•DNA Input	5 ng	3 (minimum per condition)	10 Million reads uniquely mapped	•Illumina Single End	Schmidt et al. (2009)
	Immunoprecipitation	•Control IgG					Landt et al. (2012)
ChIP-seq (broad source)	Antibody	•DNA Input	5 ng	3 (minimum per condition)	20 Million reads uniquely mapped	•Illumina Single End/ Paired end	Schmidt et al. (2009)
	Immunoprecipitation	•Control IgG					Landt et al. (2012)

Table that comprises relevant experimental design parameters, to carefully consider before applying this methodology

Table QG10.2 Available software recommendations

Software	Language/platform	Input format	Results output	Results format	Reference
FASTQC	Windows, Mac OS X, Linux, and Solaris	FASTQ, SAM, BAM	Graphical representation	Graphics	Andrews (2010)
MGA	Mac OS X, Linux, and Solaris	FASTQ	HTML	Text	Hadfield and Eldridge (2014)
Bowtie	Windows, Mac OS X, Linux, and Solaris	FASTQ	BOW/SAM/BAM (Alignment files)	Text	Langmead et al. (2009)
BWA	Windows, Mac OS X, Linux, and Solaris	FASTQ	SAI/SAM/BAM (Alignment files)	Text	Li et al. (2009)
GreyListChIP	Windows, Mac OS X, Linux, and Solaris (R/Bioconductor package)	BAM	BED	Text	Brown (2015)
MACS	Windows, Mac OS X, Linux, and Solaris	SAM/BAM	BED, XLS (actually CSV)	Tables, text	Zhang et al. (2008)
SICER	Windows, Mac OS X, Linux, and Solaris	SAM/BAM	BED, XLS	Tables, text	Zang et al. (2009)
ChipQC	Windows, Mac OS X, Linux, and Solaris (R/Bioconductor package)	BAM	HTML	Graphics, text	Carroll and Stark (2014)
DiffBind	Windows, Mac OS X, Linux, and Solaris (R/Bioconductor package)	BAM, BED	R-Data, CSV	Graphics, tables, text	Stark and Brown (2011), Ross-Innes et al. (2012)
csaw	Windows, Mac OS X, Linux, and Solaris (R/Bioconductor package)	BAM	R-Data	Graphics, tables, text	Lun and Smyth (2014)
ChIPpeakAnno	Windows, Mac OS X, Linux, and Solaris (R/Bioconductor package)	BED, GFF, MACS	R-Data, CSV	Graphics, tables, text	Zhu et al. (2010)

HOMER	Windows, Mac OS X, Linux, and Solaris	BED, tables	CSV	Graphics, tables, text	Heinz et al. (2010)
MEME	Web server, Windows, Mac OS X, Linux, and Solaris	FASTA, BEDF	HTML, motifs GFF	Graphics, tables, text	Machanick and Bailey (2011)
BETA	Web server, Windows, Mac OS X, Linux, and Solaris	BED, BAM	Tab separated text, HTML, UCSC tracks	Graphics, tables, text	Wang et al. (2013)
GREAT	Windows, Mac OS X, Linux, and Solaris	BED	Tab separated text, HTML, UCSC tracks	Graphics, tables, text	McLean et al. (2010)

Table displaying a selection of the recommended software available for the computational analysis of data yielded by this technique

References

Acevedo LG, Iniguez AL, Holster HL, Zhang X, Green R, Farnham PJ (2007) Genome-scale ChIP-chip analysis using 10,000 human cells. Biotechniques 43(6):791

Andrews S (2010) FastQC: a quality control tool for high throughput sequence data. Babraham Institute. Available at: http://www.bioinformatics.babraham.ac.uk/projects/fastqc/

Bao Y, Vinciotti V, Wit E, 't Hoen PA (2013) Accounting for immunoprecipitation efficiencies in the statistical analysis of ChIP-seq data. BMC Bioinformatics 14(1):169

Benjamini Y, Hochberg Y (1995) Controlling the false discovery rate: a practical and powerful approach to multiple testing. J R Stat Soc Ser B Methodol 57:289–300

Brown G (2015) GreyListChIP: Grey Lists – Mask Artefact Regions Based on ChIP Inputs. R package version 1.0.1. Available at: http://bioconductor.org/packages/3.1/bioc/html/GreyListChIP.html

Carroll T, Stark R (2014) Assessing ChIP-seq sample quality with ChIPQC. Available at: http://bioconductor.org/packages/3.1/bioc/html/ChIPQC.html

Carroll TS, Liang Z, Salama R, Stark R, de Santiago I (2014) Impact of artifact removal on ChIP quality metrics in ChIP-seq and ChIP-exo data. Front Genet 5:75

Cock PJ, Fields CJ, Goto N, Heuer ML, Rice PM (2010) The Sanger FASTQ file format for sequences with quality scores, and the Solexa/Illumina FASTQ variants. Nucleic Acids Res 38(6):1767–1771

ENCODE Project Consortium (2011) A user's guide to the encyclopedia of DNA elements (ENCODE). PLoS Biol 9(4):e1001046

Gentleman RC, Carey VJ, Bates DM, Bolstad B, Dettling M, Dudoit S et al (2004) Bioconductor: open software development for computational biology and bioinformatics. Genome Biol 5(10):R80

Hadfield J, Eldridge MD (2014) Multi-genome alignment for quality control and contamination screening of next-generation sequencing data. Front Genet 5:31

Heinz S, Benner C, Spann N, Bertolino E, Lin YC, Laslo P et al (2010) Simple combinations of lineage-determining transcription factors prime cis-regulatory elements required for macrophage and B cell identities. Mol Cell 38(4):576–589

Kharchenko PV, Tolstorukov MY, Park PJ (2008) Design and analysis of ChIP-seq experiments for DNA-binding proteins. Nat Biotechnol 26(12):1351–1359

Koohy H, Down TA, Spivakov M, Hubbard T (2014) A comparison of peak callers used for DNase-seq data. PLoS One 9:e96303

Kundaje A (2013) A comprehensive collection of signal artifact blacklist regions in the human genome. ENCODE [hg19-blacklist-README. doc-EBI]. Available at: https://sites.google.com/site/anshulkundaje/projects/blacklists

Kutter C, Brown GD, Gonçalves Â, Wilson MD, Watt S, Brazma A et al (2011) Pol III binding in six mammals shows conservation among amino acid isotypes despite divergence among tRNA genes. Nat Genet 43(10):948–955

Landt SG, Marinov GK, Kundaje A, Kheradpour P, Pauli F, Batzoglou S, Bernstein BE, Bickel P, Brown JB, Cayting P et al (2012) ChIP-seq guidelines and practices of the ENCODE and modENCODE consortia. Genome Res 22(9):1813–1831

Langmead B, Trapnell C, Pop M, Salzberg SL (2009) Ultrafast and memory-efficient alignment of short DNA sequences to the human genome. Genome Biol 10(3):R25

Li H, Durbin R (2009) Fast and accurate short read alignment with Burrows–Wheeler transform. Bioinformatics 25(14):1754–1760

Li H, Handsaker B, Wysoker A, Fennell T, Ruan J, Homer N et al (2009) The sequence alignment/map format and SAMtools. Bioinformatics 25(16):2078–2079

Li Q, Brown JB, Huang H, Bickel PJ (2011) Measuring reproducibility of high-throughput experiments. Annals Appl Stat 5:1752–1779

Liang K, Keleş S (2012) Detecting differential binding of transcription factors with ChIP-seq. Bioinformatics 28(1):121–122

Liu T, Ortiz JA, Taing L, Meyer CA, Lee B, Zhang Y et al (2011) Cistrome: an integrative platform for transcriptional regulation studies. Genome Biol 12(8):R83
Love MI, Huber W, Anders S (2014) Moderated estimation of fold change and dispersion for RNA-seq data with DESeq2. Genome Biol 15(12):550
Lun AT, Smyth GK (2014) De novo detection of differentially bound regions for ChIP-seq data using peaks and windows: controlling error rates correctly. Nucleic Acids Res 42(11):e95
Machanick P, Bailey TL (2011) MEME-ChIP: motif analysis of large DNA datasets. Bioinformatics 27(12):1696–1697
Massie CE, Mills IG (2012) Mapping protein–DNA interactions using ChIP-sequencing. In: Transcriptional regulation. Springer, New York, NY, pp 157–173
McCarthy DJ, Chen Y, Smyth GK (2012) Differential expression analysis of multifactor RNA-Seq experiments with respect to biological variation. Nucleic Acids Res 40:4288
McLean CY, Bristor D, Hiller M, Clarke SL, Schaar BT, Lowe CB et al (2010) GREAT improves functional interpretation of cis-regulatory regions. Nat Biotechnol 28(5):495–501
Mohammed H, Russell IA, Stark R, Rueda OM, Hickey TE, Tarulli GA et al (2015) Progesterone receptor modulates ERα action in breast cancer. Nature 523(7560):313–317
Mortazavi A, Williams BA, McCue K, Schaeffer L, Wold B (2008) Mapping and quantifying mammalian transcriptomes by RNA-Seq. Nat Methods 5(7):621–628
Murtaza M, Dawson SJ, Tsui DW, Gale D, Forshew T, Piskorz AM et al (2013) Non-invasive analysis of acquired resistance to cancer therapy by sequencing of plasma DNA. Nature 497(7447):108–112
O'Neill LP, VerMilyea MD, Turner BM (2006) Epigenetic characterization of the early embryo with a chromatin immunoprecipitation protocol applicable to small cell populations. Nat Genet 38(7):835–841
Park PJ (2009) ChIP–seq: advantages and challenges of a maturing technology. Nat Rev Genet 10(10):669–680
Pennacchio LA, Bickmore W, Dean A, Nobrega MA, Bejerano G (2013) Enhancers: five essential questions. Nat Rev Genet 14(4):288–295
Planet E, Stephan-Otto C, Reina O, Flores O, Rossell D (2012) htSeqTools: quality control, visualization and processing high-throughput sequencing data. Bioinformatics 28(4):589
Robinson MD, Oshlack A (2010) A scaling normalization method for differential expression analysis of RNA-seq data. Genome Biol 11(3):R25
Robinson MD, Smyth GK (2007) Moderated statistical tests for assessing differences in tag abundance. Bioinformatics 23(21):2881–2887
Robinson MD, McCarthy DJ, Smyth GK (2010) edgeR: a Bioconductor package for differential expression analysis of digital gene expression data. Bioinformatics 26(1):139–140
Ross-Innes CS, Stark R, Holmes KA, Schmidt D, Spyrou C, Russell R et al (2010) Cooperative interaction between retinoic acid receptor-α and estrogen receptor in breast cancer. Genes Dev 24(2):171–182
Ross-Innes CS, Stark R, Teschendorff AE, Holmes KA, Ali HR, Dunning MJ et al (2012) Differential oestrogen receptor binding is associated with clinical outcome in breast cancer. Nature 481(7381):389–393
Rye MB, Sætrom P, Drabløs F (2011) A manually curated ChIP-seq benchmark demonstrates room for improvement in current peak-finder programs. Nucleic Acids Res 39:e25
Schmidt D, Wilson MD, Spyrou C, Brown GD, Hadfield J, Odom DT (2009) ChIP-seq: using high-throughput sequencing to discover protein–DNA interactions. Methods 48(3):240–248
Schmidt D, Wilson MD, Ballester B, Schwalie PC, Brown GD, Marshall A et al (2010) Five-vertebrate ChIP-seq reveals the evolutionary dynamics of transcription factor binding. Science 328(5981):1036–1040
Stark R, Brown G (2011) DiffBind: differential binding analysis of ChIP-Seq peak data. Available at: http://bioconductor.org/packages/release/bioc/vignettes/DiffBind/inst/doc/DiffBind.pdf
Wang S, Sun H, Ma J, Zang C, Wang C, Wang J et al (2013) Target analysis by integration of transcriptome and ChIP-seq data with BETA. Nat Protoc 8(12):2502–2515
Wilbanks EG, Facciotti MT (2010) Evaluation of algorithm performance in ChIP-seq peak detection. PLoS One 5(7):e11471

Zang C, Schones DE, Zeng C, Cui K, Zhao K, Peng W (2009) A clustering approach for identification of enriched domains from histone modification ChIP-Seq data. Bioinformatics 25(15):1952–1958

Zhang Y, Liu T, Meyer CA, Eeckhoute J, Johnson DS, Bernstein BE et al (2008) Model-based analysis of ChIP-Seq (MACS). Genome Biol 9(9):R137

Zhu LJ, Gazin C, Lawson ND, Pagès H, Lin SM, Lapointe DS, Green MR (2010) ChIPpeakAnno: a Bioconductor package to annotate ChIP-seq and ChIP-chip data. BMC Bioinformatics 11(1):237

Zuo C, Keleş S (2014) A statistical framework for power calculations in ChIP-seq experiments. Bioinformatics 30:753

Chapter 11
PAR-CLIP: A Genomic Technique to Dissect RNA-Protein Interactions

Tara Dutka, Aishe A. Sarshad, and Markus Hafner

11.1 Introduction

In recent years our concept of gene expression has expanded beyond the central dogma (DNA → RNA → protein) to include post-transcriptional gene regulation (PTGR). In PTGR processes, RNA-binding proteins (RBPs) and ribonucleoprotein complexes (RNPs) control the stability, maturation, location, or translation of virtually all cellular transcripts (Keene 2007; Morris et al. 2010; Gerstberger et al. 2014). Recent estimates suggest the presence of more than 1500 genes encoding RBPs in the human genome, of which ~700 interact with and regulate mRNA, while the rest interact with other classes of RNA, such as rRNA, sn/snoRNA, and tRNA (Mattaj 1993; Mansfield and Keene 2009; Gerstberger et al. 2014). Given that approximately 10 % of RBPs have already been linked to human disease phenotypes in OMIM (http://www.ncbi.nlm.nih.gov/omim), the critical role of these proteins in cellular and organismal homeostasis cannot be overstated. However, the specific mechanistic role of most of these RBPs remains to be elucidated.

The effort to catalogue heterogeneous nuclear ribonucleoproteins (Piñol-Roma et al. 1988) led to the realization that RBPs recognize their targets via discrete RNA binding domains (RBDs) at specific structural or sequence elements, termed RNA recognition elements (RREs) (Swanson et al. 1987; Dreyfuss et al. 1988; Bandziulis et al. 1989; Query et al. 1989). Currently 75 canonical RBDs are known, including the well-characterized RRM, KH, dsrm, zf-CCHC, PAZ, and Piwi domains. While approximately 60 % of RBPs contain a single RBD, some 40 % of RBPs contain

Authors contributed equally and are listed alphabetically.

T. Dutka • A.A. Sarshad • M. Hafner (✉)
Laboratory of Muscle Stem Cells and Gene Regulation, NIAMS,
50 South Drive, Bethesda, MD 20892, USA
e-mail: markus.hafner@nih.gov

A.M. Aransay, J.L. Lavín Trueba (eds.), *Field Guidelines for Genetic Experimental Designs in High-Throughput Sequencing*,
DOI 10.1007/978-3-319-31350-4_11

either repeats or multiple different RBDs, an arrangement thought to increase affinity and sequence-specificity of an RBP (Ascano et al. 2012a).

RBPs belong to the most abundant protein classes and are expressed with low tissue-specificity, further increasing the complexity of PTGR networks by possible widespread competition and synergy. Considering that most known RBDs recognize short sequence stretches of 4–6 nt, every RBP potentially interacts with hundreds to thousands of different RNAs. RBP occupancy on a given sequence stretch of an RNA will depend on expression levels and localization of the RBP and RNA target, as well as accessibility of the target site (e.g., binding of competitors in the vicinity, secondary structure effects). These factors complicate in silico prediction of PTGR networks and their characterization thus required the development of appropriate experimental approaches (Mattaj 1993; Hafner et al. 2010; Baltz et al. 2012; Castello et al. 2012).

11.2 Techniques for Examining RNA-Protein Interactions

A variety of techniques have been employed to characterize RNA-protein interactions (Table 11.1). Here we will give a brief, noncomprehensive overview of in vitro and in vivo approaches, before describing in greater detail Photoactivatable-Ribonucleoside-Enhanced Crosslinking and Immunoprecipitation (PAR-CLIP), a method to identify RREs on a transcriptome-wide scale with nucleotide resolution.

11.2.1 In Vitro Techniques

11.2.1.1 One Protein to One RNA Species, EMSA

Several in vitro techniques allow for the investigation of the thermodynamic properties of interactions between purified RBPs and single RNA species, including isothermal titration calorimetry (Salim and Feig 2009), surface plasmon resonance (Katsamba et al. 2002; Yang et al. 2008), and filter binding (Rio 2012). Electrophoretic Mobility Shift Assay (EMSA) is one of the most widely used, and oldest, methods to determine binding affinities of RBPs and RNA. In this assay, radioactively or fluorescently labeled RNA is incubated with varying concentrations of a purified protein to permit formation of the RNPs, which are subsequently fractionated by electrophoresis under native conditions, either on an agarose or a polyacrylamide gel. The labeled RNA is then visualized and quantified using autoradiography or fluorometry. Formation of an RNP will result in higher retention of bound compared to unbound RNA and cause a “shift” in migration on the gel. Thermodynamic constants are determined using the ratio of bound versus unbound RNA at varying RBP concentrations. The structural determinants for binding can be validated by sequential mutation of target RNA or protein sequences. In the variant competition

Table 11.1 Methods to probe for RNA-protein interactions

Isothermal titration calorimetry (ITC)	Thermodynamic assessment of binding interactions of protein and RNA interactors based on measuring the heat generated or absorbed during binding.	Salim and Feig (2009)
Surface plasmon resonance (SPR)	An optical detection method that measures the changes in the refractive index near a sensor surface to analyze biomolecular interactions such as RNA and protein in real time.	Katsamba et al. (2002), Yang et al. (2008)
Filter binding	A purified protein is mixed with a purified and radioactively labeled RNA and applied to a nitrocellulose membrane. Only RNA bound to the protein is retained on the membrane confirming the interaction of the two species.	Rio (2012)
Electrophoretic mobility shift assay (EMSA)	Purified protein is mixed with a purified and radioactively/fluorescently labeled RNA and fractionated by native gel electrophoresis separating formed RNP complexes based on size, shape, and charge. Binding of RNA to protein causes a visible shift in the RNA location on the gel.	Gagnon and Maxwell (2010)
Fluorescent anisotropy/polarization	A fluorescent RNA is incubated with an RBP and binding is assed by the change in fluorescent anisotropy, which is the change in intensity of light emitted on different axes of polarization due to changes in molecular volume and/or RNA flexibility upon binding of the RBP.	Wilson (2005)
RNA mechanically induced trapping of molecular interactions (RNA-MITOMI)	An array of DNA oligos is used to transcribe and immobilize RNA on the array. RBPs are labeled with different fluorophores and incubated with the RNA pool to quantify the binding of a single RBP to many RNA variations.	Martin et al. (2012)
Systematic evolution of ligands by exponential enrichment (SELEX)	A library of in vitro transcribed ssRNA is incubated with a known RBP. The bound RNA is eluted and amplified and this process is repeated for several cycles to identify high-affinity binders.	Stoltenburg et al. (2007), Manley (2013)

(continued)

Table 11.1 (continued)

High-throughput sequencing RNA affinity profiling (HiTS-RAP)/quantitative analysis of RNA on a massively parallel array (RNA-MaP)	The principles of SELEX are adapted to the technology of Illumina Genome Analyzers such that RNA sequence variants are immobilized on a flow cell and the analyzer quantifies the association of a fluorescently tagged protein with the RNA sequences.	Buenrostro et al. (2014), Tome et al. (2014)
RNA Bind-n-seq/In vitro selection, high-throughput sequencing of RNA, and sequence-specificity landscapes (SEQRS)	An ssRNA library is incubated with multiple RBP concentrations or at a single concentration followed by less than five rounds of selection. High-throughput sequencing to determine high- and low-affinity binders follows either case.	Campbell et al. (2012), Lambert et al. (2014)
RNAcompete	A pool of ~240,000 different sequences of 30–40 nt length, with at least 16 copies of every possible RNA 9-mer sequence, is printed on a microarray, in vitro transcribed, and incubated with a tagged RBP. After recovery on the protein, relative affinity of each sequence is determined using a microarray and the RRE inferred from these sequences.	Ray et al. (2009)
Fluorescence in situ hybridization (FISH)	Detects co-localization of RNA and protein within 200 nm using fluorescent tags/probes.	Geiger and Neugebauer (2005), Tanke et al. (2005)
Fluorescence resonance energy transfer (FRET)	Detects co-localization of RNA and protein up to 1 nm using energy transfer between fluorescent tags/probes.	Selvin (2000), Ganguly et al. (2004), Huranová et al. (2009)
RNase protection	Determines the RRE of an RBP by hybridization to DNA probes followed by digestion of the area unprotected from RNA–DNA hybridization.	Günzl and Bindereif (1999), Paulus et al. (2004), Ilagan et al. (2009)
RIP-Chip	RNA immunoprecipitation coupled to microarray analysis to identify (m) RNA targets of an RBP.	Keene et al. (2006)

RIP-Seq	RNA immunoprecipitation coupled to high-throughput sequencing to identify RNA targets of an RBP.	Cloonan et al. (2008)
Crosslinking and immunoprecipitation (CLIP)	Crosslinking of RNA and interacting proteins by 254 nm UV followed by immunoprecipitation and sequencing of recovered crosslinked RNA fragments to reveal RNA target sites of an RBP. Originally, traditional Sanger sequencing was used.	Ule et al. (2003)
High-throughput sequencing-CLIP (HiTS-CLIP)	The original CLIP protocol adapted to high-throughput (Illumina) sequencing technologies.	Licatalosi et al. (2008)
Individual-nucleotide resolution CLIP (iCLIP)	Individual-nucleotide resolution CLIP takes advantage of abortive reverse transcription at the site of crosslinking during cDNA library preparation involving circularization of the truncated cDNA. The first nucleotide sequenced represents the site of crosslinking.	König et al. (2010), Huppertz et al. (2014)
UV crosslinking and analysis of cDNAs (CRAC)	CRAC replaces the IP step with purification on tandem tagged proteins on an affinity resin. This makes for even more stringent purification of RNPs.	Granneman et al. (2009)
Crosslinking, ligation, and sequencing of hybrids (CLASH)	CLASH is a high-throughput method to identify RNA-RNA interactions.	Kudla et al. (2011)
Photoactivatable-ribonucleoside-enhanced crosslinking and immunoprecipitation (PAR-CLIP)	PAR-CLIP utilizes photoreactive ribonucleoside incorporated into RNA to enhance crosslinking. Photocrosslinking of the nucleoside analogs introduces a structural change resulting in a characteristic mutation during cDNA library construction, allowing identification of the site of crosslinking.	Hafner et al. (2010)
In vivo PAR-CLIP (iPAR-CLIP)	In vivo PAR-CLIP adaptation, first demonstrated in *C. elegans*.	Jungkamp et al. (2011)

gel-shift assay, simultaneous incubation with a variety of unlabeled RNA competitors is employed to survey the relative affinities and specificities of sequence variants. It is possible to adapt EMSA assays to allow the use of extracts or partially purified proteins by incubating with an antibody against the RNP causing an additional shift in the gel band, or "supershift," which identifies the binding RBP (Gagnon and Maxwell 2010). EMSA assays rely on knowledge of possible interactions and are less able to identify kinetically weak interactions. However, some of these limitations may be overcome with crosslinking.

11.2.1.2 One Protein to Many RNA

SELEX (Systematic Evolution of Ligands by EXponential Enrichment) (Stoltenburg et al. 2007; Manley 2013) and newer variations such as RNAcompete (Ray et al. 2009) or Bind-n-seq (Lambert et al. 2014) interrogate a large sequence pool for RNAs that bind to the RBP with high affinity. In its basic iteration, SELEX involves an in vitro transcribed single stranded RNA (ssRNA) sequence library of approximately 10^{13} to 10^{15} different sequences of 20–80 randomized nucleotides flanked by constant sequences that allow reverse transcription (RT), PCR amplification, and in vitro transcription (Stoltenburg et al. 2007). This library is incubated with a target RBP, RNA molecules with low binding affinity are removed by washing, and the bound RNA is amplified by RT and PCR amplification to form a new pool as input for the next round of selection. After several cycles, generally 6–20 rounds, only RNAs that bind with high affinity remain in the amplified pool and can be identified by sequencing (Stoltenburg et al. 2007; Manley 2013). Phylogenetic analysis of the sequences reveals the optimal motif resulting in high-affinity interactions. SELEX has been used to identify high-affinity RNA ligands to proteins, cofactors, and small molecules (Stoltenburg et al. 2007). In addition, the selection conditions can be modified to isolate RNA sequences catalyzing a variety of chemical and biochemical reactions (ribozymes) (Bartel and Szostak 1993; Joyce 1994; Seelig and Jäschke 1999). Application of SELEX to a diverse set of RBPs, including Pumilio (White et al. 2001), Quaking (Galarneau and Richard 2005), and FMR1 (Chen et al. 2003), revealed candidates for their RNA recognition element (RRE) that were then used for the genome-wide prediction of RNA targets.

Newer SELEX-type experiments, such as RNA Bind-n-Seq, HiTS-RAP, and RNA-MaP (Ozer et al. 2014; Buenrostro et al. 2014; Tome et al. 2014; Lambert et al. 2014), are designed to capture lower affinity binders, which, in the context of widespread competition and synergy between RBPs, may represent equally valuable RRE candidates. For example, in RNA Bind-n-Seq, multiple RBP concentrations are used in a single binding step followed by deep sequencing and bioinformatics sequence analysis, which captures more variation than the repeated cycles of traditional SELEX.

The sophisticated imaging and fluidics capabilities of Illumina Genome Analyzers can allow the simultaneous interrogation of the binding landscapes of more than 10^7 sequences (Buenrostro et al. 2014; Tome et al. 2014). In these assays, the RNA pool is directly immobilized on the flow cell and incubated with fluorescently tagged proteins. Quantification of protein association and dissociation rates allows for the simultaneous determination of the thermodynamic binding constants for the entire sequence collection. Furthermore, these rates can be used to identify the compensatory effect of multiple sequence variations, which can be challenging for the other SELEX-type methods.

RNAcompete combines a carefully designed sequence pool with microarray-based detection methods and thus allows for an increased throughput in the identification of unstructured RREs. The RNA pool consists of ~240,000 different sequences of 30–40 nt length containing at least 16 copies of every possible RNA 9-mer sequence (Ray et al. 2009). This pool is printed on microarrays, amplified directly from these arrays, in vitro transcribed, and incubated with affinity-tagged proteins of interest in a high molar excess to ensure that at equilibrium the proportion of binding of each sequence reflects the affinity to the protein. After recovery of the protein, the enrichment of each sequence over the input pool is determined on microarrays and the RRE inferred from comparison of enrichment scores for every possible 7-mer (Ray et al. 2009). RNAcompete was used for a comprehensive survey of 193 different RBPs revealing the deep conservation of binding properties of homologous RBPs (Ray et al. 2013).

While these methods provide valuable insights into the binding specificity between RNA and proteins, they are biased towards identification of high-affinity interactions. In living cells, RBPs compete for some RNA targets and act in synergy on others. Thus, in vivo, some high-affinity interactions may be irrelevant compared to others of lower affinity due to differences in RNA abundance or localization.

11.2.2 In Vivo Techniques

11.2.2.1 Visualization of Interactions with Fluorescence

Known or suspected RNA-protein interactions can be visualized in tissues and cells by detecting co-localization using fluorescence in situ hybridization (FISH) and fluorescence resonance energy transfer (FRET) (Selvin 2000; Vyboh et al. 2012; Silahtaroglu 2014). These microscopy-based methods require labeling of the RBP, generally by expressing a chimeric protein with a fluorescent tag and incorporating fluorescently labeled RNA probes either in situ, following fixation and permeabilization, or in vivo using microinjection, streptolysin O, scrape-loading, peptide-mediated membrane transfer, or electroporation (Geiger and Neugebauer 2005; Tanke et al. 2005). While FISH indicates a possible interaction, the large resolution distance of typically 200 nm precludes definitive conclusions. The resolution can be increased to approximately 1 nm using FRET, in which a donor fluorophore excites fluorescence of an acceptor fluorophore in close proximity. This transfer causes an apparent reduction in intensity of the donor and an increase in acceptor intensity that depends on the distance between the two molecules (Selvin 2000). FRET has been successfully applied to investigate the binding of hnRNP H to its target RNA (Huranová et al. 2009), the interaction of fibrillarin and snRNA in *Giardia lamblia* (Ganguly et al. 2004), and the proteins interacting with the mutant *DMPK* gene RNA foci in Myotonic Dystrophy type 1 (Rehman et al. 2014). FISH and FRET can be useful in demonstrating in vivo interactions; however, the necessity to fluorescently label protein and RNAs limits the applicability outside cell culture systems.

11.2.2.2 Immunoprecipitation-Based Assays

Most in vivo RNA-protein interaction analyses are based on immunoprecipitation and are conceptually related to chromatin immunoprecipitation (ChIP) assays for studying DNA-protein interactions in vivo (Niranjanakumari et al. 2002).

11.2.1.1.1 RNase Protection

RNase protection assays can be used to determine binding sites and RREs on known RNA ligands for the RBPs of interest. The RNP is immunoprecipitated from cell lysates (Günzl and Bindereif 1999) and small single-stranded DNA (ssDNA) probes complementary to the suspected RRE and flanking regions are allowed to hybridize to the immunoprecipitate. Bound RBPs prevent this hybridization and protect the RRE from RNase H, which selectively degrades RNA-DNA hybrids. The extent of protection is typically quantified using Northern blotting. RNase protection has been used to examine the structure of spliceosome complexes and confirm the impact of a stem-loop structure on ribosome binding (Paulus et al. 2004; Ilagan et al. 2009).

11.2.1.1.2 RNA Immunoprecipitation Followed by Microarray Analysis (RIP-Chip) or Next-Generation Sequencing (RIP-Seq)

RNA immunoprecipitation (RIP) coupled to high-throughput methods allows for the comprehensive identification and quantification of RNA binding on a global scale. In its original form, the RNA co-immunoprecipitated with the RBP of interest was quantified using microarray analysis (RIP-Chip) (Tenenbaum et al. 2000). In place of microarrays, more recent variants involve next-generation sequencing analysis of the RNA (RIP-Seq) (Cloonan et al. 2008). RIP-Chip and RIP-Seq have been applied to multiple RBPs from a wide variety of tissues and species (see Table 1 in Morris et al. 2010).

RIP-Chip approaches gave first insights into the dynamic remodeling of RNPs in post-transcriptional gene regulatory processes. Among the important insights gained from RIP-type experiments is the understanding that RBPs typically interact with multiple (m)RNAs, with some RBPs interacting with a sizeable fraction of the transcriptome (Hogan et al. 2008). The analysis of the interacting mRNAs and the finding that they oftentimes encode functionally related proteins led to the hypothesis that RBPs are capable of coordinating so-called RNA regulons, regulatory entities conceptually related to DNA operons found in bacteria (Keene 2007). RIP-Chip assays for RBPs shuttling in and out of cytoplasmic granules under conditions of cellular stress also confirmed the dynamic nature of RNP complexes (Anderson and Kedersha 2006).

A common concern when utilizing RIP is that possible RNP reorganization during lysis and immunoprecipitation leads to a misrepresentation of the RNA target complement (Mili and Steitz 2004). This issue can be "fixed" by immobilizing the RBP onto its target RNAs with formaldehyde crosslinking before RIP (fRIP). Formaldehyde crosslinking is reversible and is compatible with cDNA library construction; however, while it allows for the recovery of more weakly bound transcripts, the 2.3–2.7 Å (Sutherland et al. 2008) crosslinking distance makes it impossible to distinguish between direct and indirect RNA-protein interactions. RNA recovered from indirectly interacting RNPs can increase background signal and further complicate the analysis and requires stringent experimental controls. However, fRIP may prove useful in analyzing RNA-protein interactions of RBPs that are refractory to other crosslinking methods (see below), as shown in a recent study of Staufen1 protein binding and function (Ricci et al. 2014).

While RIP methods are successful for dissecting the RNA content of RNPs, they do not directly identify the RRE within long RNA targets. Furthermore, computational methods are only successful in predicting RREs of high-information content from RIP-data (López de Silanes et al. 2004; Gerber et al. 2006; Zhang et al. 2007; Karginov et al. 2007; Landthaler et al. 2008). Determination of the RRE thus requires genome-wide methods with higher nucleotide resolution.

11.2.1.1.3 Crosslinking and Immunoprecipitation (CLIP)

In order to specifically isolate the RREs from RNPs, Darnell and colleagues (Ule et al. 2003) introduced Crosslinking and Immunoprecipitation (CLIP) by adapting the in vivo UV-crosslinking methods used to characterize hnRNP proteins (Dreyfuss et al. 1984). Irradiation of RNPs in living cells with 254 nm UV light leads to photoaddition of uridines to proximal (<1 Å), reactive amino acid residues of interacting proteins and nucleic acids (Kramer et al. 2014). The irreversible nature of this type of crosslinking allows for stringent purification of an RNP complex. In CLIP, the RNP of interest is immunoprecipitated and mild RNase treatment ensures that only the protected RRE bound by the RBP is recovered (Jensen and Darnell 2008). The immunoprecipitate can be further fractionated by SDS-PAGE and subsequently blotted onto nitrocellulose membranes, which helps remove contaminating, non-crosslinked RNA molecules. The RNP protein component is removed by Proteinase K and the recovered RNA is carried through small RNA cDNA library preparation protocols before sequencing.

Initially, the cDNA was cloned into bacterial vectors, followed by Sanger sequencing (Ule et al. 2003) and these sequences gave insights into the targets and functions of Nova and a handful of other proteins (Darnell 2010). However, the low throughput of traditional sequencing methods limited the comprehensive analysis of RNPs containing multiple RNA targets. Combination of CLIP with next-generation sequencing provided RREs on a genome-wide scale (Licatalosi et al. 2008). For example, for Nova protein, the number of binding sites increased from 340 RNAs with CLIP to 412,686 with HiTS-CLIP (Darnell 2010). However, the large size of the clusters, up to 1 kb, (Darnell et al. 2011), and the difficulty of distinguishing crosslinked sequences from co-purified, non-crosslinked sequences requires stringent controls and complicates the analysis.

CLIP Variants

Complete digestion of the crosslinked RBPs by proteases leaves oligopeptides attached to the RNA at the site of crosslinking, frequently resulting in abortive RT during cDNA library preparation. Thus, fragments of co-purifying, non-crosslinked RNAs will be more efficiently converted into cDNA and result in a sizeable background. Recent, careful computational analysis of HiTS-CLIP datasets indicated an increased presence of mutations within binding sites, which is potentially useful for the identification of precise binding sites. However, the nature of these transitions in HiTS-CLIP libraries is unclear and ranges from deletions and insertions to specific C-to-T mutations (Granneman et al. 2009; Zhang and Darnell 2011; Wang et al. 2012). Several newer CLIP variants try to specifically address these limitations and thereby increase nucleotide-level resolution (Table 11.2).

Individual-Nucleotide Resolution CLIP (iCLIP)

Individual-nucleotide resolution CLIP (iCLIP) achieves nucleotide-level resolution by taking advantage of the abortive RT at the crosslinking site. Instead of the introduction of primer binding sites by ligation of adapter oligonucleotides to the 5′ and 3′ end of the recovered RNA, only a 3′ adapter is ligated to the recovered RNA. The RT primer hybridizing to the 3′ adapter contains forward and reverse primer binding sites for PCR amplification (König et al. 2010; Sugimoto et al. 2012). After RT, the cDNA is circularized and purified. The circular DNA is then linearized and amplified by PCR before Illumina sequencing. Clusters of sequence reads, overlapping after mapping to the genome, are considered binding sites if they contain a sharp 5′ end, indicating the site of abortive RT (König et al. 2010; Huppertz et al. 2014). iCLIP has been performed for several intensely studied RBPs, including TDP-43 (Tollervey et al. 2011), TIA1, TIAL1 (Wang et al. 2010), hnRNP C (Zarnack et al. 2013), and hnRNP L (Rossbach et al. 2014).

Table 11.2 Overview of experimental designs for Hits-CLIP, PAR-CLIP, and iCLIP

Technique	Control library	Recommended starting material	Numbers of replicates	Sequencing depth	Recommended sequencing platforms and run	Reference
HiTS-CLIP	Optional: Cells ± RBP expression or IgG control	2–3 ml pellet	2–3 per condition	5–10 million	Illumina HiSeq 50 cycles single read	Moore et al. (2014)
PAR-CLIP	Optional: Cells ± RBP expression or IgG control	2–3 ml pellet	2–3 per condition	5–10 million	Illumina HiSeq 50 cycles single read	Hafner et al. (2010)
iCLIP	Optional: Cells ± RBP expression or IgG control	10 cm plate (~9×10^6 cells)	2–3 per condition	5–10 million	Illumina HiSeq 50 cycles single read	König et al. (2011)

11.2.1.1.4 Photoactivatable-Ribonucleoside-Enhanced Crosslinking and Immunoprecipitation (PAR-CLIP)

Photoactivatable-Ribonucleoside-Enhanced Crosslinking and Immunoprecipitation (PAR-CLIP) allows precise determination of crosslinking sites on a transcriptome-wide scale (Fig. 11.1). It relies on the incorporation of photoactivatable thioribonucleoside analogs, such as 4-thiouridine (4SU) and 6-thioguanosine (6SG), into nascent RNAs. Thioribonucleoside-labeled RNA is efficiently crosslinked to interacting RBPs using UV of 310–365 nm (UVA and UVB). RBP IP, RNA recovery, and cDNA library construction are performed analogous to other CLIP procedures and the cDNA library is sequenced using next-generation sequencing methods. While PAR-CLIP offers similar or greater efficiency of crosslinking and thus RNA recovery compared to other methods, one unique hallmark of PAR-CLIP is a structural change in the thioribonucleoside induced by the photoaddition reaction, which leads to specific misincorporation of bases in the RT reaction. The reverse transcriptase incorporates a T instead of a C when using 6SG, or a G instead of A in the case of 4SU. This misincorporation allows for specific mapping of the site of interaction by evaluating these characteristic mutations. In addition to revealing the site of crosslinking, the transition also enables the efficient removal of any non-crosslinked, background sequences. PAR-CLIP has been applied to identify the binding sites and specificities of dozens of RNA-binding proteins in various cell lines (reviewed in Ascano et al. 2012a) and in model organisms such as yeast and *C. elegans* (Creamer et al. 2011; Jungkamp et al. 2011).

11.2.3 Practical Considerations for PAR-CLIP

In the following section, we will describe detailed considerations for the setup and analysis of a PAR-CLIP experiment.

11.2.3.1 Scale of the Experiment

The scale of a PAR-CLIP experiment will depend on the expression levels of the RBP of interest. In our experience, for RBPs expressed at high copy numbers of 50,000–100,000 copies per cell, 10–50 million cells are sufficient for PAR-CLIP. It may be necessary to adjust the cell and tissue lysis conditions according to the RBP examined. Most cytoplasmic and nucleocytoplasmic shuttling RBPs are amenable to NP-40 lysis; however, chromatin-associated RBPs may require more specialized cell lysis conditions to generate the extract for immunoprecipitation

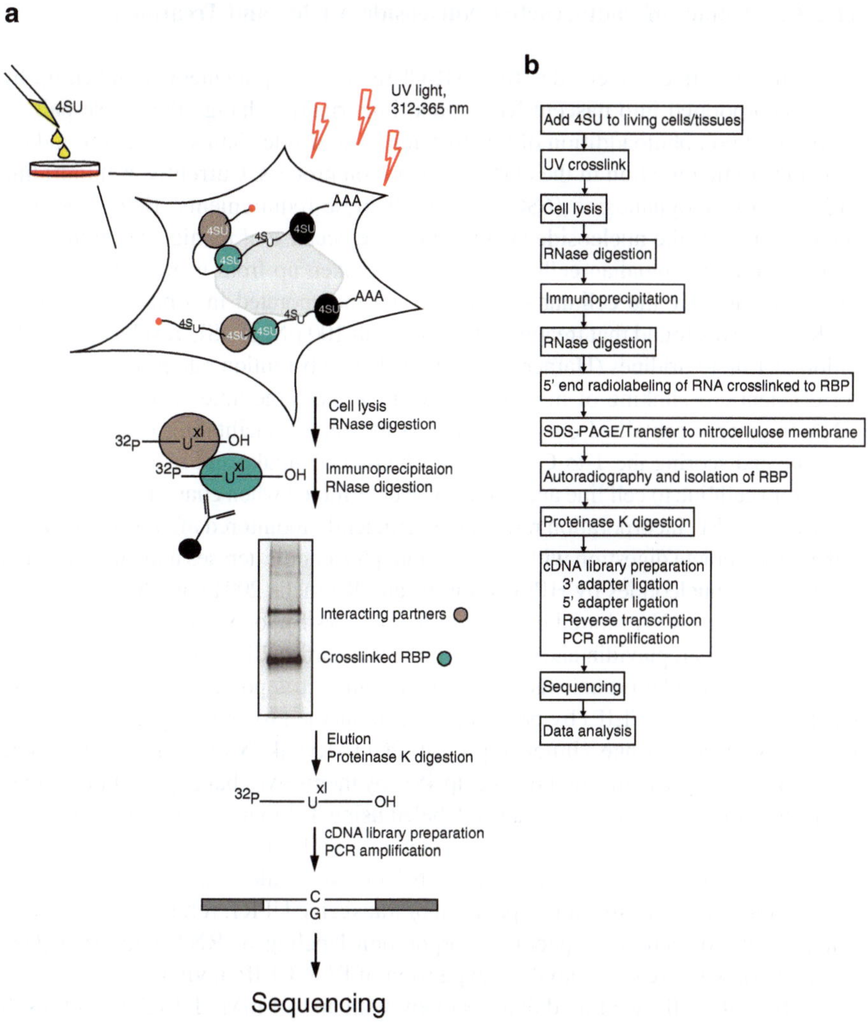

Fig. 11.1 Schematic depiction of PAR-CLIP workflow. (**a**) A photoreactive nucleoside analog, 4-thiouridine (4SU) or 6-thioguanosine (6SG, not shown), is directly added to the growth medium of cultured cells. Cells are irradiated with 312 nm (UVB) or 365 nm (UVA) UV light followed by lysis and RNase treatment to isolate RNP complexes. The RBP of interest (*green oval*) is immunoprecipitated, and the crosslinked and protected RNA further trimmed using RNases, and finally the RNP is further purified by SDS-PAGE fractionation. Other, directly interacting RBPs (*brown oval*) will also be visualized on an autoradiograph of the SDS-PAGE. The crosslinked RNA is recovered and carried through a small RNA cDNA library preparation protocol and deep sequenced. Photocrosslinking induces a structural change on the thioribonucleoside, which causes a characteristic mutation in the corresponding position in the cDNA library (T-to-C for 4SU, or G-to-A when using 6SG) and can be used to filter RBP binding sites from the background of fragments from abundant cellular RNAs. 4SU denotes 4-thiouridine, U^{xl} denotes crosslinked uridines, and ^{32}P denotes P-32 labeled 5′ phosphate. (**b**) PAR-CLIP workflow

11.2.3.2 Choice of Photoreactive Nucleoside Analog and Treatment

Ideal photoreactive nucleosides for PAR-CLIP will be spontaneously taken up by cells, incorporated into nascent RNAs, and importantly, change their base-pairing properties upon photoaddition of reactive amino acid side chains in order to induce a characteristic mutation in the cDNA preparation process. Currently, 4-thiouridine (4SU) and 6-thioguanosine (6SG) satisfy all these requirements. Below, we will focus on 4SU as the nucleoside analog of choice because of its high reactivity.

In cultured mammalian cells, 4SU is readily taken up from growth medium, triphosphorylated by the cellular machinery, and incorporated into nascent RNA. In HEK293, it was found that treatment for 16 h with 100 μM of 4SU results in a substitution of 1 in 40 uridines (Hafner et al. 2010). This substitution rate proved sufficient for efficient crosslinking of most RBPs, while at the same time ensuring that most RNA fragments contained only a single U substitution, facilitating mapping to the genome and scoring the T-to-C mutation. The rate of uptake and incorporation may vary from cell line to cell line and needs to be determined when characterizing a novel cell system. 4SU incorporation rates can be efficiently monitored after RNA recovery either by complete digestion with snake venom phosphodiesterases and quantification of individual nucleotides by HPLC (Andrus and Kuimelis 2001) or, alternatively, by derivatization of RNA blotted onto nylon membranes by iodoacetamido-biotin reaction with HRP-streptavidin and comparison to standards (Rädle et al. 2013).

For simple model organisms such as *C. elegans*, it is possible to perform PAR-CLIP in vivo (iPAR-CLIP) by providing 4SU to larvae grown in liquid cultures and harvesting at adult stage (Jungkamp et al. 2011; Rybak-Wolf et al. 2014). Some organisms, e.g., yeast, that do not take up 4SU by themselves but express uracil phosphoribosyltransferases (UPRT) can be labeled using 4-thiouracil, which is converted by UPRT into 4SU. Heterologous expression of UPRT has been used in *Drosophila melanogaster* to label newly synthesized RNA in vivo (Miller et al. 2009). Expanding on this concept, tissue-specific expression of transgenic UPRT, referred to as TU tagging, allows for cell type specific incorporation labeling of RNA with 4SU (Gay et al. 2014), with great potential for expansion of PAR-CLIP in vivo.

In HEK293 cells, we found that treatment with up to 1 mM of 4SU did not result in a noticeable change in the mRNA profile, indicating low toxicity at working concentration. However, a recent study reported effects of prolonged 4SU treatment at high concentrations on rRNA processing, demonstrating the need to monitor for possible toxicity of the employed photoreactive analog in the cell system of choice (Burger et al. 2013).

11.2.3.3 Crosslinking and Immunoprecipitation

Upon irradiation with UVA (365 nm) and UVB light (312 nm), 4SU labeled RNA forms photoadducts with reactive amino acid side chains of interacting RNA-binding proteins, as well as RNA or DNA. For cells grown in monolayers, such as HEK293, an energy dose of 0.15–0.5 J/cm^2 was found to be sufficient for efficient crosslinking of the majority of examined RBPs. For cells grown in suspension and for model organisms with varying opacity, it may be necessary to determine the optimal energy dose, e.g., 2 J/cm^2 was used for iPAR-CLIP in *C. elegans* (Jungkamp et al. 2011).

A key step in the PAR-CLIP protocol is the immunoprecipitation, which needs to be both comprehensive and as specific as possible. Quality control of the antibody used is necessary prior to performing PAR-CLIP experiments. Such assays can include probing lysate from RBP-knockout or depleted cells to document specificity. In addition, lysates from cells expressing epitope-tagged transgenic versions of the RBP can be used to monitor IP efficiency. If no suitable antibody for the RBP is available, we routinely generate stable cell lines inducibly expressing FLAG/HA-tagged versions of the RBP of interest. Note that inappropriate placing of such epitope tags may interfere with RBP function, e.g., changing the C-terminus of Argonaute proteins abolishes their capacity of binding small RNAs. In most cases, antibodies can be immobilized on Protein G coated magnetic beads, allowing convenient buffer exchanges and other experimental manipulations. Prior blocking of the matrix with BSA or heparin can further minimize background introduced by unspecific binding of RNA and proteins. For optimal capture of the studied RNP the final amount of antibody and matrix used in the PAR-CLIP experiments should be adjusted based on RBP expression levels as well as the affinity of antibody.

11.2.3.4 RNase Digestion

The main motivation to use PAR-CLIP approaches to study an RBP is to gain insights into its binding sites (or RREs) on target RNAs at nucleotide resolution. Thus, cross-linked and co-immunoprecipitated RNA needs to be trimmed to reveal the footprint of the RBP using RNases. We suggest titrating the amount of RNases used to ensure that the length of the recovered RNA distributes between 20 and 40 nt. At these lengths, >90 % of nonrepetitive sequences uniquely map to the human genome, and thus allow unambiguous determination of target sites. Furthermore, limiting the length of resulting clusters of overlapping sequence reads helps pinpointing RREs and eliminates confounding contributions of other proteins binding in close proximity. Finally, small RNA cDNA libraries can be cost-efficiently sequenced in a standard Illumina sequencing run of 50 cycles. We routinely use RNase T1 for its high activity and specificity of cutting after guanosines, which affords an additional layer of quality control for sequenced reads by requiring their genomic mapping directly after G. Nevertheless, multiple other RNases have been used for PAR-CLIP, including micrococcal nuclease and RNase I (Kishore et al. 2011; Munschauer et al. 2014).

11.2.3.5 Labeling of RNA Molecules and Denaturing SDS-PAGE

Trimming of RNA with RNases generally leaves the RNA with 5′ hydroxyl and 2′3′ cyclic phosphate or 2′ or 3′ phosphate termini, which are not compatible with the small RNA cDNA library preparation procedure (Hafner et al. 2012). Thus, the RNA needs to be treated with phosphatases to remove these termini and then labeled with γ-^{32}P-ATP to facilitate downstream detection of the RNP. After radiolabeling the immunoprecipitated RNP complex is fractionated by denaturing SDS polyacrylamide electrophoresis to allow isolation of RNA specifically interacting with the RBP of interest and to remove contributions of other possibly interacting RBPs. Transferring the fractionated RNP onto nitrocellulose helps further remove non-crosslinked RNA molecules. RNPs are visualized by autoradiography and the relative intensity of the radioactive bands provides a measure of the occupancy of the RBP in vivo. Finally, the crosslinked RNA is recovered by excision of the radioactive band corresponding to the RBP of interest and removal of the RBP with Proteinase K. Note the importance of using maximally active proteinases to minimize the length of oligopeptides remaining covalently bound at the site of crosslinking, which may otherwise interfere with the cDNA preparation.

11.2.3.6 cDNA Library Preparation for Sequencing

Recovered RNA with 5′ phosphate and 3′ hydroxyl termini is carried through a small RNA cDNA library preparation protocol (Hafner et al. 2012). The first step typically involves ligation of an oligonucleotide adapter to the 3′ end of the sample RNA using T4 RNA ligases to allow RT priming and subsequent PCR. To avoid undesired side reactions, such as circularization and concatamerization of the 5′ phosphorylated RNAs, we recommend using truncated and mutated T4 RNA ligase 2 (Rnl2(1–249)K227Q) and preadenylated 3′ adapters with chemically blocked 3′ ends for the reaction (Lau 2001). The next step consists of joining the 3′-OH of the 5′-adapter oligonucleotide to the 5′ end of the 3′-adapter ligation product. Side reactions are of no concern because the 3′ end of the 3′-adapter ligation product is chemically modified, and the 5′ adapter does not have a reactive 5′-phosphate. After each adapter ligation step the reaction products containing the desired RNA of 20 to ~40 nt are size-selected by denaturing urea PAGE to minimize co-purification of adapter-adapter ligation products formed by the vast excess of adapters over input RNA. To maximize RT across crosslinking sites, thermostable RT enzymes, such as the SuperScript family, are preferably used. Finally, a PCR reaction is required to amplify the cDNA as input for Illumina sequencing.

11.2.3.7 Computational Analysis

Current depths of Illumina sequencing reach >200 million sequence reads per sample and data analysis requires sophisticated approaches to identify binding sites. A number of biocomputational pipelines as well as databases for the analysis of PAR-CLIP datasets have been made available (Table 11.3).

The exogenous adapter sequence is trimmed off before aligning sequenced reads to the genome, allowing for at least one error (substitution, insertion, or deletion) to

Table 11.3 Available software for CLIP-based data analysis[a]

MicroMummie	A model for predicting miRNA binding sites using PAR-CLIP data	Majoros et al. (2013)
PARma	Software for analyzing PAR-CLIP targets	Erhard et al. (2013)
PARalyzer	Identifies high-confidence interaction sites from PAR-CLIP data based on a kernel-density estimate from T-to-C conversion frequency and sequence read density	Corcoran et al. (2011)
doRiNA	Repository of miRNA and RBP target sites	Anders et al. (2012)
wavClusteR	Defines clusters at high resolution based on binding site (clusters) identification algorithm	Sievers et al. (2012)
CLIPZ	Defines binding sites from CLIP-based methods at the genomic and individual transcript levels	Khorshid et al. (2011)
PIPE-CLIP	Galaxy-based tool for CLIP, HiTS-CLIP, PAR-CLIP, and iCLIP data analysis	Chen et al. (2014)
starBase	Repository of published CLIP data	Yang et al. (2011)
CLIPdb	Repository of published CLIP data	Yang et al. (2015)
Piranha	Algorithm identifying binding sites from CLIP-based methods	Uren et al. (2012)
dCLIP	Database including quantitative comparative analysis of published CLIP-seq	Wang et al. (2014)
miRTarCLIP	Tool defining miRNA target sites from RBP CLIP data	Chou et al. (2013)

[a]Results are aligned sequence reads as SAM/BAM files and target clusters as csv files

capture reads with crosslinking-induced mutations. Overlapping sequence reads are grouped, taking into account the frequency of crosslinking-induced mutations. To allow insights into the RBP's binding preferences, these groups of overlapping sequence reads can be then mapped against the transcriptome to annotate and categorize them as derived from exonic regions of mRNA (5′ untranslated region (UTR), coding sequence (CDS), 3′UTR), introns, rRNA, long noncoding RNAs, tRNAs, etc. Note that the presence of a single site within a cluster containing T-to-C (or G-to-A when using 6SG as photoreactive nucleoside) can occasionally be misinterpreted as a crosslinking event, if it is derived from sequence polymorphisms or sequencing errors. Recently, a repository of sequences contaminating CLIP-based experiments in human cell lines has been created and can be used to further refine the analysis (Friedersdorf and Keene 2014). The frequency of the T-to-C (or G-to-A) mutations allows ranking of the groups to predict those RBP-RNA interactions with the highest functional impact. In addition, the top-ranked groups provide a useful set of sequences as input into motif-finding programs to determine the underlying RRE. Some of the algorithms listed in Table 11.3, such as PARalyzer, take advantage of the frequency and distribution of crosslinking-induced mutations to predict the shortest possible region of interaction between RBP and RNA that harbors the RRE. Several programs initially developed for the analysis of transcription-factor binding sites on DNA are available to calculate the common sequence motifs of the

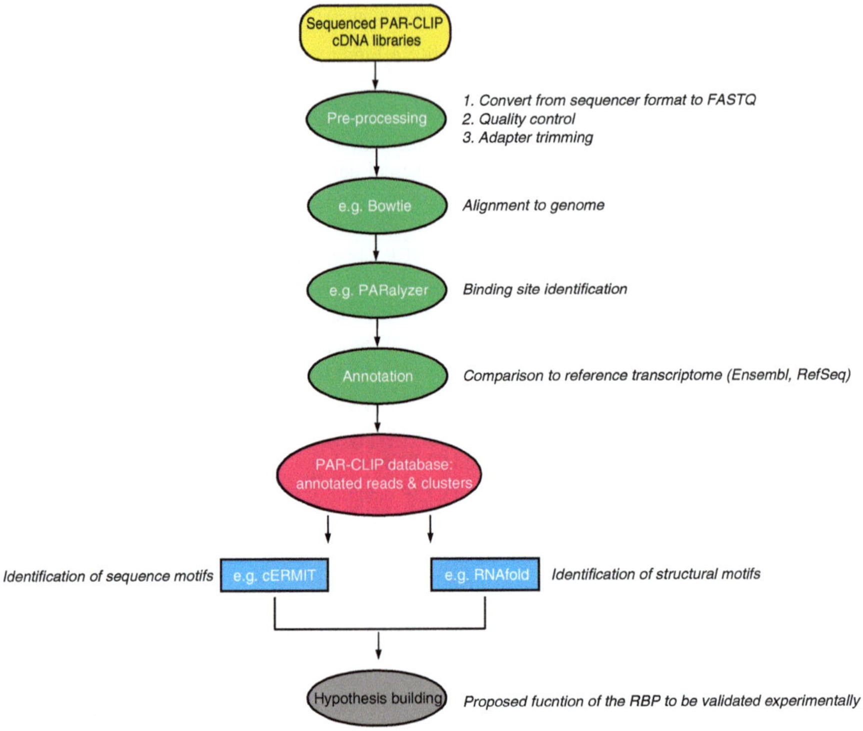

Fig. 11.2 PAR-CLIP computational analysis workflow. The programs listed are examples of software that could be used for these analysis steps. Other programs exist and each algorithm will yield slightly different results. The RBP functions hypothesized based on these analyses must be validated by follow-up experiments

RRE, including MEME (Bailey 2002), MDScan (Liu et al. 2002), cERMIT (Georgiev et al. 2010), and Gimsan (Ng and Keich 2008). For the identification of structured RREs (e.g., hairpins), secondary structure prediction algorithms such as RNAfold (Hofacker and Stadler 2006) or Mfold (Zuker 2003) may be useful and should be coupled to analysis of evolutionary conservation of binding sites to identify signatures of base-pair covariation (Eddy and Durbin 1994) (Fig. 11.2).

11.2.3.8 Follow-Up Experiments

PAR-CLIP substantially furthers the understanding of the in vivo binding preferences and specificity of an RBP. Follow-up experiments are necessary to couple information from the thousands to tens of thousands of binding sites to the regulatory function of an RBP.

The in vitro methods described in the first section of this chapter can be used in order to experimentally validate the RRE predicted from the identified binding sites.

These methods will yield dissociation constants for synthetic sequences representing the RREs alone, the top-ranked binding sites, or RRE/binding sites with introduced mutations. Thermodynamic constants for various sequence elements can then be related to their enrichment in RIP-type experiments (Ascano et al. 2012b; Mukherjee et al. 2014).

Many RNA-binding proteins regulate RNA stability, turnover, and splicing. In these instances, the effect of an RBP on its targets can be conveniently studied by perturbing the RBP of interest and quantifying its RNA targets, e.g., on mRNA microarrays or by RNAseq. In cultured cells, RBP levels can be easily manipulated by overexpression from plasmids of the normal RBP and/or a mutant form lacking RNA binding ability (Teplova et al. 2013), by knockdown using siRNA and shRNAs (Hafner et al. 2010), or by knockout using the emerging CRISPR-Cas system (Ran et al. 2013; Doudna and Charpentier 2014). Ideally, these results would be related to RNA quantification from model organisms with and without the RBP knocked-out in the relevant cells or tissues to verify possible functionality in vivo.

Recently developed high-throughput proteomics methods based on isotopic labeling, such as SILAC and iTRAQ, allow for the analysis of the regulatory impact of an RBP on target gene product levels (Lebedeva et al. 2011; Hafner et al. 2013; Graf et al. 2013). To monitor subtle, cumulative effects on translation efficiency, next-generation sequencing-based ribosome profiling may prove to be a useful alternative (Ingolia et al. 2009; Guo et al. 2010).

More customized follow-up experiments may be necessary to understand the influence of RBPs on other post-transcriptional processes, such as RNA transport or localization. For example, to dissect the regulatory roles of the MBNL1 protein, RNAseq from multiple cellular compartments was performed (Wang et al. 2012). In addition, individual transcripts may be tracked using FISH methods or using reporter systems.

11.3 Conclusion

The role of PTGR in basic cellular function and human disease has become increasingly appreciated over time. Each new discovery in this field has been supported by novel methods to test if, how, and where protein-RNA interactions occur and to relate these interactions to a cellular and organismal function. CLIP-based methods, including PAR-CLIP, dissect RNA-protein interaction sites on a genome-wide scale with nucleotide resolution and have already vastly increased our knowledge of the extent of PTGR networks. Rapid advances in single-cell genomics and single-molecule imaging technologies will provide further granularity in the dissection of PTGR networks. In conjunction with the emerging data of genomic and transcriptomic variation between individuals (1000 Genomes Project Consortium 2010), these novel approaches will provide insights into the role of PTGR in development and disease.

Annex: Quick Reference Guide

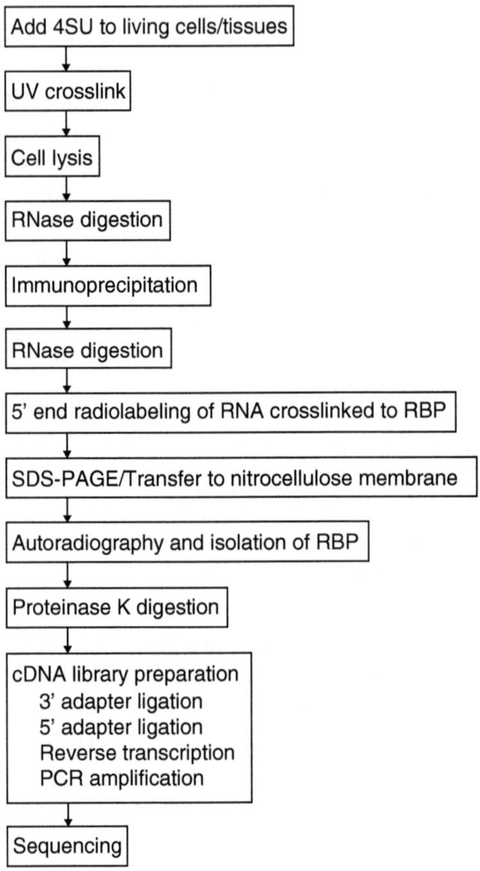

Fig. QG11.1 Representation of the wet-lab procedure workflow

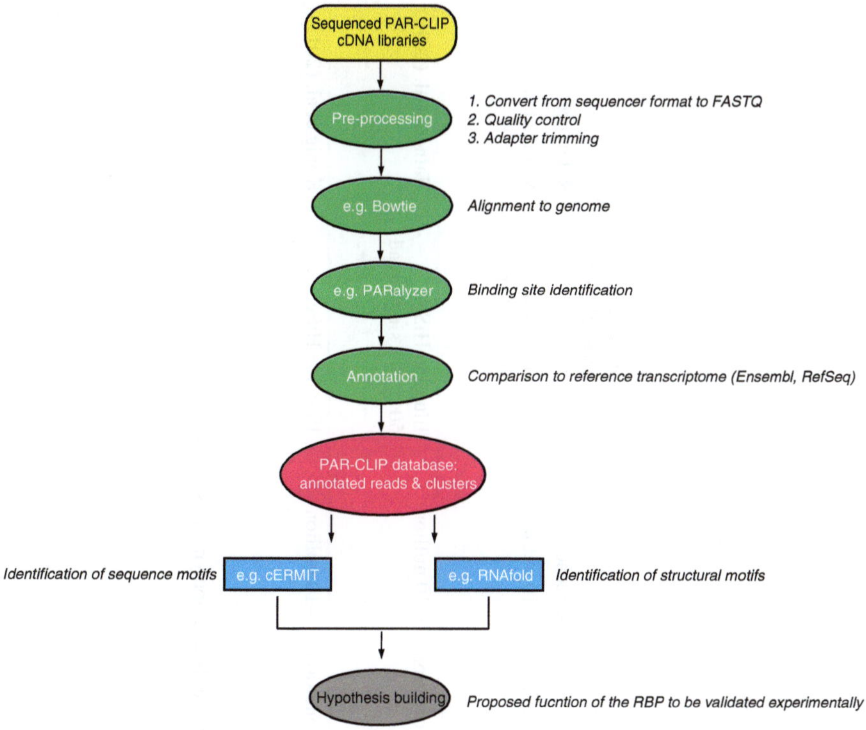

Fig. QG11.2 Main steps of the computational analysis pipeline

Table QG11.1 Experimental design considerations

Technique	Control library	Recommended starting material	Numbers of replicates	Sequencing depth	Recommended sequencing platforms and run	Reference
HiTS-CLIP	Optional: Cells ± RBP expression or IgG control	2–3 ml pellet	2–3 per condition	5–10 million	Illumina HiSeq 50 cycles single read	Moore et al. (2014)
PAR-CLIP	Optional: Cells ± RBP expression or IgG control	2–3 ml pellet	2–3 per condition	5–10 million	Illumina HiSeq 50 cycles single read	Hafner et al. (2010)
iCLIP	Optional: Cells ± RBP expression or IgG control	10 cm plate (~9×10^6 cells)	2–3 per condition	5–10 million	Illumina HiSeq 50 cycles single read	König et al. (2011)

Table that comprises relevant experimental design parameters, to carefully consider before applying this methodology

Table QG11.2 Available software recommendations

MicroMummie	A model for predicting miRNA binding sites using PAR-CLIP data	Majoros et al. (2013)
PARma	Software for analyzing PAR-CLIP targets	Erhard et al. (2013)
PARalyzer	Identifies high-confidence interaction sites from PAR-CLIP data based on a kernel-density estimate from T-to-C conversion frequency and sequence read density	Corcoran et al. (2011)
doRiNA	Repository of miRNA and RBP target sites	Anders et al. (2012)
wavClusteR	Defines clusters at high resolution based on binding site (clusters) identification algorithm	Sievers et al. (2012)
CLIPZ	Defines binding sites from CLIP-based methods at the genomic and individual transcript levels	Khorshid et al. (2011)
PIPE-CLIP	Galaxy-based tool for CLIP, HiTS-CLIP, PAR-CLIP, and iCLIP data analysis	Chen et al. (2014)
starBase	Repository of published CLIP data	Yang et al. (2011)
CLIPdb	Repository of published CLIP data	Yang et al. (2015)
Piranha	Algorithm identifying binding sites from CLIP-based methods	Uren et al. (2012)
dCLIP	Database including quantitative comparative analysis of published CLIP-seq	Wang et al. (2014)
miRTarCLIP	Tool defining miRNA target sites from RBP CLIP data	Chou et al. (2013)

Table displaying a selection of the recommended software available for the computational analysis of data yielded by this technique

References

1000 Genomes Project Consortium, Abecasis GR, Altshuler D et al (2010) A map of human genome variation from population-scale sequencing. Nature 467:1061–1073. doi:10.1038/nature09534

Anders G, Mackowiak SD, Jens M et al (2012) doRiNA: a database of RNA interactions in post-transcriptional regulation. Nucleic Acids Res 40:D180–D186. doi:10.1093/nar/gkr1007

Anderson P, Kedersha N (2006) RNA granules. J Cell Biol 172:803–808. doi:10.1083/jcb.200512082

Andrus A, Kuimelis RG (2001) Base composition analysis of nucleosides using HPLC. Curr Protoc Nucleic Acid Chem Chapter 10:Unit 10.6–10.6.6. doi:10.1002/0471142700.nc1006s01

Ascano M, Hafner M, Cekan P et al (2012a) Identification of RNA-protein interaction networks using PAR-CLIP. Wiley Interdiscip Rev RNA 3:159–177. doi:10.1002/wrna.1103

Ascano M, Mukherjee N, Bandaru P et al (2012b) FMRP targets distinct mRNA sequence elements to regulate protein expression. Nature 492:382–386. doi:10.1038/nature11737

Bailey TL (2002) Discovering novel sequence motifs with MEME. Curr Protoc Bioinformatics Chapter 2:Unit 2.4–2.4.35. doi:10.1002/0471250953.bi0204s00

Baltz AG, Munschauer M, Schwanhäusser B et al (2012) The mRNA-bound proteome and its global occupancy profile on protein-coding transcripts. Mol Cell 46:674–690. doi:10.1016/j.molcel.2012.05.021

Bandziulis RJ, Swanson MS, Dreyfuss G (1989) RNA-binding proteins as developmental regulators. Genes Dev 3:431–437. doi:10.1101/gad.3.4.431

Bartel DP, Szostak JW (1993) Isolation of new ribozymes from a large pool of random sequences [see comment]. Science 261:1411–1418

Buenrostro JD, Araya CL, Chircus LM et al (2014) Quantitative analysis of RNA-protein interactions on a massively parallel array reveals biophysical and evolutionary landscapes. Nat Biotechnol 32:562–568. doi:10.1038/nbt.2880

Burger K, Mühl B, Kellner M et al (2013) 4-thiouridine inhibits rRNA synthesis and causes a nucleolar stress response. RNA Biol 10:1623–1630. doi:10.4161/rna.26214

Campbell ZT, Bhimsaria D, Valley CT et al (2012) Cooperativity in RNA-protein interactions: global analysis of RNA binding specificity. Cell Rep 1:570–581. doi:10.1016/j.celrep.2012.04.003

Castello A, Fischer B, Eichelbaum K et al (2012) Insights into RNA biology from an atlas of mammalian mRNA-binding proteins. Cell 149:1393–1406. doi:10.1016/j.cell.2012.04.031

Chen B, Yun J, Kim MS et al (2014) PIPE-CLIP: a comprehensive online tool for CLIP-seq data analysis. Genome Biol 15:R18. doi:10.1186/gb-2014-15-1-r18

Chen L, Yun S-W, Seto J et al (2003) The fragile X mental retardation protein binds and regulates a novel class of mRNAs containing U rich target sequences. Neuroscience 120:1005–1017. doi:10.1016/S0306-4522(03)00406-8

Chou C-H, Lin F-M, Chou M-T et al (2013) A computational approach for identifying microRNA-target interactions using high-throughput CLIP and PAR-CLIP sequencing. BMC Genomics 14(Suppl 1):S2. doi:10.1186/1471-2164-14-S1-S2

Cloonan N, Forrest ARR, Kolle G et al (2008) Stem cell transcriptome profiling via massive-scale mRNA sequencing. Nat Methods 5:613–619. doi:10.1038/nmeth.1223

Corcoran DL, Georgiev S, Mukherjee N et al (2011) PARalyzer: definition of RNA binding sites from PAR-CLIP short-read sequence data. Genome Biol 12:R79. doi:10.1186/gb-2011-12-8-r79

Creamer TJ, Darby MM, Jamonnak N et al (2011) Transcriptome-wide binding sites for components of the Saccharomyces cerevisiae non-poly(A) termination pathway: Nrd1, Nab3, and Sen1. PLoS Genet 7, e1002329. doi:10.1371/journal.pgen.1002329

Darnell JC, Van Driesche SJ, Zhang C et al (2011) FMRP stalls ribosomal translocation on mRNAs linked to synaptic function and autism. Cell 146:247–261. doi:10.1016/j.cell.2011.06.013

Darnell RB (2010) HITS-CLIP: panoramic views of protein-RNA regulation in living cells. Wiley Interdiscip Rev RNA 1:266–286. doi:10.1002/wrna.31

Doudna JA, Charpentier E (2014) Genome editing. The new frontier of genome engineering with CRISPR-Cas9. Science 346:1258096. doi:10.1126/science.1258096

Dreyfuss G, Choi YD, Adam SA (1984) Characterization of heterogeneous nuclear RNA-protein complexes in vivo with monoclonal antibodies. Mol Cell Biol 4:1104–1114

Dreyfuss G, Swanson MS, Piñol-Roma S (1988) Heterogeneous nuclear ribonucleoprotein particles and the pathway of mRNA formation. Trends Biochem Sci 13:86–91

Eddy SR, Durbin R (1994) RNA sequence analysis using covariance models. Nucleic Acids Res 22:2079–2088

Erhard F, Dölken L, Jaskiewicz L, Zimmer R (2013) PARma: identification of microRNA target sites in AGO-PAR-CLIP data. Genome Biol 14:R79. doi:10.1186/gb-2013-14-7-r79

Friedersdorf MB, Keene JD (2014) Advancing the functional utility of PAR-CLIP by quantifying background binding to mRNAs and lncRNAs. Genome Biol 15:R2. doi:10.1186/gb-2014-15-1-r2

Gagnon KT, Maxwell ES (2010) Electrophoretic mobility shift assay for characterizing RNA–protein interaction. In: Nielsen H (ed) RNA. Humana, Totowa, NJ, pp 275–291

Galarneau A, Richard S (2005) Target RNA motif and target mRNAs of the Quaking STAR protein. Nat Struct Mol Biol 12:691–698. doi:10.1038/nsmb963

Ganguly S, Ghosh S, Chattopadhyay D, Das P (2004) Antisense molecular beacon strategy for in situ visualization of snRNA and fibrillarin protein interaction in Giardia lamblia. RNA Biol 1:48–53. doi:10.4161/rna.1.1.928

Gay L, Karfilis KV, Miller MR et al (2014) Applying thiouracil tagging to mouse transcriptome analysis. Nat Protoc 9:410–420. doi:10.1038/nprot.2014.023

Geiger JA, Neugebauer KM (2005) Fluorescent detection of nascent transcripts and RNA-binding proteins in cell nuclei. In: Bindereif A, Schön A, Westhof E, Hartmann RK (eds) Handbook of RNA biochemistry. Wiley-VCH Verlag GmbH, Weinheim, Germany, pp 729–736

Georgiev S, Boyle AP, Jayasurya K et al (2010) Evidence-ranked motif identification. Genome Biol 11:R19. doi:10.1186/gb-2010-11-2-r19

Gerber AP, Luschnig S, Krasnow MA et al (2006) Genome-wide identification of mRNAs associated with the translational regulator PUMILIO in Drosophila melanogaster. Proc Natl Acad Sci U S A 103:4487–4492. doi:10.1073/pnas.0509260103

Gerstberger S, Hafner M, Tuschl T (2014) A census of human RNA-binding proteins. Nat Rev Genet 15:829–845. doi:10.1038/nrg3813

Graf R, Munschauer M, Mastrobuoni G et al (2013) Identification of LIN28B-bound mRNAs reveals features of target recognition and regulation. RNA Biol 10:1146–1159. doi:10.4161/rna.25194

Granneman S, Kudla G, Petfalski E, Tollervey D (2009) Identification of protein binding sites on U3 snoRNA and pre-rRNA by UV cross-linking and high-throughput analysis of cDNAs. Proc Natl Acad Sci U S A 106:9613–9618. doi:10.1073/pnas.0901997106

Guo H, Ingolia NT, Weissman JS, Bartel DP (2010) Mammalian microRNAs predominantly act to decrease target mRNA levels. Nature 466:835–840. doi:10.1038/nature09267

Günzl A, Bindereif A (1999) Oligonucleotide-targeted RNase H protection analysis of RNA-protein complexes. In: Haynes SR (ed) RNA-protein interaction protocols. Humana, New Jersey, pp 93–103

Hafner M, Landthaler M, Burger L et al (2010) Transcriptome-wide identification of RNA-binding protein and microRNA target sites by PAR-CLIP. Cell 141:129–141. doi:10.1016/j.cell.2010.03.009

Hafner M, Max KEA, Bandaru P et al (2013) Identification of mRNAs bound and regulated by human LIN28 proteins and molecular requirements for RNA recognition. RNA 19:613–626. doi:10.1261/rna.036491.112

Hafner M, Renwick N, Farazi TA et al (2012) Barcoded cDNA library preparation for small RNA profiling by next-generation sequencing. Methods 58:164–170. doi:10.1016/j.ymeth.2012.07.030

Hofacker IL, Stadler PF (2006) Memory efficient folding algorithms for circular RNA secondary structures. Bioinformatics 22:1172–1176. doi:10.1093/bioinformatics/btl023

Hogan DJ, Riordan DP, Gerber AP et al (2008) Diverse RNA-binding proteins interact with functionally related sets of RNAs, suggesting an extensive regulatory system. PLoS Biol 6:e255. doi:10.1371/journal.pbio.0060255

Huppertz I, Attig J, D'Ambrogio A et al (2014) iCLIP: protein-RNA interactions at nucleotide resolution. Methods 65:274–287. doi:10.1016/j.ymeth.2013.10.011

Huranová M, Jablonski JA, Benda A et al (2009) In vivo detection of RNA-binding protein interactions with cognate RNA sequences by fluorescence resonance energy transfer. RNA 15:2063–2071. doi:10.1261/rna.1678209

Ilagan J, Yuh P, Chalkley RJ et al (2009) The role of exon sequences in C complex spliceosome structure. J Mol Biol 394:363–375. doi:10.1016/j.jmb.2009.09.019

Ingolia NT, Ghaemmaghami S, Newman JRS, Weissman JS (2009) Genome-wide analysis in vivo of translation with nucleotide resolution using ribosome profiling. Science 324:218–223. doi:10.1126/science.1168978

Jensen KB, Darnell RB (2008) CLIP: crosslinking and immunoprecipitation of in vivo RNA targets of RNA-binding proteins. Methods Mol Biol 488:85–98. doi:10.1007/978-1-60327-475-3_6

Joyce GF (1994) In vitro evolution of nucleic acids. Curr Opin Struct Biol 4:331–336

Jungkamp A-C, Stoeckius M, Mecenas D et al (2011) In vivo and transcriptome-wide identification of RNA binding protein target sites. Mol Cell 44:828–840. doi:10.1016/j.molcel.2011.11.009

Karginov FV, Conaco C, Xuan Z et al (2007) A biochemical approach to identifying microRNA targets. Proc Natl Acad Sci U S A 104:19291–19296. doi:10.1073/pnas.0709971104

Katsamba PS, Park S, Laird-Offringa IA (2002) Kinetic studies of RNA-protein interactions using surface plasmon resonance. Methods 26:95–104. doi:10.1016/S1046-2023(02)00012-9

Keene JD (2007) RNA regulons: coordination of post-transcriptional events. Nat Rev Genet 8:533–543. doi:10.1038/nrg2111

Keene JD, Komisarow JM, Friedersdorf MB (2006) RIP-Chip: the isolation and identification of mRNAs, microRNAs and protein components of ribonucleoprotein complexes from cell extracts. Nat Protoc 1:302–307. doi:10.1038/nprot.2006.47

Khorshid M, Rodak C, Zavolan M (2011) CLIPZ: a database and analysis environment for experimentally determined binding sites of RNA-binding proteins. Nucleic Acids Res 39:D245–D252. doi:10.1093/nar/gkq940

Kishore S, Jaskiewicz L, Burger L et al (2011) A quantitative analysis of CLIP methods for identifying binding sites of RNA-binding proteins. Nat Methods 8:559–564. doi:10.1038/nmeth.1608

König J, Zarnack K, Rot G et al (2010) iCLIP reveals the function of hnRNP particles in splicing at individual nucleotide resolution. Nat Struct Mol Biol 17:909–915. doi:10.1038/nsmb.1838

König J, Zarnack K, Rot G, et al (2011) iCLIP—transcriptome-wide mapping of protein-RNA interactions with individual nucleotide resolution. J Vis Exp:2638. doi:10.3791/2638

Kramer K, Sachsenberg T, Beckmann BM et al (2014) Photo-cross-linking and high-resolution mass spectrometry for assignment of RNA-binding sites in RNA-binding proteins. Nat Methods 11:1064–1070. doi:10.1038/nmeth.3092

Kudla G, Granneman S, Hahn D et al (2011) Cross-linking, ligation, and sequencing of hybrids reveals RNA-RNA interactions in yeast. Proc Natl Acad Sci U S A 108:10010–10015. doi:10.1073/pnas.1017386108

Lambert N, Robertson A, Jangi M et al (2014) RNA Bind-n-Seq: quantitative assessment of the sequence and structural binding specificity of RNA binding proteins. Mol Cell 54:887–900. doi:10.1016/j.molcel.2014.04.016

Landthaler M, Gaidatzis D, Rothballer A et al (2008) Molecular characterization of human Argonaute-containing ribonucleoprotein complexes and their bound target mRNAs. RNA 14:2580–2596. doi:10.1261/rna.1351608

Lau NC (2001) An abundant class of tiny RNAs with probable regulatory roles in Caenorhabditis elegans. Science 294:858–862. doi:10.1126/science.1065062

Lebedeva S, Jens M, Theil K et al (2011) Transcriptome-wide analysis of regulatory interactions of the RNA-binding protein HuR. Mol Cell 43:340–352. doi:10.1016/j.molcel.2011.06.008

Licatalosi DD, Mele A, Fak JJ et al (2008) HITS-CLIP yields genome-wide insights into brain alternative RNA processing. Nature 456:464–469. doi:10.1038/nature07488

Liu XS, Brutlag DL, Liu JS (2002) An algorithm for finding protein-DNA binding sites with applications to chromatin-immunoprecipitation microarray experiments. Nat Biotechnol 20:835–839. doi:10.1038/nbt717

López de Silanes I, Zhan M, Lal A et al (2004) Identification of a target RNA motif for RNA-binding protein HuR. Proc Natl Acad Sci U S A 101:2987–2992. doi:10.1073/pnas.0306453101

Majoros WH, Lekprasert P, Mukherjee N et al (2013) MicroRNA target site identification by integrating sequence and binding information. Nat Methods 10:630–633. doi:10.1038/nmeth.2489

Manley JL (2013) SELEX to identify protein-binding sites on RNA. Cold Spring Harb Protoc 2013(2):156–163. doi:10.1101/pdb.prot072934

Mansfield KD, Keene JD (2009) The ribonome: a dominant force in co-ordinating gene expression. Biol Cell 101:169–181. doi:10.1042/BC20080055

Martin L, Meier M, Lyons SM et al (2012) Systematic reconstruction of RNA functional motifs with high-throughput microfluidics. Nat Methods 9:1192–1194. doi:10.1038/nmeth.2225

Mattaj IW (1993) RNA recognition: a family matter? Cell 73:837–840. doi:10.1016/0092-8674(93)90265-r

Mili S, Steitz JA (2004) Evidence for reassociation of RNA-binding proteins after cell lysis: implications for the interpretation of immunoprecipitation analyses. RNA 10:1692–1694. doi:10.1261/rna.7151404

Miller MR, Robinson KJ, Cleary MD, Doe CQ (2009) TU-tagging: cell type-specific RNA isolation from intact complex tissues. Nat Methods 6:439–441. doi:10.1038/nmeth.1329

Moore MJ, Zhang C, Gantman EC et al (2014) Mapping Argonaute and conventional RNA-binding protein interactions with RNA at single-nucleotide resolution using HITS-CLIP and CIMS analysis. Nat Protoc 9:263–293. doi:10.1038/nprot.2014.012

Morris AR, Mukherjee N, Keene JD (2010) Systematic analysis of posttranscriptional gene expression. Wiley Interdiscip Rev Syst Biol Med 2:162–180. doi:10.1002/wsbm.54

Mukherjee N, Jacobs NC, Hafner M et al (2014) Global target mRNA specification and regulation by the RNA-binding protein ZFP36. Genome Biol 15:R12. doi:10.1186/gb-2014-15-1-r12

Munschauer M, Schueler M, Dieterich C, Landthaler M (2014) High-resolution profiling of protein occupancy on polyadenylated RNA transcripts. Methods 65:302–309. doi:10.1016/j.ymeth.2013.09.017

Ng P, Keich U (2008) GIMSAN: a Gibbs motif finder with significance analysis. Bioinformatics 24:2256–2257. doi:10.1093/bioinformatics/btn408

Niranjanakumari S, Lasda E, Brazas R, Garcia-Blanco MA (2002) Reversible cross-linking combined with immunoprecipitation to study RNA-protein interactions in vivo. Methods 26:182–190. doi:10.1016/S1046-2023(02)00021-X

Ozer A, Pagano JM, Lis JT (2014) New technologies provide quantum changes in the scale, speed, and success of SELEX methods and aptamer characterization. Mol Ther Nucleic Acids 3, e183. doi:10.1038/mtna.2014.34

Paulus M, Haslbeck M, Watzele M (2004) RNA stem-loop enhanced expression of previously non-expressible genes. Nucleic Acids Res 32:e78. doi:10.1093/nar/gnh076

Piñol-Roma S, Choi YD, Matunis MJ, Dreyfuss G (1988) Immunopurification of heterogeneous nuclear ribonucleoprotein particles reveals an assortment of RNA-binding proteins. Genes Dev 2:215–227. doi:10.1101/gad.2.2.215

Query CC, Bentley RC, Keene JD (1989) A common RNA recognition motif identified within a defined U1 RNA binding domain of the 70K U1 snRNP protein. Cell 57:89–101. doi:10.1016/0092-8674(89)90175-x

Ran FA, Hsu PD, Wright J et al (2013) Genome engineering using the CRISPR-Cas9 system. Nat Protoc 8:2281–2308. doi:10.1038/nprot.2013.143

Ray D, Kazan H, Chan ET et al (2009) Rapid and systematic analysis of the RNA recognition specificities of RNA-binding proteins. Nat Biotechnol 27:667–670. doi:10.1038/nbt.1550

Ray D, Kazan H, Cook KB et al (2013) A compendium of RNA-binding motifs for decoding gene regulation. Nature 499:172–177. doi:10.1038/nature12311

Rädle B, Rutkowski AJ, Ruzsics Z, et al. (2013) Metabolic labeling of newly transcribed RNA for high resolution gene expression profiling of RNA synthesis, processing and decay in cell culture. J Vis Exp (78):e50195. doi:10.3791/50195

Rehman S, Gladman JT, Periasamy A et al (2014) Development of an AP-FRET based analysis for characterizing RNA-protein interactions in myotonic dystrophy (DM1). PLoS One 9:e95957. doi:10.1371/journal.pone.0095957

Ricci EP, Kucukural A, Cenik C et al (2014) Staufen1 senses overall transcript secondary structure to regulate translation. Nat Struct Mol Biol 21:26–35. doi:10.1038/nsmb.2739

Rio DC (2012) Filter-binding assay for analysis of RNA-protein interactions. Cold Spring Harb Protoc 2012:1078–1081. doi:10.1101/pdb.prot071449

Rossbach O, Hung L-H, Khrameeva E et al (2014) Crosslinking-immunoprecipitation (iCLIP) analysis reveals global regulatory roles of hnRNP L. RNA Biol 11:146–155. doi:10.4161/rna.27991

Rybak-Wolf A, Jens M, Murakawa Y et al (2014) A variety of dicer substrates in human and C. elegans. Cell 159:1153–1167. doi:10.1016/j.cell.2014.10.040

Salim NN, Feig AL (2009) Isothermal titration calorimetry of RNA. Methods 47:198–205. doi:10.1016/j.ymeth.2008.09.003

Seelig B, Jäschke A (1999) A small catalytic RNA motif with Diels-Alderase activity. Chem Biol 6:167–176. doi:10.1016/S1074-5521(99)89008-5

Selvin PR (2000) The renaissance of fluorescence resonance energy transfer. Nat Struct Biol 7:730–734. doi:10.1038/78948

Sievers C, Schlumpf T, Sawarkar R et al (2012) Mixture models and wavelet transforms reveal high confidence RNA-protein interaction sites in MOV10 PAR-CLIP data. Nucleic Acids Res 40:e160. doi:10.1093/nar/gks697

Silahtaroglu A (2014) Fluorescence in situ hybridization for detection of small RNAs on frozen tissue sections. Methods Mol Biol 1211:95–102. doi:10.1007/978-1-4939-1459-3_9

Stoltenburg R, Reinemann C, Strehlitz B (2007) SELEX—a (r)evolutionary method to generate high-affinity nucleic acid ligands. Biomol Eng 24:381–403. doi:10.1016/j.bioeng.2007.06.001

Sugimoto Y, König J, Hussain S et al (2012) Analysis of CLIP and iCLIP methods for nucleotide-resolution studies of protein-RNA interactions. Genome Biol 13:R67. doi:10.1186/gb-2012-13-8-r67

Sutherland BW, Toews J, Kast J (2008) Utility of formaldehyde cross-linking and mass spectrometry in the study of protein-protein interactions. J Mass Spectrom 43:699–715. doi:10.1002/jms.1415

Swanson MS, Nakagawa TY, LeVan K, Dreyfuss G (1987) Primary structure of human nuclear ribonucleoprotein particle C proteins: conservation of sequence and domain structures in heterogeneous nuclear RNA, mRNA, and pre-rRNA-binding proteins. Mol Cell Biol 7:1731–1739

Tanke HJ, Dirks RW, Raap T (2005) FISH and immunocytochemistry: towards visualising single target molecules in living cells. Curr Opin Biotechnol 16:49–54. doi:10.1016/j.copbio.2004.12.001

Tenenbaum SA, Carson CC, Lager PJ, Keene JD (2000) Identifying mRNA subsets in messenger ribonucleoprotein complexes by using cDNA arrays. Proc Natl Acad Sci U S A 97:14085–14090. doi:10.1073/pnas.97.26.14085

Teplova M, Hafner M, Teplov D et al (2013) Structure-function studies of STAR family Quaking proteins bound to their in vivo RNA target sites. Genes Dev 27:928–940. doi:10.1101/gad.216531.113

Tollervey JR, Curk T, Rogelj B et al (2011) Characterizing the RNA targets and position-dependent splicing regulation by TDP-43. Nat Neurosci 14:452–458. doi:10.1038/nn.2778

Tome JM, Ozer A, Pagano JM et al (2014) Comprehensive analysis of RNA-protein interactions by high-throughput sequencing-RNA affinity profiling. Nat Methods 11:683–688. doi:10.1038/nmeth.2970

Ule J, Jensen KB, Ruggiu M et al (2003) CLIP identifies Nova-regulated RNA networks in the brain. Science 302:1212–1215. doi:10.1126/science.1090095

Uren PJ, Bahrami-Samani E, Burns SC et al (2012) Site identification in high-throughput RNA-protein interaction data. Bioinformatics 28:3013–3020. doi:10.1093/bioinformatics/bts569

Vyboh K, Ajamian L, Mouland AJ (2012) Detection of viral RNA by fluorescence in situ hybridization (FISH). J Vis Exp. doi:10.3791/4002

Wang ET, Cody NAL, Jog S et al (2012) Transcriptome-wide regulation of pre-mRNA splicing and mRNA localization by muscleblind proteins. Cell 150:710–724. doi:10.1016/j.cell.2012.06.041

Wang T, Xie Y, Xiao G (2014) dCLIP: a computational approach for comparative CLIP-seq analyses. Genome Biol 15:R11. doi:10.1186/gb-2014-15-1-r11

Wang Z, Kayikci M, Briese M et al (2010) iCLIP predicts the dual splicing effects of TIA-RNA interactions. PLoS Biol 8:e1000530. doi:10.1371/journal.pbio.1000530

White EK, Moore-Jarrett T, Ruley HE (2001) PUM2, a novel murine puf protein, and its consensus RNA-binding site. RNA 7:1855–1866

Wilson GM (2005) RNA folding and RNA-protein binding analyzed by fluorescence anisotropy and resonance energy transfer. In: Geddes CD, Lakowicz JR (eds) Reviews in fluorescence 2005. Springer US, Boston, MA, pp 223–243

Yang J-H, Li J-H, Shao P et al (2011) starBase: a database for exploring microRNA-mRNA interaction maps from Argonaute CLIP-Seq and Degradome-Seq data. Nucleic Acids Res 39:D202–D209. doi:10.1093/nar/gkq1056

Yang Y, Wang Q, Guo D (2008) A novel strategy for analyzing RNA-protein interactions by surface plasmon resonance biosensor. Mol Biotechnol 40:93. doi:10.1007/s12033-008-9066-3

Yang Y-CT, Di C, Hu B et al (2015) CLIPdb: a CLIP-seq database for protein-RNA interactions. BMC Genomics 16:51. doi:10.1186/s12864-015-1273-2

Zarnack K, König J, Tajnik M et al (2013) Direct competition between hnRNP C and U2AF65 protects the transcriptome from the exonization of Alu elements. Cell 152:453–466. doi:10.1016/j.cell.2012.12.023

Zhang C, Darnell RB (2011) Mapping in vivo protein-RNA interactions at single-nucleotide resolution from HITS-CLIP data. Nat Biotechnol 29:607–614. doi:10.1038/nbt.1873

Zhang L, Ding L, Cheung TH et al (2007) Systematic identification of C. elegans miRISC proteins, miRNAs, and mRNA targets by their interactions with GW182 proteins AIN-1 and AIN-2. Mol Cell 28:598–613. doi:10.1016/j.molcel.2007.09.014

Zuker M (2003) Mfold web server for nucleic acid folding and hybridization prediction. Nucleic Acids Res 31:3406–3415

Chapter 12
Metagenomic Design and Sequencing

William L. Trimble, Stephanie M. Greenwald, Sarah Owens, Elizabeth M. Glass, and Folker Meyer

12.1 Introduction

The technological advances in sequencing technology in the recent decade have made determining the sequence nucleic acid polymers affordable and widespread. While the study of variations in the genomes of model organisms including humans is a rich and fruitful area of investigation, microorganisms have vastly greater numbers and sequence diversity than macroorganisms. Sequencing of DNA from environmental samples has become a fast-growing application of sequencing technology.

Metagenomics is the analysis of genetic material extracted from environmental samples or extracted from mixtures of organisms. Two general approaches are available, targeted gene sequencing and random "shotgun" sequencing. Targeted sequencing amplifies gene fragments of interest using PCR primers corresponding to conserved regions of selected genes. Subsets of the sequence of the prokaryotic rRNA 16S subunit, the internal transcribed spacer (ITS) in fungi, functional genes of interest to oxidation–reduction metabolism (NifH, AmoA), and conserved non-RNA phylogenetic marker genes are all suitable. The 16S rRNA gene has proven most popular for surveying the composition of microbial communities, and as one of the genes that has been under investigation for the longest, its primers and sequences have been the most studied and have the largest number of database sequences, and protocols for high-throughput sample preparation are available (Caporaso et al. 2012).

W.L. Trimble (✉) • S.M. Greenwald
Institute for Genomics and Systems Biology, 9700 S. Cass Ave, Argonne, IL 60439, USA
e-mail: trimble@anl.gov; smoormann@anl.gov

S. Owens • E.M. Glass • F. Meyer
Argonne National Laboratory, 9700 S. Cass Ave, Argonne, IL 60439, USA
e-mail: Sarah.Owens@anl.gov; marland@anl.gov; folker@anl.gov

A.M. Aransay, J.L. Lavín Trueba (eds.), *Field Guidelines for Genetic Experimental Designs in High-Throughput Sequencing*,
DOI 10.1007/978-3-319-31350-4_12

Random "shotgun" sequencing provides unaligned samples from each organism's thousands of genes, rather than amplifying a single gene per organism or organismal type. This increases the complexity of the sequencing data product several thousandfold, and as a result, much greater per-sample sequencing effort is required. This higher sequencing effort has meant that environmental shotgun sequencing has been enabled disproportionately by low-cost sequencing technologies, and as a consequence the total amount of shotgun metagenomic sequence data has been rising rapidly. The Sequence Read Archive has (as of May 2015) 34 Terabases of sequence data tagged as metagenomic in origin; IMG/M and MG-RAST claim to have 4.5 and 76×10^{12} bp, respectively. Most metagenomic shotgun datasets at present have between a gigabase and a few-tens-of-gigabases sequencing effort. Sequencing single samples to depths of a hundred gigabases or greater have been uncommon but not unheard of.

12.2 Design

Generally, researchers are interested in the effect of external (non-sequence derived) variables on the composition of microbial communities. For both the targeted-gene and shotgun approaches, a vector of inferred relative taxonomic abundances is produced. For shotgun sequencing, the sequences can be further interpreted as relative abundances of fragments from different functional classes of genes. Analytical approaches that use additional information (from comparative genomics, or from chemical reaction networks) to extend the inferred profiles are in current use.

Finally, we can confidently recommend engaging the specialists in the wetlab and in computational analysis early in the sequencing process; many steps along the sample- and data-handling path have different sensitivities and different efficiencies for different sorts of target data; DNA handling and DNA processing technicians can only help if they are informed about the purpose of the experiment, the type of experimental design, and the relevant sampling characteristics.

12.2.1 Sample Replicates

Experimental design for metagenomic sampling is similar to that for RNA-seq experiments, where block experimental design and at least fivefold biological replication are recommended. Threefold biological replication is tolerated, but may not be forever. Biological samples are much more valuable than technical samples in supporting the detection of significant differences between treatments.

Auer (Auer and Doerge 2010) and Williams (Williams et al. 2001) have suggested using barcodes for blocked experimental designs that control for per-lane technical effects. It has been our experience that the technical repeatability within platforms is very good, and that the principal source of technical variability lies between different types of sequencing (different read lengths, ABI SOLiD vs. 454 vs. Illumina) and different protocols for library preparation (use of different fragmentation techniques, use of different PCR parameters for low-content samples, or use of different read lengths or sequencing platforms). Block designs to balance technical variation are better spent on the factors of the experiment, randomizing treatments to sequencing runs or batches of sample processing than to hedge against the effect of lanes or barcodes.

Sequencing samples sometimes fail; when sequencing many libraries at once, the failure of some of the samples becomes likely. A single lane or a single barcode can fail, producing insufficient quality or quantity data while other samples at the same time produce good sequence. The principal benefit of a design that spreads samples across several lanes is that this design provides insurance against a technical failure that is confined to a single lane. If one lane fails, a loss of one eighth of the sequencing depth is less disruptive to experimental design that the loss of data for one eighth of the samples. Block randomization is clearly indicated, however, if the sequencing protocol, whether extraction, template construction or purification, sequencing chemistry, or platform is changed during an experimental campaign, or when there are so many samples that batch changes in the sequencing protocol could confound the results.

12.2.2 Sequencing Options

The target sequencing depth—the number of sequences and base pairs to collect per sample—is the next design parameter. For experiments using shotgun sequencing, the relationship between number of samples and sequencing depth per sample is seen as the principal design constraint (Auer and Doerge 2010). While there is evidence of diminishing returns on RNA-seq sampling in excess of ten million tags (2 gigabases with 2×100 cycle sequencing reads) for eukaryotic RNA-seq (Wang et al. 2011), shotgun metagenomic samples typically target 10 gigabases per sample. This allows one or two samples per MiSeq flowcell (seven million spots at 400 bp per spot) and four samples per HiSeq flowcell at 2×101.

The large complexity difference between shotgun and targeted gene surveys and the availability of protocols to multiplex more than 600 samples in a single sequencing run, invite researchers to sequence large numbers of samples with just a single gene, and to apply shotgun sequencing to selected samples. Another sequencing option is to sequence one or a small number of samples to much greater depth than the others. This approach is not recommended, as it (by definition) consumes large amounts of sequencing effort that would usually be better applied to more samples to permit characterization of the within-group variability of sequence signals.

Unlike RNA-seq experiments, metagenomic shotgun experiments suffer when individual read lengths are less than 150 bp. Individual metagenomic reads bear the burden of identifying which organism they come from and which biochemical entity they represent. The reads must do this individually, since each random fragment

Table 12.1 Sequencing platforms suitable for metagenomic sequencing

Platform	Read length	Read number	Raw data yield	Error rate %	Targeted	Shotgun
Iontorrent PGM 318	200,400	5M	1–2 Gb/cell	2	OK	
Iontorrent Proton	200	10M	2 Gb/cell	3	OK	
Illumina MiSeq	2×100–2×300	16M	3–5 Gb/cell	1	[a]	[a]
Illumina NextSeq	2×150	120M	110 Gb/cell	1	[a]	[a]
Illumina HiSeq	2×100, 2×150	160M×8	40 Gb/lane	1	[a]	[a]
ABI SOLiD	50	1.4G	70G	5		
PacBio	6000	50k	300M	15		Supplemental

M millions
[a]Recommended platforms

may or may not be from the same organism. This makes longer high-quality reads—reads in the range of 150–450 bp—more valuable for exploitation than even overwhelming numbers of short (<75 bp) reads. ABI SOLiD has been successfully applied to metagenomics (Iverson et al. 2012), but the short read lengths (ca. 50 bp) present a challenge both to assembly and annotation. Iontorrent has been applied successfully to targeted-gene metagenomic analysis, but is not recommended for shotgun metagenomics because it has similar read lengths and costs to Illumina, but has poorer error characteristics. On the other side of the read-length continuum, some instruments produce very long (>3 kbase) reads with very poor sequence quality—base call error rates above 10 % (Pacific Bio-sciences, Oxford Nanopore). The anonymous nature of individual reads makes these poor choices for metagenomics unless complemented with Illumina data with high base accuracy. The simultaneous inference of the organism and the corrected sequence is not currently feasible with only long-read low-quality data except, perhaps, in the lowest complexity samples (Table 12.1).

12.2.3 Library Types

There are two main metagenomic library types/kits that we have tested thoroughly and can confidently recommend for metagenomic sequencing. These are the TruSeq and the Nextera, both from Illumina, Inc. (San Diego, California). These two library types differ in their approach in two key components of metagenomic library generation: the fragmenting or shearing of the input material and the ligation of sequencing adapters and sample-identifying barcodes. TruSeq libraries have been on the market longer, so there are more kits and biotechnology companies that cater to their creation. Nextera libraries are newer and, to date can only be made with Illumina reagent kits. TruSeq library generation uses mechanical shearing in a sonicator to fragment the DNA and ligates adapters separately. Nextera library generation uses an engineered transposase enzyme to simultaneously fragment and ligate adapters to the input material. The TruSeq and Nextera approaches differ considerably in the amount of input material needed. TruSeq libraries require 500–1000 ng of input DNA, while Nextera needs only 50 ng. This makes Nextera libraries is particularly helpful with low biomass samples. Because of the use of sonication instead of enzymatic incubation, TruSeq libraries give the user greater control over the insert size of library fragments.

12.2.4 Sample Requirements

Input DNA quantities for library preparation kits range from 1 ng to 1 μg of material. It is important to make sure that the amount of genetic material available for library preparation falls within the range given by the kit's protocol. Because library creation depends on creating fragments in size ranges that work well with the sequencing technology, and because fragments in the wrong size range can be filtered out during library creation, the quality of the input nucleic acids has a large effect on library success. Even if a researcher has ample genetic material, if the material is not of good quality a robust library often cannot be made. Sample quality, referring to the survival of high-molecular-weight nucleic acids, depends on the circumstances of extraction and storage as well as properties of the sampling environment; samples taken from hot or acidic environments tend to have lower nucleic acid quality compared to samples from cold or more neutral environment.

12.3 Wetlab Protocol

12.3.1 Storage

The proper storage of a sample also plays a role in overall sample quality. Storage variables such as delay before storage, storage temperature, and storage time can drastically affect relative abundances of microorganisms. Systematic studies have shown that samples stored at room temperature and at –4 °C show loss of 16S diversity and storage-associated microbial composition biases (Rubin et al. 2013). We recommend storing samples at −80 °C as soon as possible after collection, avoiding free-thaw cycles, and consistent extraction following storage to reduce storage associated community shits.

We have experience with the MoBio PowerSoil DNA isolation Kit.

12.3.2 *Quantification*

DNA quantification is an essential step at many places in the library creation workflow because some of the steps in library preparation and sequencing are concentration-dependent.

For assessing the quantity of a sample after extraction we recommend Invitrogen's Qubit Fluorometer. The Qubit utilizes a fluorescent dye that binds to nucleic acids to determine the starting concentrations. We recommend avoiding the NanoDrop, as it consistently overestimates nucleic acid concentrations. Unlike the NanoDrop, the Qubit Fluorometer can discriminate between DNA and contaminants, such as RNA. To assess the quality of the genetic material we recommend using Agilent's 2100 Bioanalyzer or an agarose gel. Generally, high quality nucleic acids destined for metagenomic research will be free of any fragments below 100–200 bp. If the number of fragments smaller than 200 bp outnumbers the rest this is an indication of overfragmentation or low input quality.

Equipment
1. Invitrogen Qubit Fluorometer
2. Covaris S-series system
3. Wafergens Apollo 324 system
4. Magnetic Stand or Rack (holds 1.5 ml or 96 well plates)
5. Thermocycler
6. Sage Sciences BluePippin Prep
7. Agilent 2100 Bioanalyzer

12.3.3 *Positive and Negative Controls*

There are numerous controls utilized throughout metagenomic library preparation in order to ensure quality data. The first of these controls is the extraction blank, a negative control. When extracting DNA from metagenomic samples researchers should include 1–3 extraction blanks with the sample set. The researcher will then compare the quality and quantity of the extraction blank to the samples and if a sample is found to match the extraction blank it will be discarded as a false positive. The second of these controls is the library blank and is used in the same manner as the extraction blank. Water will be run through the library preparation process in tandem with the samples and used to remove false positives from the set. Use of a negative control during library preparation is more common during the PCR step and several negative controls will often be included. This is because primer-dimers will often be generated by PCR along with the amplified libraries. Researchers will use the nucleic acid concentration of the negative controls to determine the concentration of primer-dimers in any given sample, often called the background noise. Researchers also employ positive controls in metagenomic preparation. These positive controls consist of sequences of DNA that are of high quality, well studied, and explicitly known. The most common positive control is called PhiX. These positive controls can either be spiked into the samples as an internal control or they can be run by themselves, separately barcoded, as an external control. The positive controls are then compared to the individual samples to help determine the quality of the library and the effectiveness of the library preparation method.

12.3.4 DNA Quantification

We recommend starting with 500 ng of high-quality DNA for TruSeq metagenomic library prep, although lower quality and concentrations may be used. By contrast, the Nextera protocol is optimized for exactly 50 ng, and samples should be diluted to that level.

We recommend the following protocol:

1. Make a Qubit working solution by diluting the Qubit DNA reagent 1:200 in Qubit DNA buffer using a sterile plastic tube.
2. Load 190 μL of Qubit working solution into tubes labeled standard 1 and 2.
3. Add 10 μL of standard 1 solution and standard 2 solution to the appropriate tube and mix by vortexing for 2–3 s.
4. Note: These are positive and negative controls used to calibrate the instrument.
5. Load 198 μL of Qubit working solution into each individual assay tube.
6. Add 2 μL of DNA to each assay tube and mix by vortexing for 2–3 s. The final volume of this solution should equal 200 μL.
7. Note: The amount of sample and working solution added to each assay tube can vary depending on concentration of the sample. The sample can vary between 1 and 20 μL, and the working solution can vary between 199 and 180 μL with the final volume equaling 200 μL. It is recommended to use 2 μL of sample to produce the most accurate results.
8. Allow all the tubes to incubate at room temperature for 2 min.
9. Select DNA assay on the Qubit Fluorometer. Select run a new calibration.
10. Insert the tube containing Standard 1, close lid and press read.
11. Remove standard 1 and repeat step 8 for standard 2.
12. Insert sample, close lid and press read.
13. Calculate concentration using dilution calculation on Qubit Fluorometer by selecting original volume of sample added to the assay tube.
14. Repeat steps 10 and 11 until all samples have been quantified.

12.3.5 *TruSeq Metagenomic Library Prep*

12.3.5.1 Insert Size Determination

Insert size determination is an important consideration for all Illumina libraries. Due to the enzymatic shearing of Nextera libraries, the ratio of DNA to enzyme and the enzymatic cut sites will determine the size distribution of a Nextera library. For TruSeq library prep, however, the user has more control over the size distribution. It is critically important to determine what library insert size will work best for your downstream analysis. Often, bioinformaticians will have a preference. We recommend consulting with the bioinformaticians that will be analyzing data before making your libraries.

The current generation of sequencing platforms produces reads with error rates that vary as a function of position in the read. The deterioration of sequence quality results from imperfect extension reactions that cause the sequencing signal to fade in strength and contrast, in part due to contributions from nonsynchronized populations of template molecules. Paired-end sequencing, which initiates synthesis from primers on opposite ends of the sequencing template, allows the high-quality bases to be drawn from both ends of the templates.

Careful selection of the size of the template molecules further permits reads to overlap. Libraries constructed so that the end of the first read overlaps with the end of the second are called "overlapping" libraries and allow the construction of longer composite reads, where the low-quality parts of each reads are complemented by redundant sequencing. Read merging is computationally inexpensive compared to assembly. When applied to well constructed libraries, more than 90 % of paired reads can be found to overlap. Variations in the overlap fraction between different samples likely result from differences in template length distribution, and to the extent that this affects annotation, this may be one of the sources of library-construction biases that occur in annotation output as batch effects.

Getting 90 % overlap requires careful control of the insert size. Templates that are too short result in more overlap (and less resulting sequence) than expected, reducing sequencing yield. Templates that are too long result in nonoverlapping sequences, or mixtures of nonoverlapping and overlapping sequences. When templates are much too short, shorter than the read length, the sequencer sequences the template and a piece of the normally unsequenced adapter on each end—resulting in reads that overlap for most of the beginning of the sequences, but that have unrelated artificial barcode sequences at their ends. These sequences can be recovered bioinformatically, but are of lower value than optimally overlapping sequences.

For paired-end sequencing, insert sizes fall in several qualitatively different regimes, illustrated in Fig. 12.2. A 160–180 base pair insert (270–280 bp including adapters) will result in overlapping reads on a 2 × 100 bp HiSeq run, with 20–40 bp of overlap. A 250 base pair insert (350 bp with adapters) will result in overlapping reads on a 2 × 150 HiSeq run. A 500 base pair insert (600 bp with adapters) will result in no overlap on a 2 × 100 or 2 × 150 HiSeq run. Finally, a 350–450 base pair insert (450–550 bp with adapters) will result in no overlap with a known distance between the reads for single genome assembly.

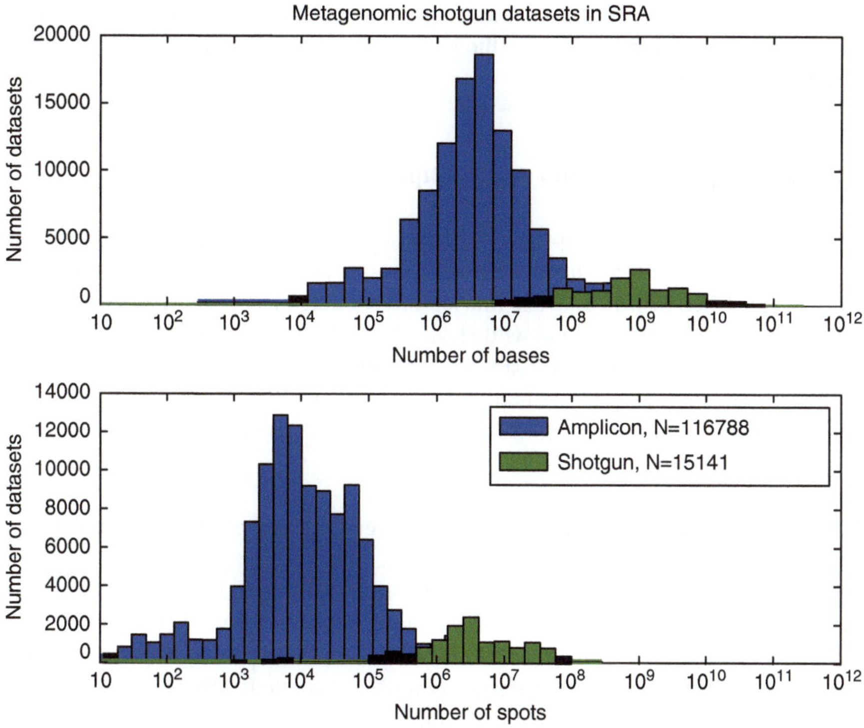

Fig. 12.1 Histogram of dataset sizes for metagenomic datasets in the Sequence Read Archive as of June 2015. Shotgun datasets have much greater sequencing requirements, and as a consequence, targeted-gene datasets outnumber them by a factor of 7

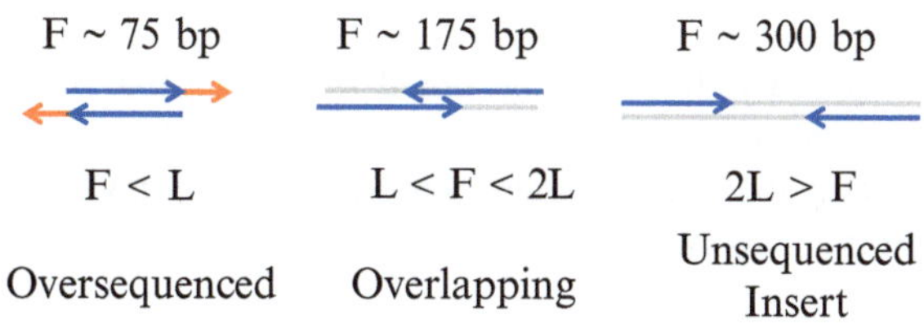

Fig. 12.2 Cartoons of possible insert sizes, with attention to overlap between paired reads. The *blue lines* indicate DNA from the library and the *orange lines* indicate the forward and reverse adapters for 2 × 100 paired-end sequencing.

12.3.5.2 Shearing of Libraries

For TruSeq libraries we recommend using the Covaris S-series system for mechanical shearing. The following instructions pertain to the S2 system but can be easily adapted to the S1 series. We recommend setting the water bath between 6 and 8 °C and using a minimum of 500 ng sample in 50–100 μL. It is important to use no less than 50 μL of sample as the Covaris relies on surface area to appropriately shear the material. If there is not at least 500 ng in 100 μL, we recommend using Agencourt

Ampure XP Beads to concentrate the sample down to a smaller volume using a 1.8× beads ratio. The conditions set on the Covaris are directly related to the preferred insert size of the final library.

12.3.5.3 Choosing Barcodes and Multiplexing

When multiplexing, it is important to choose mixtures of barcodes that result in complementary color mixtures within the same pool/sequencing lane. Illumina MiSeq and HiSeq instruments use four-color encoding, and bases A and C are principally found in the red channel while bases T and G are read out in the green channel. Mixtures of balanced color signals for each base—including the bases in the barcode—help the software maintaining high data quality. We recommend choosing indexes for samples that allow for at least one base in each channel per pool.

12.3.5.4 End Repair, A-Tailing, and Adapter Ligation on the Apollo 324

After samples have been sheared, there are several different kits that will perform end-repair and ligation of A-tails and adapters. We recommend Illumina's TruSeq PCR free Sample Prep, Illumina's Nano DNA Sample Prep Kits, Kapa Biosystems Library Amplification kits, or Wafergens PrepX Complete ILMN DNA library Kits. Each of these kits uses the same basic pipeline of end-repair, A-tail ligation, and adapter ligation and each will produce high quality libraries. The kits differ by the amount of input material they can handle, the insert size ranges they can produce, the time investment needed to complete the protocol, and the price. We prefer the Wafergen PrepX Complete ILMN DNA library kit as it has the fastest completion time and it is completed on an automated system allowing for less human error and increased reproducibility. It should be noted that the Kapa Biosystems Library Amplification kits can also be used on Wafergens Apollo 324 system. The following protocol pertains only to using Wafergens PrepX Complete ILMN DNA library Kits on Wafergens Apollo 324 system.

12.3.5.5 PCR and Size Selection

PCR and further size selection is not always necessary. For some applications the wide size distribution generated during library prep is sufficient. If the libraries are at least 2 nM concentration, then PCR is unnecessary. If size selection is unnecessary start this protocol at step 9. If PCR is necessary, we recommend using Bio-O Scientifics NEXTflex DNA Barcodes and PCR mixture. We recommend 10–15 cycles of PCR to achieve at least 2 nM concentration. If a tighter size distribution is necessary, we recommend further size selection with the BluePippin Prep (Sage Science, Inc., Beverly, MA), agarose gels, or E-Gels. Each of these methods vary in the amount of input material they can handle, the insert size ranges they can produce, the time investment needed to complete the protocol, and the price. We prefer the BluePippin Prep due to its ability to produce tighter sized libraries. The protocol below applies to the BluePippin Prep only. It is important to remember that we have accounted for the approximately 100 bp of adapter length to the library. For instance, a 180 bp insert must be thought of as a 280 bp library. Thus, we will size-select for 100 bp larger than the given insert size to accommodate for the adapters. We recommend the following protocol:

1. Choose the appropriate cassette to the given insert size and library size.
   (a) 3 % cassette ranges from 90 to 200 bp.
   (b) 2 % cassette ranges from 100 to 600 bp.
   (c) 1.5 % cassette ranges from 250 bp to 1.5 kb.
   (d) 0.75 % cassette ranges from 1 to 50 kb.
2. Program the Pippin.
   (a) In the BluePippin software go to the Protocol Editor tab.
   (b) Click on the Cassette folder that matches the appropriate cassette for the given library size.
   (c) Select either range or tight and enter in the given base pair range or peak.
   (d) Click the use internal standards button.
3. Calibrate the Optics.
   (a) Place the calibration fixture in the optical nest, close the lid and hit calibrate.
   (b) Continue only if it passes, if it does not pass, try again.
4. Load the Cassette.
   (a) Inspect the cassette from bubbles, breakage of agarose column, and equal buffer levels.
   (b) Dislodge any bubbles from the elution chamber.
   (c) Place the cassette into the optics nest.
   (d) Fill the sample well to the top with buffer.
   (e) Remove any buffer from the elution well and fill it with 40μL of fresh buffer.
   (f) Place a seal over the elution wells to keep them from overflowing during the run.
   (g) Run a continuity test and continue only if it passes. Try again if it fails.
5. Mix the library and dye.
   (a) Mix at least 30 μL of library with 10 μL of dye. If there is less than 30 μL use nuclease-free water to dilute the libraries to 30 μL.
   (b) Vortex the libraries and dye well and spin the mixture down.
6. Load the samples.
   (a) Remove 40 μL of buffer from the sample well and replace it with the 40 μL mixture of sample and dye.
   (b) Repeat for each sample.
   (c) Close the lid and hit the start button.

7. The BluePippin will run for 30–56 min depending on the given program.
8. Open the lid, remove the samples from the elution wells and place into a collection tube.
9. Check the concentration of the samples with a DNA HS assay on the Qubit Fluorometer as referenced above.
10. Use the Qubit concentration and estimated library size (100 bp + insert size) to calculate the molarity of the sample with the following equation (with X = ng/μL concentration and Y = estimated size of fragment in bp): Molarity in nM $=[X/1\times10^{-6}]/[Y\times660]$.
11. If the estimated molarity is less the 2 nM then proceed to PCR in step 12. If it is 2 nM or higher proceed to final library quantification.
12. PCR using Bio-O Scientifics NEXTflex™ DNA Barcodes and PCR mixture.

    (a) Mix 7.5 μL of the library, 29.5 μL of nuclease-free water, 12 μL of NEXTflex PCR master mix, and 2 μL NEXTflex Primer Mix in a well of a PCR strip tube or plate.
    (b) Set a pipette to 50 μL and mix by pipetting up and down ten times.
    (c) PCR on a thermocycler under the following settings.

    - 2 min at 98 °C
    - 10–15 cycles of: 30 s at 98 °C, 30 s at 65 °C, 60 s at 72 °C
    - 4 min at 72 °C

    (d) Add 44 μ L of AMPure XP Beads.
    (e) Incubate at room temperature for 15 min. During incubation, prepare an 80 % ethanol solution.
    (f) Place the tubes or plate on the magnetic stand at room temperature for at least 5 min, until the liquid appears clear.
    (g) Remove and discard the supernatant from each tube. Do not disturb the beads.
    (h) With the samples still on the magnetic stand, add 200 μ L of freshly prepared 80 % ethanol to each sample, without disturbing the beads.
    (i) Incubate at room temperature for at least 30 s while still on the magnetic stand, then remove and discard all of the supernatant from each tube. Again, do not disturb the beads.
    (j) Repeat steps 6 and 7 one more time for a total of two 80 % ethanol washes.
    (k) Allow the tubes to air-dry on the magnetic stand at room temperature for 15 min or until the beads no longer appear wet.
    (l) Add 15 μL of nuclease-free water to each tube.
    (m) Thoroughly resuspend the beads by gently pipetting ten times.
    (n) Incubate the tubes at room temperature for 2 min.
    (o) Place the tubes back onto the magnetic stand at room temperature for at least 5 min, until the liquid appears clear.
    (p) Transfer the clear supernatant from each tube to an appropriate collection tube. Leave at least 1 μL of the supernatant behind to avoid carryover of magnetic beads.

13. Proceed to Final Library Quantification.

12.3.6 Nextera Metagenomic Library Prep

For metagenomic library prep of low biomass samples, we recommend using Illumina's Nextera DNA kit. It is important that exactly 50 ng of sample is used as this protocol is optimized for exactly 50 ng. The sample should be in a 20 μL volume at a concentration of 2.5 ng/μL. If the sample has 50 ng but is in a volume that is larger than 20 μL, a 1.8× ratio of Agencourt Ampure XP Beads can be used to bring the sample to the appropriate volume. Please note that all of the abbreviations in this protocol refer to abbreviations used to describe reagents in the Illumina Nextera DNA kit.

12.3.6.1 Tagmentation of Genomic DNA

In this step, the transposome fragments the DNA while adding adapter sequences to the ends, allowing it to be amplified by PCR in later steps; our protocol includes a 5 min incubation at 55 °C.

12.3.6.2 Cleanup of Tagmented DNA

This step is critical because without it the Nextera transposome can bind tightly to the DNA and will interfere with downstream processing. We recommend using ZymoTM Purification Kit (ZR-96 D NA clean and Concentrator TM-5) for this protocol.

12.3.6.3 Choice of Barcodes

Nextera libraries are dual-indexed, meaning that each sample has two barcodes (an i7 and i5 index) ligated on opposite adapter/primers. It is important to ensure that no two samples in the same pool have the exact same combination of indexes. We recommend arranging samples in a 96 well plate and to assign each column an i7 index and each row an i5 index when working with moderate numbers of samples.

When multiplexing, it is also important to choose barcodes for individual samples that will be color-complementary with the barcodes of other samples in a given pool, avoiding mixtures of barcodes with extreme signals in the green and red channels. Illumina MiSeq and HiSeq instruments use four-color encoding, and bases A and C are principally found in the red channel while bases T and G are read out in the green channel. For example, if only samples with index 701(TAAGGCGA) and 704(TCCTGAGC) were in a pool, during the first read of the index the machine would only detect samples in the green channel (base T for index 1 and base T for index 2). This deprives the machine of the color contrast that it requires to identify clusters and issue confident base calls. We recommend choosing indexes for samples that allow for at least one base in each channel per pool. For Nextera libraries, it is also important to achieve red-green channel balance for both the i7 and i5 index mixtures separately.

12.3.6.4 PCR Amplification

It is critical to use the full amount of recommended input DNA at this step to ensure libraries that produce high quality sequencing results.

12.3.7 Final Library Quantification

The molarity and library size are critical for successful clustering and sequencing. Figure 12.3 shows examples of Bioanalyzer scans of completed library types. Illumina recommends that completed libraries achieve a molarity of at least 2 nM or greater in order to be sequenced with quality results. It is important to remove any primer dimers that may be present. Primer-dimers will be visible on a Bioanalyzer electropherogram between bases 0–100 depending on the length of PCR primers you are using. If primer-dimers are present, use a 1× ratio of AMPure XP Beads to remove them. To assess the quality of the completed library, we recommend the following protocol:

1. Use the Qubit Fluorometer to determine the concentration of libraries in ng/μL. As referenced above.
2. Use the Agilent 2100 Bioanalyzer to determine the library insert size and length, as referenced above.
3. Use the concentration from the Qubit and peak base pair size generated by the Bioanalyzer to calculate the molarity of the sample with the equation provided above.
4. The libraries are considered complete and ready for Illumina sequencing if the molarity is 2 nM or greater.

12.4 Analysis

12.4.1 Sequence Complexity

Nucleic acid sequences determined from environmental samples are difficult to interpret for a variety of reasons, and as a result exploitation of metagenomic shotgun data is computationally expensive compared to the study of model organisms. Some parts of microbial genomes evolve quickly and make detection of similarity technically difficult. Many environmental microbes and microbial genes lack close relatives in cultured organisms, resulting in large fractions of many environmental

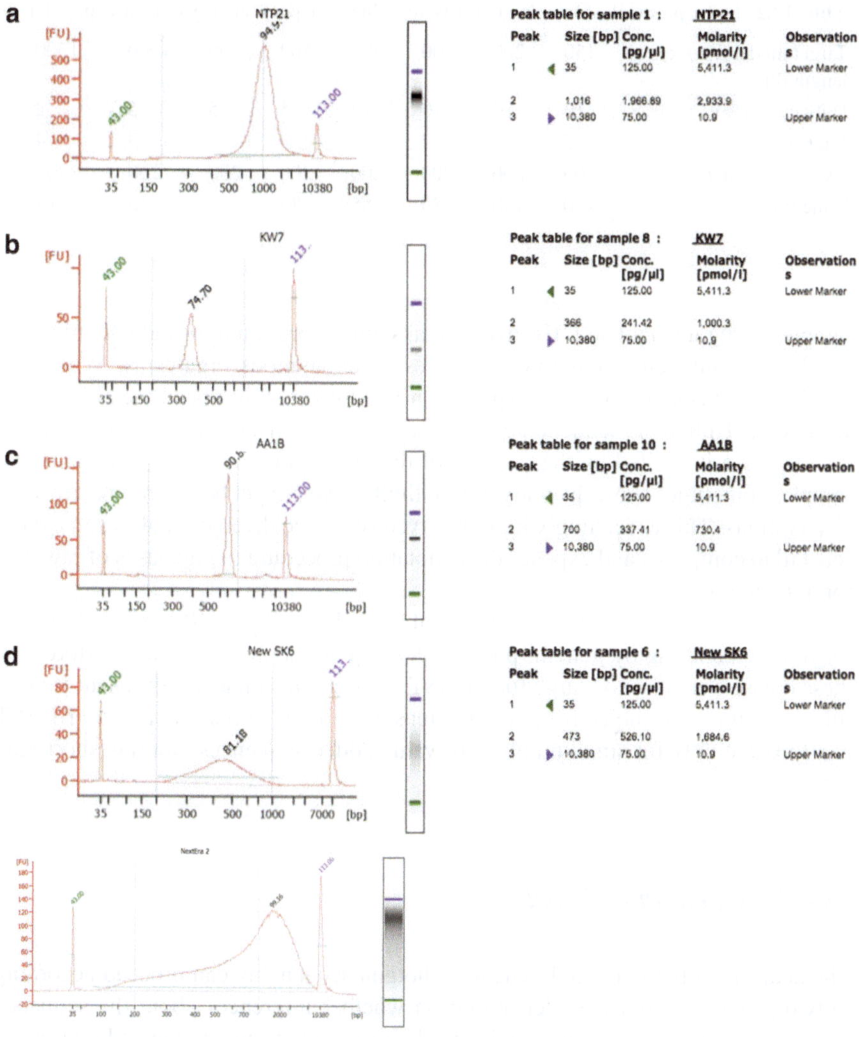

Fig. 12.3 Bioanalyzer traces of complete libraries. Panel **a** shows a TruSeq library with a narrow range peaking around 1 kb. Panel **b** shows a TruSeq library with a narrow range peaking around 360 bp. Panel **c** shows a TruSeq library with a narrow range peaking around 700 bp. Panel **d** shows a TruSeq library with a broad range peaking around 450 bp. Panel **e** shows a Nextera library peaking around 1600 bp

Table 12.2 Input protocols for a S2 Covaris to achieve a peak at a given base pair length

Target modal fragment length (bp)	150	200	300	400	500	800	1000	1500
Duty cycle (%)	10	10	10	10	5	5	5	2
Intensity	5	5	4	4	3	3	3	4
Cycles per burst	200	200	200	200	200	200	200	200
Time (s)	430	180	80	55	80	55	40	15

samples going unannotated. Environmental samples present a formidable inference problem of unraveling unknown mixtures of anonymous organisms. When this problem has been approached, expectation-maximization has been the algorithmic workhorse. Unlike the sequencing of reference organisms, where the complexity of the sequence is limited by the genome of the underlying organism, environmental samples sometimes show diversity whose limits have not yet been circumscribed by observations. This exceedingly high observed sequence diversity makes some datasets fail to compress, and exposes the annotation procedure to gigabases of raw data for annotation.

High-throughput sequencing datasets of typical size (10^8 to 10^{11} bp) are too large for routine handling by general-purpose desktop and laptop computers. Moreover, these datasets are also too large for BLAST. Faster, presumably less sensitive algorithms are the only choice for searching tens of millions of reads at a time; BLAST is affordable only for small numbers of value-added sequences, not raw short-read data.

12.4.2 *Open and Closed*

The analysis of both targeted-gene and shotgun sequencing can proceed according to two general approaches, depending on whether inferences about the sequence content of the samples depend on the databases used for comparison and interpretation. These approaches are called closed-reference and open-reference. Open-reference approaches are presumably more powerful, but involve unknown sample-dependent biases that cause the completeness of the analytical representation of the sequences to vary.

Comparing new sequence data to a database of sequences or sequence signatures is called "closed-reference annotation." Closed-reference annotation has the advantages that datasets annotated using the same procedure can be reliably compared because the space of possible annotations is limited and can be known in advance. Experience has shown that DNA recruitment of environmental samples to the genomes of all cultivated organisms often explains low (10–50 %) fractions of the dataset, leaving 50–90 % of environmental shotgun sequences without recognizable similarity to database sequences.

Constructing sequence hypotheses from the data and performing a database-comparison annotation on value-added sequences is called open reference annotation. For targeted gene sequencing, the sequence hypotheses are clusters constructed from the observed data; for shotgun sequencing hypotheses are usually the products of sequence assembly of the shotgun data. Unlike closed-reference annotation, open-reference annotation can discover and describe sequence patterns present in the data but not in the database. Open-reference annotation is more technically tricky and suffers from uncharacterized biases in the sequence construction phase, and difficulty in interpretation of the results. The sequences resulting from assembly, called contigs, are longer and can contain both complete genes and chains of genes from the same organism, permitting better resolution when comparing to databases and allowing analysis of synteny in metagenomic data.

The increased value of the sequences in open-reference annotation, however, comes with added analytical complexity. The collection of all the assembled contigs is always an incomplete summary of the metagenomic dataset, and contigs are not of equal importance in light of the sequence data. Contigs vary both in length and in depth, and the effects of this heterogeneity, which depends on uncontrolled properties of the sample and its biological diversity, on analysis are as yet unexplored.

The growing nature of the set of reference sequences in open-reference (assembly-based) analysis of shotgun metagenomic data means that sequences are typically analyzed in batches using defined sets of reference sequences, and comparisons of sample sets between lots with different sets of references are not straightforward.

12.4.3 Analysis Workflow Overview

To address artifacts associated with sequencing technology and to improve ultimate signal-to-noise, metagenomic data are subjected to a number of sequence-level filters before assembly or annotation. These preprocessing steps remove uninformative sequences, correct low-level errors, and discard sequence subsets enriched in errors. Removal of known sequence contaminants or positive control spikes is computationally straightforward when the contaminating sequences are known. Samples of host-associated microbes may contain varying amounts of host DNA, and the varying host content of the samples (or perhaps other host characteristics) represents an unwanted, potentially confounding signal in the genetic analysis of microbial community composition. Reads are compared to the reference genome with a fast read aligner (bwa and bowtie2 are the current state of the art) and reads that match are excluded from further analysis. Fecal samples from humans and animals, samples of wounds, and plant-associated sampling are all subject to this sort of confounding from host-organism contamination.

Current sequencing technologies all have platform-specific artificial sequences which are part of the sequencing technology. These include PCR primers, barcodes, and linker sequences that are ligated onto the sequences of interest. For some protocols, these sequences are intended to appear in the output, but in the standard Illumina single-end and paired-end protocols, adapter sequences in the sequence output are a symptom of poorly executed sequencing library preparation, either bad sequence size selection or survival of primer-dimers to the sequencing stage. These contaminants can include as few as 30 bp and as much as 150 bp of distinct sequence. Removing these "adapter" sequences is not computationally expensive, but there are no generally accepted guidelines on how much contamination is acceptable. Contamination ranges from minimal, affecting less than 10^{-5} of reads in a dataset, to as much as half of some sequence datasets; 10^{-3} is typical. Adapter sequences are a bigger problem for assembly than for recruitment or annotation.

12.4.4 The Human Factor

Just as with laboratory technicians, bioinformatic data processing requires people with specialized skills. The bioinformatic handling of any sort of sequencing data requires computational competency, including familiarity with transport, storage, and format conversion of large data files; management of maintainable workflows; ability to navigate sequence archives for sets of relevant reference sequences; and the ability to replicate computational workflows described in the literature, which requires installing and troubleshooting software.

Researchers usually get better results by sharing research goals, hypotheses, and prior information with the specialists, both in the wetlab and on the computational end. In order to suggest or apply procedures in the wetlab or in the computer lab to attenuate unwanted, contaminating DNA or sequences, technicians need to know what signals are interesting, and what likely uninformative signals look like. These specialists cannot help you if you give them DNA and sample numbers and no further instructions.

12.5 Reporting

The output from an Illumina Next Generation sequencing run is ultimately one or more FASTQ files (Cock et al. 2010). Metagenomes will be analyzed using the available online resources (e.g., IMG/M, MGRAST, CAMERA, EBIs Metagenomics portal) providing annotation by comparing transcripts to different functional gene databases (e.g., using BLAST to assign functions against M5NR, SFams, and SEED). For more detailed descriptions of potential functional pipelines and analyses of these data see (Meyer et al. 2008; Thomas et al. 2012; Wilke et al. 2015).

The results of closed-reference annotation are per-sample "feature vectors" representing the number of observations of biological molecules of a given type. The number of dimensions of this vector can range from a handful (classifying reads merely by estimated domain) to millions (counting each database sequence as a distinct potential unit of observation), and in general the number of observable features far exceeds the number of samples.

Raw shotgun metagenomics datasets range from hundreds of megabytes to hundreds of gigabytes in size. Since 2009, the NCBI's Sequence Read Archive has archived raw data from sequencing runs with mandatory metadata on protocol, sampling, and sequencing. The archive issues accession numbers for individual samples, individual instrument runs, collections of runs with the same protocol, collections of runs with the same purpose but different protocols, and collections of sequencing experiments with different samples, and has some features for accessioning analysis products. In addition to the public archives, some annotation services (MG-RAST, iMicrobe, and IMG/M) host metagenomic sequence data and annotation results and allow making public raw and value-added sequence data. NCBI's Whole Genome Shotgun archive accepts assembled contigs in FASTA format if sufficient metadata are provided, and is one option for making contigs available for later use for comparative study. Some consortia have published their value-added data products (for example annotation tables, results from assemblies) separately from the public sequence archives.

The accession numbers from depositions of the raw read data and of assemblies derived from the data must be included and associated with sample names when publishing results using metagenomic sequencing.

Annex: Quick Reference Guide

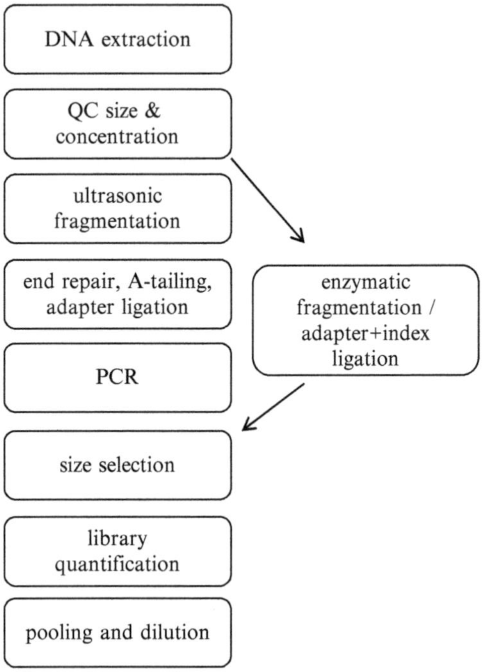

Fig. QG12.1 Representation of the wet-lab procedure workflow

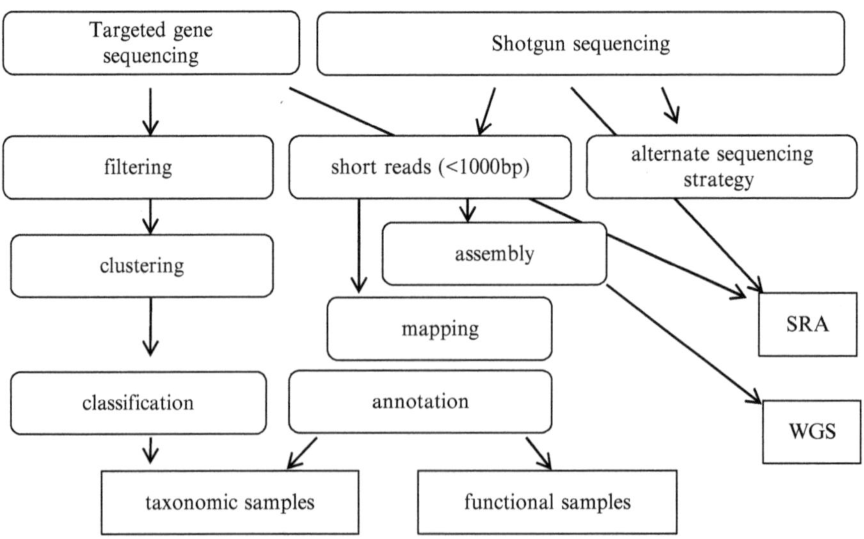

Fig. QG12.2 Main steps of the computational analysis pipeline

Table QG12.1 Experimental design considerations

Technique	Platform	Multiplexing	Target depth
Metagenomic/meta-transcriptomic shotgun	HiSeq 2000, 2500	4 samples/lane	40 M ~170 bp reads (2×100)
	MiSeq	1 sample/flowcell	15 M ~400 bp reads (2×250)
	NextSeq	3 samples/flowcell	30 M ~250 bp reads (2×150)
Targeted-gene amplicon	MiSeq preferred	700 samples/flowcell	>10,000 reads; size depends on primers
	Iontorrent	96 samples/flowcell	>10,000 reads / sample; size depends on primers
Design hints on matching samples to platforms			

Table that comprises relevant experimental design parameters, to carefully consider before applying this methodology

Table QG12.2 Available software recommendations

Software	Method	Language/platform	Input format
HUMAnN	tblast	Independent	fastq
CAMERA/RAMMCAP	Clustering + similarity	Independent (454-size data)	fastq
MG-RAST	Clustering + similarity	Online; API	fastq
EBI metagenomics portal	Similarity	Online	fastq
IMG-M	Similarity	Online	contig fasta
RAST	Genome guided similarity	Online; API	contig fasta

Table displaying a selection of the recommended software available for the computational analysis of data yielded by this technique. Pipelines for functional analysis of shotgun data. Because of the high computational burden relative to known-genome similarity searching, most general-purpose analysis seems to go through online portals. Note that none of the existing pipelines automate sequence assembly

References

Auer PL, Doerge RW (2010) Statistical design and analysis of RNA sequencing data. Genetics 185(2):405–416

Caporaso JG, Lauber CL, Walters WA, Berg-Lyons D, Huntley J, Fierer N, Owens SM, Betley J, Fraser L, Bauer M, Gormley N, Gilbert JA, Smith G, Knight R (2012) Ultra-high-throughput microbial community analysis on the Illumina HiSeq and MiSeq platforms. ISME J 6(8): 1621–1624

Cock PJA, Fields CJ, Goto N, Heuer ML, Rice PM (2010) The Sanger FASTQ file format for sequences with quality scores, and the Solexa/Illumina FASTQ variants. Nucleic Acids Res 38(6):1767–1771

Iverson V, Morris RM, Frazar CD, Berthiaume CT, Morales RL, Armbrust EV (2012) Untangling genomes from metagenomes: revealing an uncultured class of marine euryarchaeota. Science 335(6068):587–590

Meyer F, Paarmann D, D'Souza M, Olson R, Glass EM, Kubal M, Paczian T, Rodriguez A, Stevens R, Wilke A, Wilkening J, Edwards RA (2008) The metagenomics RAST server—a public resource for the automatic phylogenetic and functional analysis of metagenomes. BMC Bioinformatics 9(1):386–388

Rubin BE, Gibbons SM, Kennedy S, Hampton-Marcell J, Owens S, Gilbert JA (2013) Investigating the impact of storage conditions on microbial community composition in soil samples. PLoS One 8(7):e70460

Thomas T, Gilbert J, Meyer F (2012) Metagenomics—a guide from sampling to data analysis. Microb Inform Exp 2(1):3

Wang Y, Ghaffari N, Johnson CD, Braga-Neto UM, Wang H, Chen R, Zhou H (2011) Evaluation of the coverage and depth of transcriptome by RNA-seq in chickens. BMC Bioinformatics 12(Suppl 10):S5

Wilke A, Bischof J, Harrison T, Brettin T, D'Souza M, Gerlach W, Matthews H, Paczian T, Wilkening J, Glass EM, Desai N, Meyer F (2015) A RESTful API for accessing microbial community data for MG-RAST. PLoS Comput Biol 11(1):e1004008

Williams AG, Thomas S, Wyman SK, Holloway AK (2001) RNA-seq data: challenges in and recommendations for experimental design and analysis. Curr Protoc Hum Genet Unit 11.13(Suppl 83):11.13.1–11.13.20

Chapter 13
A Hitchhiker's Guide to Metatranscriptomics

Mariana Peimbert and Luis David Alcaraz

13.1 Transcriptomics, Metatranscriptomics, and Bacterial RNA Complications

Transcriptomics is defined as the complete set of RNA molecules produced in a cell (Güell et al. 2011). Metatranscriptomics is the assessment of environmental gene expression, be it in a population or a whole community. The rapid advance in sequencing technologies has allowed to rapidly increase the environmental genomics related works. At the beginning of the Next Generation Sequencing (NGS) about some 10 years ago from now, most of the works were only able to describe microbial taxonomic diversity by means of amplicon sequencing (16S/18S rRNA sequences), and then the introduction of 454 pyrosequencing led to multiple groups start working with Whole Genome Shotgun (WGS) metagenomics. Although the work was merely descriptive at the beginning of WGS metagenomics, it threw light on both taxonomic and functional diversity of the studied environments. Within the functional diversity, metagenomics is only describing the potential outcome, but to test the functional profile of a microbial community further methodologies for the expression (metatranscriptomics), and translation (metaproteomics) are required. The race for cheaper sequencing is still going on, and there is no such thing as a universal and unique best solution platform in the market but there are several technologies leading the competition like is the case for Illumina®, and several

M. Peimbert
Departamento de Ciencias Naturales, Universidad Autónoma Metropolitana Unidad Cuajimalpa, Av. Vasco de Quiroga 4871a, Col. Santa Fe, 05348 Cuajimalpa, Cd. Mx., México

L.D. Alcaraz (✉)
Departamento de Ecología de la Biodiversidad, LANCIS, Instituto de Ecología, Universidad Nacional Autónoma de México, AP 70-275, Ciudad Universitaria, UNAM, 04510 Coyoacán, Cd. Mx., México
e-mail: lalcaraz@iecologia.unam.mx

A.M. Aransay, J.L. Lavín Trueba (eds.), *Field Guidelines for Genetic Experimental Designs in High-Throughput Sequencing*,
DOI 10.1007/978-3-319-31350-4_13

newcomers are still on its way with promising technologies like nanopores, and solid state based solutions.

The challenge of describing genome wide expression has been done historically by means of microarray chips, and they have the advantage of describing overall gene expression, but previous knowledge about the genomic sequence of the organism is mandatory. A previous NGS technique, to describe microbe's transcripts, is expressed sequence tags (ESTs); the current transcriptome sequencing strategies are just an up-scaling of ESTs. While microarrays have been proved as an effective tool for describing the expression profiles for model organisms, they are not still a major player in metatranscriptomics. The cause of the microarrays relegated role in metatranscriptomics is that for complex environments with high diversity there would be the need to sequence the metagenome, then select representative gene clusters, and print them into the microarray, making it expensive and laborious. Although it would be possible to design environmental microarrays looking for some particular genes (pathogenesis, virulence, etc.) or particular species, this would be limited when comparing to current RNA-seq approaches (Westermann et al. 2012).

The main advantage of current NGS metatranscriptome is that is possible to associate gene expression patterns of even unknown genes, thus showing light that the unknown gene is transcribed under a particular condition. Hence, metatranscriptomics aids to identify novel genes related with environmental functions, with no necessary previous knowledge about any particular gene present in the sample (so no probe or primer design needed). The main drawback of environmental NGS metatranscriptomics is that most, some times >95 %, of the environmental RNA isolated under any situation corresponds to ribosomal RNA (rRNA), and the prokaryotes do not have a polyA track in the 3′ end of mRNA which is central for the transcriptome sequencing of eukaryotes, because it allows to start reverse transcription from the terminal polyA track and consequently the cDNA is almost exclusively formed by mRNAs (Sorek and Cossart 2010). Although rRNA is useful to determine community structure and having by PCR an unbiased picture of the active taxonomic diversity out there (by identifying, and annotating 16S rRNA fragments), when trying to define the community functional profile, getting rid of rRNA could be a challenge. However, with the current NGS technologies, it is feasible to think of having less than 5 % of mRNAs in the total sample, and still have thousands of cDNAs to tell a story about, but nevertheless cleaning the rRNA is required.

There has been an active development for technologies trying to enrich the amount of total mRNA and they could be divided in the following four main strategies: (1) Ribosomal RNA capture (rRNA hybridization), (2) 5–3′ exonuclease degrading processed RNAs, (3) adding polyA to mRNAs by means of polyA polymerase (from *Escherichia coli*), and (4) antibody capture of mRNAs interacting with selected proteins (Sorek and Cossart 2010). The polyA and antibody capture methods are highly biased, thus not recommended. The cDNAs enrichment is a major issue when designing the overall strategy and experiments.

A crucial factor in transcriptomics is whether you have a reference genome sequence to map the transcripts against or you will be performing de novo transcript assembly. It is the same situation with metatranscriptomics, if you have or not a reference metagenome obtained at the very same time to map against. The major advantage of having a reference metagenome is that you can see if there is correspondence between raw gene abundance, and its expression levels. There are plenty of options to map NGS sequencing data against references like BWA, bowtie, and tophat (Langmead et al. 2009; Li and Durbin 2009), and at the end of the day you could build count tables with each transcript abundance, and mapping Single Nucleotide Polymorphisms (SNPs) for each of the transcripts. If you are just interesting to sequence the metatranscriptome without metagenomic reference you should assemble the reads first using some NGS assemblers like SOAPdenovo, Velvet, Celera, and then perform ORF prediction with some tool like Glimmer, or Metagenemark (see Table QG13.2).

Up to date there are plenty of resources to address a metatranscriptome study. This work intention is to give an overall view of the metatranscriptomics process, experiments and analysis, and put the spotlight in the plenty of guides, tutorials and resources that have been systematically ordered for this purpose. Methodologically, the metatranscriptome uses the very same techniques and analytical tools as is single species precursor, the transcriptome.

13.2 Get to Know the Basics on Transcription Before Going Further

Previous work on systematizing the huge amount of information related to RNA-seq experiments in microorganisms has been done, and we strongly recommend to check out the biological and technical information before getting into the experimental design. A great starting point for understanding our current knowledge about bacteria transcription could be assessed in two excellent reviews the first by Sorek and Cossart 2010, and then read the review by Güell and collaborators (2011), both works on Nature Reviews Microbiology. Some previous protocols on metatranscriptomics are available as well (Gilbert and Hughes 2011), though thinking on a virtually retired technology (454 pyrosequencing), but all principles are still valid. The literature recommendations are based on first have a general outlook of what we know about bacteria gene regulation and how this is being enriched by transcriptomics. We also recommend to check some of the works to see the final publication output of metatranscriptomics, and how this is reported (Benítez-Páez et al. 2014; Frias-Lopez et al. 2008; Gilbert et al. 2008; Gosalbes et al. 2011; Hewson et al. 2009; Franzosa et al. 2014).

13.3 Experimental Design

If you want to try metatranscriptomics, the first thing would be choosing what kind of experimental approach is correct to your needs, and budget. Basically, there are two great first approaches to it, a qualitative or quantitative (Fig. 13.1).

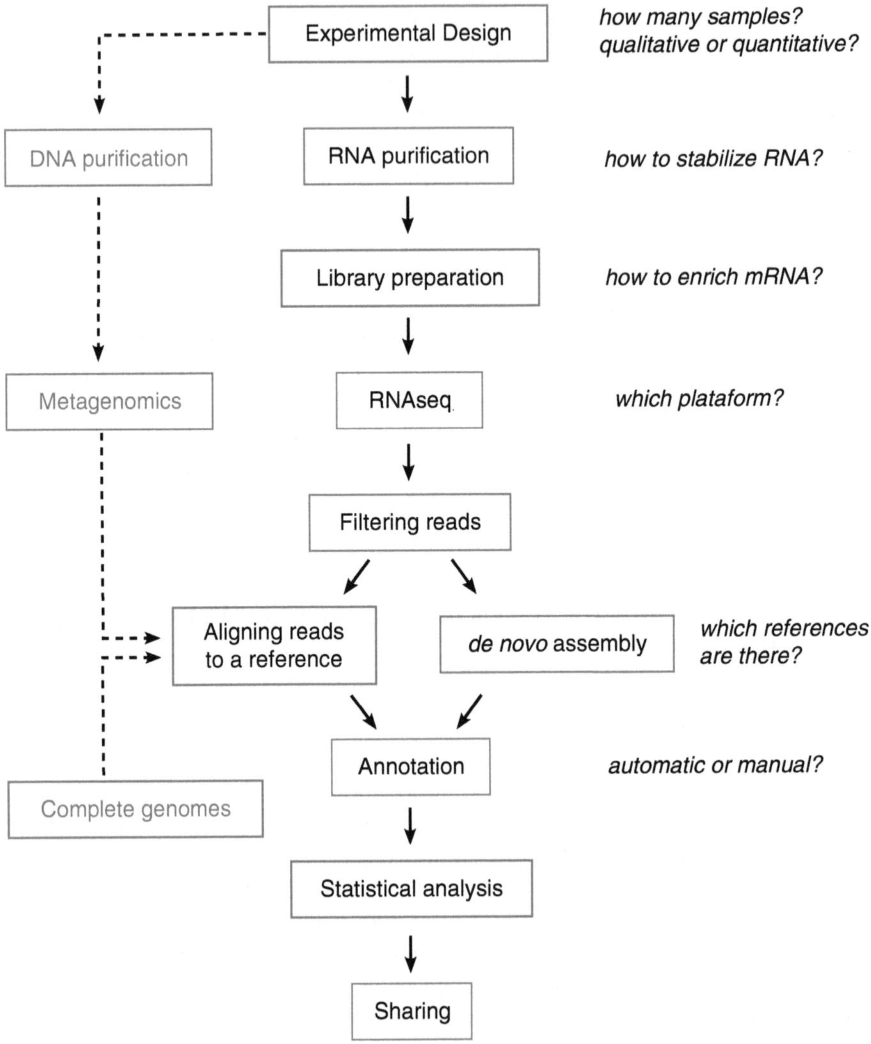

Fig. 13.1 The overall metatranscriptome process. Here are shown the main steps from the initial stages of a metatranscriptome study, in the *central* part are the mandatory steps. In the *left* part of the schema are shown in *grey* some of the optional steps. On the *right*, the main questions to address in order to perform a successful workflow

For metatranscriptomics using RNAseq, the qualitative approach is highly valuable, because even the high amount of rRNA obtained, this could be used to describe community structure and describe the metabolically active members of it. The databases with 16S rRNA are still the best repositories for bacteria taxonomic diversity out there, and even though it may not be possible to perform the tasks done with

PCR microbiome amplicons like multiple alignments, and diversity metrics derived from them (like Unifrac, Phylogenetic Distance methods, etc.), it is possible to identify by homology each of the sequenced reads, and use some tools like the RDP classifier or Greengenes to classify the overall active bacteria diversity (Lozupone et al. 2011; Cole et al. 2009; Schloss 2010; DeSantis et al. 2006). The rRNA classification for a metatranscriptome has the additional advantages of not biasing the diversity due to primer election, and PCR amplification effects. Moreover, the expected 5 % of mRNA helps to identify expressed genes in the community, some of the genes are going to have known homologs in the databases and they will be annotated accordingly but for the orphan genes (with no homologs in DBs) we will have information about them being expressed under the tested situation, something not into reach with metagenomes and so the importance of knowing previously the tested variables and the metadata that will be available for future comparisons.

The quantitative approach is the most used when doing transcriptomes on single organisms. This is because this approach allows us to detect significant differences between the overall gene expression in contrasting situations. Single organism transcriptomes in several contrasting experimental conditions had been proved to be a powerful tool when looking for Differential Gene Expression (DGEs). The success of getting DGEs depends on several factors like the number of conditions tested (biotic, abiotic), the number of biological replicas, sequencing coverage, read length. The sequencing coverage and read length could be easily planned if there is a reference genome. If there is no such thing like a reference genome one rule would be to dedicate equal sequence coverage for each of the replicas (i.e., if using Illumina HiSeq 2500® dedicate a single sequencing lane to each replica).

If you are planning to conduct a metatranscriptome it would help a lot if you have some preliminary data on helping you to answer the basic how many sequences do I need? This could be the result of pilot studies on 16S rRNA amplicon diversity, a previous metagenome, or even diversity estimates from related systems of what you are currently studying. There are several tools aiding with the design and replica number in RNA-seq experiments, like EDDA (Experimental Design in Differential Abundance analysis) which is available like an R's Bioconductor package (EDDA), or as a web server (Luo, et al. 2014). Within EDDA you can upload some pilot data you might have and test about the experimental design. The key questions are: How many replicates should I use? How much sequencing depth? Is the experimental design helping out to capture biological variation?

One rule of the thumb would be to use the same number of replicas for each condition tested and a minimum number of two replicates per condition to gain insight into the biological variance. Thus, considering one treatment and one control groups would be the simplest, and most widely used experimental design (See Table QG13.1). If you are trying a nested experimental design the number of replicates would increase dramatically but this is out of the reach of this chapter, please refer to experimental design guides, a good starting point for this was provided by Knight and collaborators (2012).

13.4 What Sequencing Platform Is the Best for Metatranscriptomics?

This is the most frequent question for most of researchers entering into the meta-omics world. There is no easy answer for this, as expected. The main trade-off would be between the overall cost, the read length, and the sequencing depth of each platform. The current major used platform is Illumina® due to its overall cost-benefit, though it has several possible configurations (MiSEQ, HiSEQ, etc.), the major platform used not that long ago was 454, and now is practically retired from metatranscriptomics, the message here is that the market is still far from being stable and new players are coming all days into it. The actual major players are: Illumina's HiSeq (X, 3000/4000, NexSeq, High-Output), and MiSeq, Life Technologies (PGM, Proton), Pacific Biosciences (RS), and the former 454. The sequence read length spans from 50 bp (Illumina) to 1.5 kb (PacBio), and the cost per Mb goes from USD$ 0.06 (Illumina) to USD$8.72 (454), and the output yield goes from ~40 Mb (PacBio) to ~300 Gb (Illumina). There are some recent works that show that in overall the gene expression profiles are similar across platforms and the main differences are the costs for detecting splice variants (Li et al. 2014). But keep in mind that the price is rather limiting but not the only variable to consider, please take into account the quality of the data, the support for the available technology (aligners, assemblers) and compare the options offered by different providers, there are some places like http://allseq.com and https://genohub.com where you can quote multiple providers all at once. Also keep in mind that you can mix two strategies, i.e., Illumina's deep coverage mixed with PacBio long reads to aid in the assembly process. The main questions are: How many samples do you want to sequence? What is your desired read length? How many reads do you need per sample? How much money do you have?

13.5 Sequencing Depth or the Number of Aligned Reads Required for a Reliable Analysis

A bacterial genome is considered complete when it has an 8× coverage depth. For an average 5 Mb genome it would be necessary to sequence at least 40 Mb to have that amount of coverage. When talking about a metatranscriptome in the ideal scenario one would have previous data about the studied system, like the species abundance with 16S rRNA amplicon sequencing. Lets say that a given environment hosts 700 species and assuming a 5 Mb genome per species one would need at least 28 Gb of sequencing to have an 8× coverage depth. This is assuming some unrealistic situations like having equal abundances for each species and genes, and that they are all the same genome size. This is not an easy task, but with Illumina's deep sequencing it is expected to generate up to 300 Gb of sequencing that would be equivalent to a 85× coverage for each species of this hypothetical scenario, and considering that not every gene is always being expressed we can have an ultra-deep coverage of the metatranscriptome that can even be multiplexed.

Most of the meta-omics analysis are highly biased to over-represented features, even of ultra deep sequencing we cannot be certain that we are not recovering rare species, or genes because of the sequencing effort. The rule of the thumb for sequencing depth is to be equitable for each condition and replicate tested.

13.6 General Considerations for Wet-Lab

When working with RNA is important to pay close attention to cleanliness of the bench working area, equipment and reagents. All living cells and all cell types produce intracellular and extracellular RNases. RNases are essential for the regulation of gene expression and are an important part of the immune system; that is the reason why there are several types of these enzymes, some of which are very resistant to inactivation treatments. Some RNases have several disulfide bridges so even after frozen or denatured they can be reactivated. RNase contamination main sources are the skin, saliva, hair, perspiration, clothing, fungi, bacteria, mites, plant, or any living cell (Sambrook and Russell 2012). This is why you should always take the following precautions:

1. Always wear gloves.
2. Change gloves frequently. Every time you touch the phone, the handle of the fridge, your face, skin, etc. you should change gloves.
3. Wear clean gown. The lab coat protects the experiment from dust on the clothes.
4. Use RNase-free tips and tubes. Providers indicate when their products meet this quality criterion. Bags and boxes must remain closed otherwise they are no longer RNase-free.
5. Work in a specific clean area with low traffic and free from air currents.
6. Use RNase-free reagents. We recommend using commercial kits, and reagents designed to work with RNA. Remember, tubes and bottles must be handled with gloves and must be closed as long as possible.
7. Clean every material to be used in a way that is free of RNase (see Sect. 13.6.1).

   Some labs still take extra precautions such as:
8. Use filter tips to avoid aerosols that could contaminate the sample.
9. Have a unique set of pipettes to work with RNA.
10. Aliquot reagents to reduce handling.
11. Use an RNase-free fumehood or cabinet.
12. Have a clean room equipped.

13.6.1 Treatments for RNase Cleaning

Contrary to common sense the autoclave does not inactivate all RNases.

All the water in contact with the RNA must be free of RNase. The most commonly used protocol for this is treatment with DEPC (diethyl pyrocarbonate). DEPC covalently modifies the secondary amines inactivating RNases permanently. However, it also modifies RNA so it must be destroyed before use. For this treatment, a 0.1 % DEPC solution is prepared and incubated for 12 h at 37 °C. Then, the solution is autoclaved for 15 min for DEPC degradation. Buffers and other reagents with amines (Tris, MOPS) should not be incubated with DEPC. To prepare these buffers water is first treated and then reagents are dissolved.

All nondisposable material should be treated. Glassware should be washed and baked at 240 °C for 4–16 h. Another protocol is to dip the glassware in water with 0.1 % DEPC for 12 h at 37 °C and then autoclaved for 15 min to remove DEPC. It is important to wrap with foil glassware before putting it in the oven or autoclave. It is also recommended to have a clean area for all reagents and materials to be used.

Electrophoresis tank must also be treated; it should be washed with detergent, rinsed with RNase-free water, and finally rinsed with ethanol.

Some companies sell DEPC alternatives that do not require autoclave. RNase inhibitors are commercially available, inhibitors are high affinity proteins specific for RNase type A. RNase inhibitors are expensive, and it is recommended only to preserve the purified sample.

13.6.2 RNA Purification

Using commercial kits is recommended, mainly because they ensure that the solutions are RNase-free. Please pay attention to the amount of sample that is recommended by the supplier as excess can result in very low efficiencies. RNA purification is divided into the next steps: sampling, RNA stabilization, cell lysis, RNA isolation and treatment with DNAse I. Here we describe various protocols for each of these steps.

13.6.3 Sampling

The samples should be acquired quickly and aseptically. The sample should be processed immediately or snap-frozen. Generally, samples are frozen directly on the field in either liquid nitrogen, or dry ice/acetone to stop metabolism without damaging cell structures, however when samples are thawed RNases will be active. When planning your sampling you should anticipate how to stabilize RNA because usually this is done before freezing (see below).

13.6.4 Stabilization

As previously mentioned, all cells have intracellular RNase, the mRNA in bacteria generally have a few minutes life span so RNA can be degraded while purified. Moreover, transport and purification can induce the synthesis of new mRNA changing expression profiles. Several reagents may serve to inactivate endogenous RNase. The simplest is to add to the sample a 1:10 solution of 5 % phenol in ethanol. Another option is to start with the isolation process before freezing adding guanidinium thiocyanate–phenol–chloroform solution, commercially known as TRIzol®, Qiazol®, or TRi®. One of the most popular stabilizers is RNAlater® containing EDTA, sodium citrate, and ammonium sulfate, it is used for all cell types and has been tested in bacteria. RNAprotect® is a stabilizer designed for bacteria; this works for gram-negative and -positive bacteria.

13.6.5 Cell Wall Lysis

The three most popular methods to lyse the cell wall are: mechanical disruption (bead beater), enzymatic lysis (lysozyme or lysostaphin) and proteinase K digestion. In axenic cultures lysate efficiency is important for the total amount of RNA but when it comes to communities, lysis will also affect RNA distribution, as some bacteria are more sensitive to some treatments. If the aim is a qualitative study, it probably is best to mix all methods of lysis, to obtain as many as possible RNA, but if you want to make a quantitative study, you would better use a mechanical method that can lyse all bacterial types and is the most reproducible one.

When working with soil communities is important to consider the contamination with humic acids, as they inhibit further PCR reactions. PowerSoil® kit is specially designed to deal with humic acids. If you do not have access to the kit, we recommend washing the cells several times with phosphate buffer and follow a purification protocol with CTAB.

13.6.6 DNase I Treatment

RNA samples often have trace contamination of genomic DNA, so the final step is to treat the samples with DNAse I, and its subsequent inactivation. DNAse I can interfere with the following steps, if not inactivated. Once again RNA can be purified by extraction and precipitation or by silica columns. The RNeasy® kit allow using DNAse when the RNA is bound to the column, which prevents the second purification.

To prevent freezing and thawing we suggest to aliquot pure RNA samples. *Store samples at −80 °C before and after purification.*

13.6.7 RNA Quality Determination

There are three factors to consider in determining the quality of a RNA sample: concentration, purity, and integrity. These three factors are important in deciding whether to continue the experiment or if repurification are necessary. We always advise to perform an UV absorbance spectrum (220–350 nm), NanoDrop® instrument allows to measure small volumes from 1 μL; absorbance at 260 nm indicates the concentration of nucleic acids, absorbance at 280 nm allows to estimate the protein concentration; while 230 nm absorbance indicates the presence of humic acids salts or compounds that were used for purification. The disadvantage of this method is that it cannot determine if the RNA is degraded and this not either distinguishes DNA contaminations. It is generally considered good quality samples when the 260/280 ratio is greater than 1.8 and the 260/230 ratio is greater than 1.7. If the sample is not pure, the concentration may be overestimate as contaminants also absorb at this wavelength (Fig. 13.2). Fluorescent dyes detect lower RNA concentrations, and these only emit in the presence of nucleic acids, so RNA concentration is more reliable. Fluorescent dyes, generally, do not discriminate between different nucleic acids and this technique cannot determine the purity and integrity of the sample. The agarose gel electrophoresis allows knowing RNA integrity; the criterion for determining that the RNA is intact is to observe 23S and 16S rRNA bands in a 1.8:1 ratio. The presence of genomic DNA can be identified in agarose gel because its size is much greater than 23S, but it do not allow us to estimate other kinds of contamination. One of its great advantages is that it is an inexpensive method that can be done in most laboratories; nevertheless, it is a qualitative method. The 2100-Bioanalyzer® is a quantitative method that uses cartridges ready to use for capillary electrophoresis. This equipment generates electropherograms and includes software that integrates the peaks to determine the RNA integrity number (RIN; Fig. 13.2). The big disadvantage of Bioanalyzer equipment and cartridges is their price, this method also allows to determine sample concentration.

13.6.8 Enrichment of mRNA

One of the most complicated steps in studying bacterial transcriptomes and metatranscriptomes is mRNA enrichment; in eukaryotes the problem is trivial by the presence of the polyA tail. The two most popular strategies to enrich the mRNA are rRNA hybridization, and degradation of processed RNA. rRNA hybridization is based on magnetic microbeads and oligo mixtures which hybridize with 16S and 23S (MICROBExpress™, and Ribo-Zero™). The hybridization method is the most popular because RNA integrity is not required. This approach is sequence specific and does not eliminate all bacteria rRNA, for example those from high GC content. Another limitation is that oligos can also hybridize with some mRNA. Degradation of processed RNA requires a 5′ monophosphate exonuclease for the removal of

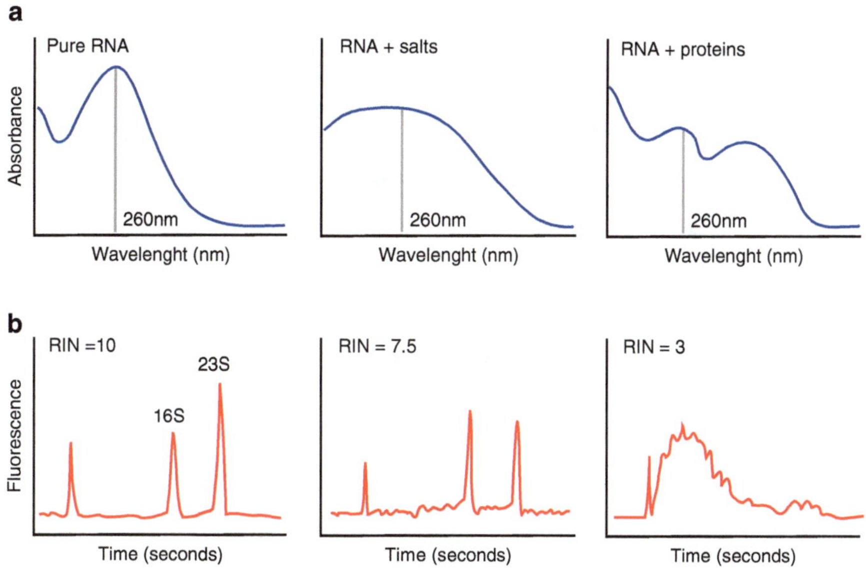

Fig. 13.2 Assessing RNA quality. (**a**) NanoDrop®'s absorbance UV spectrums, in the *left* plot an ideal sample with Pure RNA is shown, in the *middle* and *right* plots possible contaminations are shown. (**b**) Bioanalyzer® electropherogram profiles showing in the *left* plot the best case scenario with pure RNA, in the *middle* a plot of a partially degraded sample, and in the *right* a shred sample

rRNA (mRNA-Only™). Most mRNAs carry 5′-end triphosphates therefore are not degraded. 5′ monophosphate may be created by pyrophosphatase or endonuclease cuts. The advantage of this method is that sample diversity does not interfere; however, it requires very pure RNA as exonuclease is susceptible to inhibition by impurities; this also requires high RNA integrity (RIN > 8) otherwise exonuclease degrades both rRNA and mRNA (Fig. 13.3).

There are other strategies that enable deeper sequencing such as immunoprecipitations or duplex-specific nuclease digestion (DSN), these type of approaches only makes sense for specific experiments since strong bias is introduced. If your interest is to work with small RNA, these can be purified from an agarose gel. Specific biotinylated primers can be designed to eliminate other sequences, whether rRNA which are not recognized by hybridization kits or some other dominant messenger in the sample (Li et al. 2013).

Transcriptomic analyses are based on cDNA synthesis so the polarity (5′–3′) information is lost. The polarity of the transcripts can give important information for antisense RNA and novel transcripts identification. If your interest is to know the polarity, there are protocols that incorporate dUTP in the synthesis of the second strand, allowing subsequent removal by uracil-DNA-Glycosylase (UDG) treatment (Parkhomchuk et al. 2009).

The rapid development of sequencing technologies, and larger sequencing yields soon will make possible that rRNA would only need to be filtered in silico.

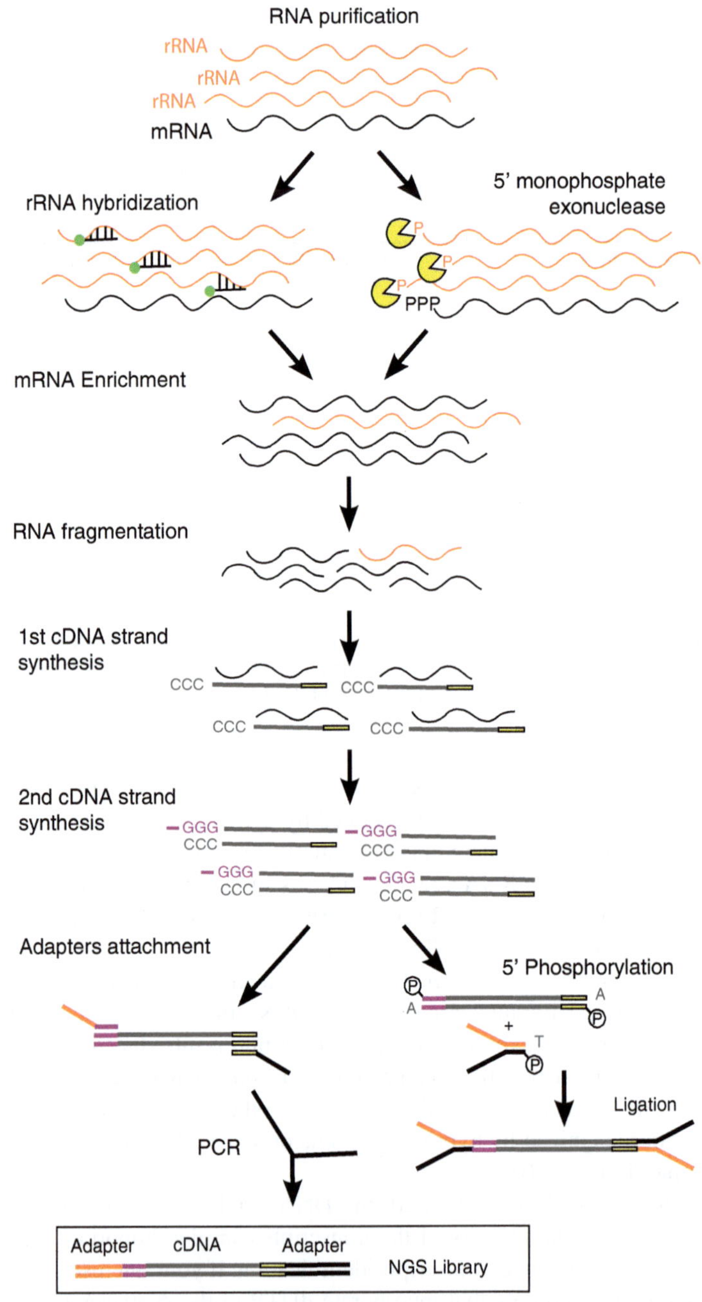

Fig. 13.3 The metatranscriptomics library preparation process. The main two strategies for mRNA enrichment are shown, first using rRNA separation by means of hybridization with 16S and 23S rRNA probes, and the second one is a depletion of rRNAs by means of a 5′-exonuclease. Then, first strand of cDNA is synthesized by means of reverse transcriptase using random hexamers. Second strand of cDNA is synthesized by a DNA polymerase. Finally, sequencing adapters need to be attached to the cDNA strands, and this could be done either by PCR or by ligation

13.6.9 Library Preparation

Regardless of sequencing platform that will be used, the general idea is the same: to produce cDNA of a certain size (50–400 bp) that is flanked by adapters. So library preparation requires fragmenting the RNA, first strand synthesis, second strand synthesis, coupling adapters, and validating the library. Sequence service providers can perform the library preparation.

cDNA should be of a certain size to optimize sequencing, depending on the platform is the size fragments must be. Fragmentation can be done with enzymes, metals, heat or sonication. Incubation times for fragmentation must be optimized for each case, as the integrity of each sample is usually different.

The synthesis of the first cDNA strand is performed by a reverse transcriptase and generally random hexamer primers are used. The synthesis of the second strand of DNA is done with a DNA polymerase. In this case, primers with guanines at 3′ are generally used since reverse transcriptase leaves a polyC overhang (Fig. 13.3).

Sequencing adapters include a region for binding to platform support and a region for primer hybridization. Additionally, they can include a barcode that serves to identify the sample if several samples are mixed in the same run (multiplexing). Depending on the used adapter kit is how many samples can be multiplexed. Illumina allows sequencing of the complementary strand, which allows for longer reads (pair-end). The adapters can be attached by a PCR or ligation reaction (Fig. 13.3).

Currently the most widely used platform is Illumina for which there are kits like TruSeq® and SMARTer®. The superiority of the former is that it allows multiplexing up to 96 samples while SMARTer® allows only 16 samples. The advantage of the latter is that you can start with 1 ng of enriched RNA whereas TruSeq® requires at least 100 ng (Alberti et al. 2014).

The last step is to validate the library. DNA concentration and size can be determined by the 2100-Bioanalyzer® coupled to a DNA chip like Agilent DNA 1000. We recommend contacting your sequencing provider, they have proven experience doing NGS on a daily basis, and they can assist you in fine-tuning the details about your samples. Sometimes your providers would even suggest some new sequencing platforms you have not noticed yet with higher yields at lower costs.

13.7 Bioinformatic Analyses

The metatranscriptome analysis involves a conceptual and technical challenge when dealing with huge amounts of multivariate information. There is an intermediate level of computing knowledge required to be able to deal with this data, and we want to provide some basic steps previous to metatranscriptome analysis that should be fulfilled if you do not have bioinformatics experience, the overall bioinformatic analysis process is summarized in Fig. 13.4.

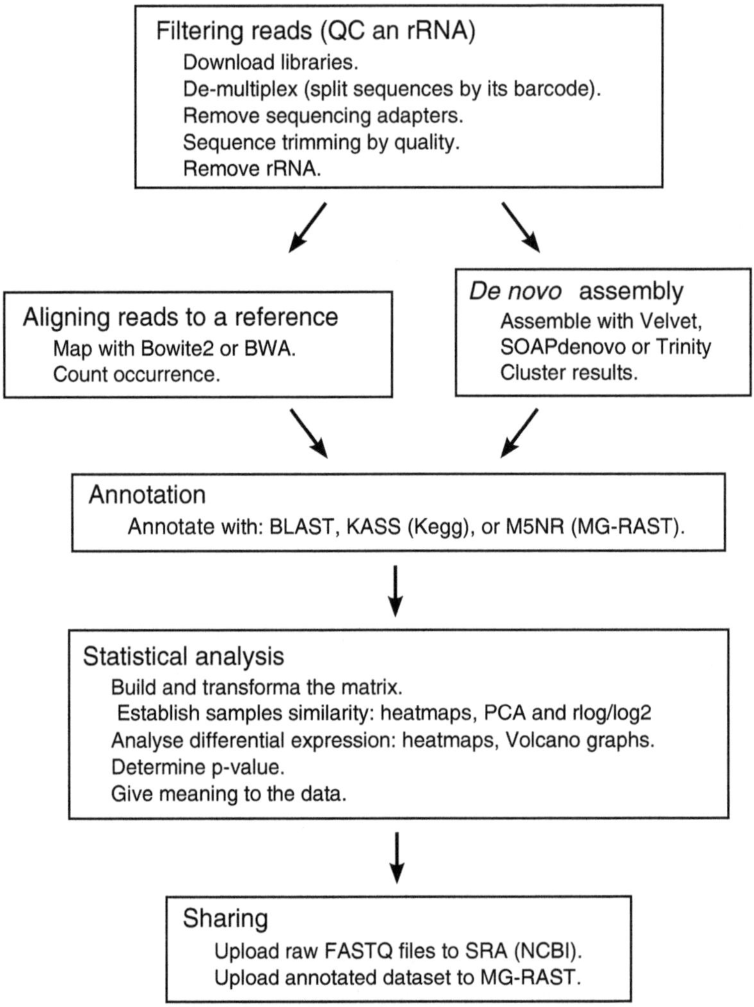

Fig. 13.4 The metatranscriptomics bioinformatic overall process. The main steps are: Filtering reads, choosing between aligning to reference sequences and performing de novo assembly, annotation, statistical analysis, and uploading the raw, assembled, and annotated data sets to the appropriate repositories

First basic steps:

1. Use the terminal (Linux, UNIX, Mac OS) or if you are running on Windows, immediately switch to Linux and learn how to use it. Use Ubuntu as it is the most supported Linux out there. And then, look for a Linux command line interface tutorial. Completing this exercise is highly recommended (see Table QG13.2).
2. Download your brand new transcriptome files from your provider FTP or provided URL. The file is normally a FASTQ, which is a text that contains both the

sequence read and the base calling, encoded in ASCII characters (non directly human readable). Use any web browser, the web browser negotiates different transfer protocols (FTP, HTTP) in a Graphic User Interface (GUI), or you could automate this with Linux/UNIX's commands like wget, rsync, curl, and ftp.
3. Unzip and manipulate the files only on the terminal, this means in the Command Line Interface (CLI, also known as terminal). If you are using your mouse and clicking the files to open/unzip them, you will be out of your computer resources pretty soon.
4. Install the compilers (transforms source code to an executable), this is mandatory to install software from source, on Ubuntu's terminal type:

```
$sudo apt-get install build-essential
```

For Mac OSX google: "Install the Command Line C Compilers in OS X" and follow the instructions.

5. Download your first program (fastx_toolkit, see Table QG13.2), and follow the install instructions.
6. Install R (see Table QG13.2).
7. Install Bioconductor (see Table QG13.2).
8. If you manage to do all the above tasks you are ready to install, and run almost any existent tools on Linux/UNIX.

If you do not want to improve your CLI skills, there are Graphical User Interfaces (GUIs) designed to cope with most of the sequence files processing like the Galaxy Server (see Table QG13.2). If you manage to do a local installation, you are doing it right. This is for the basic processing of the data, QC filtering, and trimming. Also, this is manageable by most of modern personal computers.

The overall process could be divided in the following stages: (1) Quality Control (QC), (2) Mapping against reference sequences, (3) de novo assembly, (4) annotation, (5) statistical analysis, (6) sharing your results. Each stage is described with useful hints at every step:

13.7.1 Sequences Quality Control

1. Split the libraries into individual files, this is also known as de-multiplexing, if you are using barcodes to mix several samples in a single run. Here the samples are split based on its barcode sequence.
2. Remove sequencing adapters. Removing this sequences that were used as templates for the sequencing is important and could help to further steps of mapping or assembly.
3. Quality Control, sequence trimming (and grooming). Each sequenced base has its own quality value, which is known as Phred score. Phred score serves as a proxy probability calculator, a Phred value of 30 accounts for 1 error every 1000 bases, or a 99.9 % of accuracy. This is a good standard to make a cut-off. Visualize the overall quality of your sequences via boxplots.

4. Filter rRNA. A quick way to do this step can be done with an rRNA DB and MegaBLAST (Altschul et al. 1997). There are other strategies using Interpolated Markov Models like Infernal and SSU-align and will help at this stage (Nawrocki et al. 2009; Nawrocki 2009).

The fast_toolx is a relatively easy way to perform the QC steps, plot qualities, and manipulate FASTA/FASTQ files. If command line is not an option, you should try the Galaxy servers to perform de-multiplexing, trimming adapters, and quality control (NGS QC and manipulation). The trade-off between working on the cloud or locally is the speed and fine tweaking of the pipelines, which are better controlled in our own computers. There are plenty of tutorials helping beginners to become familiar with Galaxy (see Table QG13.2; Kosakovsky et al. 2009).

13.7.2 Mapping Against Reference Sequences

1. Mapping against the reference metagenome/genomes. Use short read aligners. If there is no reference sequence(s), go to Sect. 13.7.3.

Here the standard options for short read mapping are Bowtie2 (Langmead et al. 2009), and BWA (Li and Durbin 2009). All of the mentioned programs are freely available online to be installed in CLI. There is also a cloud option provided by Galaxy under NGS Mapping. You should provide reference sequences, index the references if you are running this locally, and your metatranscriptome fastq files. After the alignment, you need to take the SAM/BAM resulting file and count the occurrence of each gene model (if available). The counting of each gene could be accomplished with R. R is a computer language intended for statistical computing and graphics, and the main recommended tool for downstream analysis (R Development Core Team 2004). For this purpose, use the libraries Rsamtools, summarizeOverlaps, and featureCounts of BioConductor (Huber et al. 2015).

13.7.3 De Novo Assembly

1. De novo assembly of metatranscriptome. This step applies if you do not have a reference, or you can do this step with the reads that were not aligned to it.

You can perform de novo assembly if you do not have reference sequences, keep in mind that the most frequent limiting factor during assembly is the amount of RAM memory of your computer. The amount of time required for assembly could last from minutes to days depending on the amount of sequences, and its complexity (repeats, SNPs, transcript forms, etc.). The most frequent choices are Velvet (Zerbino and Birney 2008), SOAPdenovo (Li et al. 2009), and Trinity (Grabherr et al. 2011). There is no clear better option when talking about assembly, you can try each one of them and can cluster the overall results at the end (with CD-HIT-est; Huang et al. 2010). All the mentioned programs are freely available online ready to be installed in your CLI. Trinity, has a cloud Galaxy based service that you could give a try (see Table QG13.2), this is recommended if you do not have enough computational resources locally.

13.7.4 Annotation

1. Annotate each transcript. If you have a metagenomic/genomic dataset already annotated, the coordinates could help you. Otherwise search by homology must be done. If there are no homolog sequences, you can try to use some RNA structural tools.

For the annotation, a hierarchical schema is suggested. If you know the species you are comparing and there are available annotated genome sequences for them, you could perform BLAST searches directly to them (Altschul et al. 1997). Then, for the sequences without homologs, go up to the next hierarchy a bacterial DB (see Table QG13.2). If there are still not homologs, try the largest DB, the NCBI's NR (see Table QG13.2). This could be tricky if you do not have the computational resources or the skills to perform it. Don't panic, there are some other cloud-based solutions like the KAAS, which is the KEGG's Automatic Annotation Server, where you can upload your assembled transcripts and annotate them, this is the most fast annotation tool that we are aware of (Moriya et al. 2007).

The other main web-server solution is MG-RAST, which has the most elegant DB design which is named M5NR (Wilke et al. 2012). M5NR merges information from plenty of Databases in a nonredundant way like the annotation ontologies COG, SEED, eggNOG, KEGG, UniProt, IMG, Patric, RefSeq, SwissProt, TrEMBL, GO, and the NCBI's NR (Tatusov et al. 2000; Overbeek et al. 2014; Powell et al. 2014; Kanehisa and Goto 2000; UniProt Consortium 2008; Markowitz et al. 2008; Wattam et al. 2014; Pruitt et al. 2005; The Gene Ontology Consortium 2014), all this information is accessible through the metagenomics analysis server (MG-RAST; Glass and Meyer 2012). This is the source to have the most cost-effective annotation pipeline for a regular wet-lab, though you will not learn any bioinformatic skill with this. The MG-RAST accepts uploads of FASTQ or regular FASTA files but be aware that you will need to upload some experiment metadata, the data remains private until you ask the MG-RAST system to release it to the public, so it also serves as a sequence repository.

If no homolog is present in your DB, you could use some tools like tRNA-SCAN (Lowe and Eddy 1997), and RNAFold (Denman 1993) to find out if there is a chance to classify your sequences by its secondary structure (i.e., hairpins, loops). The structural look at your data is demanding in computational and human resources to inspect the results. This approach could be useful if you are looking for particular class or regulatory elements (sRNAs, riboswitches). An excellent overview on annotation that should be reviewed, to understand the complexity of using multiple evidences to annotate, was done by Yandell and Ence (2012).

13.7.5 Statistical Analysis

1. Build a count matrix. This could be done by counting the mapped reads or to cluster the sequences of all the experimental conditions by its identity and count the number of occurrences in each sample/experiment. This step is required for parsing the annotation data to the Data Analysis pipeline. If you have processed your datasets on MG-RAST there is an option to export the whole dataset in BIOM format (http://biom-format.org/). The BIOM format is an acceptable input to R. There are ways to switch from BIOM to plain tabulator separated file with biom-convert tools. If you do not feel like using BIOM matrix, you could build a "table" where each row represents each individual gene and each column accounts for each sample/replica, save the file in plain text would work fine for R's input. In R, be sure to read the data as matrix.
2. Transform your matrix. There are several methods to accomplish this, one is the regularized-logarithm transformation (rlog), when measuring distances and sample similarities, and other normalizations like DESeq, which uses a negative binomial distribution, are preferred for differential expression. The log 2 and regularized logarithm transformation, also known as r-log, are the usual choice. This works to normalize your data between experiments, samples, and replicas, diminishing the importance and dependence of mean values. To perform this we recommend to use the R's Bioconductor package DESeq2 and its function RNAseqGene (Love et al. 2014).
3. Assess sample/treatment similarity, using heatmaps, Principal Component Analysis and calculating the distance on the r log/log 2 transformed data. With the transformed matrix, we can now describe the dissimilarity between samples/replicates/experiments by means of clustering analysis. The preferred option is to use heatmaps and Principal Component Analysis (or whatever ordination method you feel comfortable with). For this purpose we recommend to use the packages heatmap.2 and the function plotPCA, part of DESeq2 package.
4. Perform the differential expression analysis. In this point, you need to calculate the log 2 fold changes between your treatments (control vs. experiment). Here you will have to calculate the mean, log 2 fold change, its standard error, and test the null hypothesis that there is no change between treatments on each gene and, thus, reported as a *p*-value. For this step of the process, you could employ plenty of available tools some of the most used ones are: edgeR, DESeq, baySeq, NOISeq, and Cuffdiff (Trapnell et al. 2013; Tarazona et al. 2011; Hardcastle and Kelly 2010; Anders and Huber 2010; Robinson et al. 2010). The differences between the tools are based on what tests and assumptions they are based upon: Fisher's, negative binomial, parametric or nonparametric methods.

5. The p-value of RNA-seq is not what you are used to. You need to perform multiple testing correction, to calculate the amount of false discovery rate (FDR), or in other words the amount of false positives, and then assess the significance of the adjusted p-value. Remember that this is to answer how much of false positives could be accepted. There are multiple tools to calculate FDR and corrected p-values like metagenomeSeq which is available as part of Bioconductor and a standalone webserver (metastats), thus just working for pairwise control and experiment comparisons. This can also be done with DESeq2 package and its p-adjusted (p-adj) values.
6. Visualize the amount of significant differentially expressed genes. You can do this by means of Volcano plots, and heatmaps. If you are running a pairwise comparison, one way to accomplish this is by means of Volcano plots (log 2 fold changes versus significance), or an *MA* plot (M=log ratios, A=average). This is done also by R's Bioconductor.
7. Connect the most abundant features with its annotation. To this purpose is extremely helpful to use an ontology. An ontology is a controlled dictionary about gene functions, organized in hierarchical way like: SEED, COG, GO, KO. After determining the overall significant differentially expressed genes, usually they are coded with an identifier to reduce the amount of data loaded into R. A new table with the DE-genes and its annotations is extremely useful. To build that table the use or relational databases (MySQL, PostgreSQL) makes this an easy task.
8. Make sense of the known and annotated genes to direct new working hypothesis about their gene expression under the tested circumstances. The whole dataset of significant genes derived from the previous steps could be divided into two main groups: genes with known functions, and genes with unknown functions. Most of the functional analysis will focus on the known annotated genes, and it is the easier part of the dataset to explain but most probably a large amount of the data from your metatranscriptome will be transcripts of unknown function and thus are suitable candidates to design further experiments to discover their function (mutants, heterologous expression, etc.). An expressed gene is better than a total hypothetical predicted gene. For the genes with a known function, a process of data mining will be necessary to get the most about the functions and processes involving their participation. There are several starting points for gene function data mining like the Protein Data Bank, UniProt, Pfam, InterPro, EcoCyc, STRING, and KEGG (Berman 2000; Finn et al. 2008; Karp et al. 2002; Szklarczyk et al. 2011; Kanehisa and Goto 2000; Hunter et al. 2012). The main advantages of using those starting points is to gain insights about the current knowledge of the proteins and access to the overall information like if there are any available crystal structures, the phylogenetic distribution, known and predicted interactions. The main resource to integrate the information would be spending hours searching PubMed for related literature, and connecting it on new associations something that not machine, for the moment, could not do better than our brains.

13.7.6 Sharing Your Results

1. Upload your RNA-seq experiments to appropriate databases and repositories. To upload your datasets the main repository is NCBI's Short Read Archive (SRA) where you need to register your project and then upload your raw FASTQ files to it. To upload your assemblies there is the Transcriptome Shotgun Assembly Sequence DB (see Table QG13.2). The suggested way to share the annotated dataset is through the MG-RAST server, this also assures you to have up-to-date annotations, and it becomes available to be compared with other publicly available datasets.

13.8 Final Remarks

Metatranscriptomics as its relative metagenomics is attracting newcomers from multiple disciplines. The potential outcome to study both the environmental genome and its expression under certain conditions is a promising tool to describe the taxonomic and functional diversity out there. There is a hype about the meta-omics everywhere now, and everyone is trying to sequence; this is great and opens new opportunities to learn from a myriad of scientific perspectives. We just want to recommend to be cautious before getting into the omics fashion trend, and be aware that you need some prerequisites before getting into the adventure: a well-established and -equipped molecular biology laboratory, some computational hardware, and the most valuable asset of trained people on both experimental and analytical aspects. Take your time to plan the experimental design before getting started; do not be part of a growing disappointed crowd that ventures without any experimental design/controls and thus not able to get trustworthy biological meaningful data, or people with large experimental background but lacking the required analytical skills to tackle millions of multivariate data. Recognize your strengths and weaknesses, and go for successful collaborations; welcome to and good luck in the vibrant meta-omics road.

Acknowledgments This work was supported by the UNAM-DGAPA-PAPIIT grant IA200514, and the CONACyT Ciencia Básica grant 0237387.

Annex: Quick Reference Guide

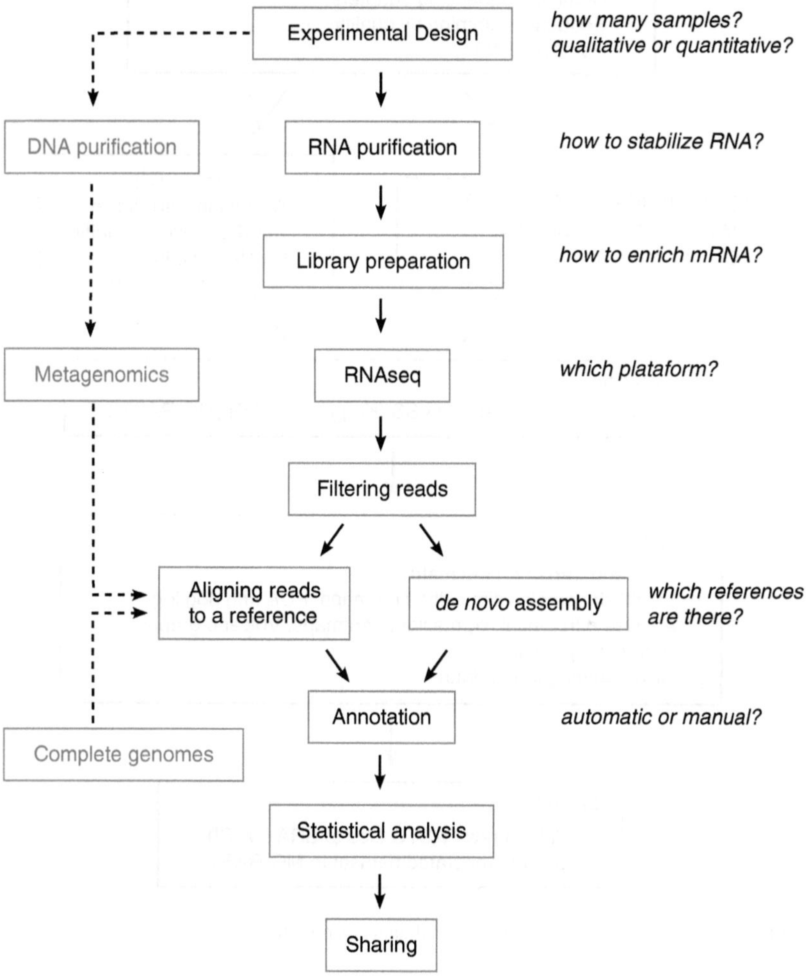

Fig. QG13.1 Representation of the wet-lab procedure workflow

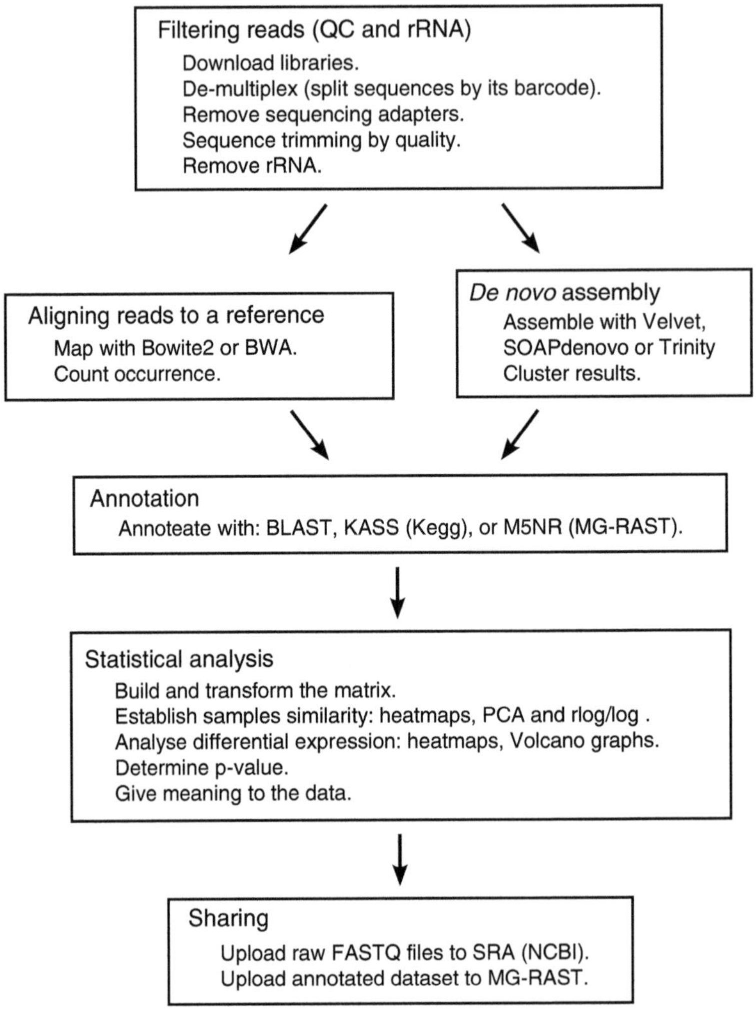

Fig. QG13.2 Main steps of the computational analysis pipeline

Table QG13.1 Experimental design considerations

Technique	Protocol	Control library	Recommended starting material (ng)	Number of replicates	Sequencing depth	Recommended sequencing platform and run
Metatranscriptomics	RNAseq	cDNA Input	Minimum of 1 ng (TruSeq®)	2 minimum for each library, for being able to estimate variances	If possible, estimate species diversity (16S rRNA amplicons). Then, calculate expected average genome sizes, then calculate an 8× minimum coverage	Illumina's MiSEQ 2× 300 bp (0.3–15 Gb, HiSeq 2500 (10–1000 Gb)
						First check with your provider to be able to use the latest technology
						Considerations:
						Quality
						Coverage
						Read length
						Sample number
		Control	Typically 100 ng cDNA			Budget
			RNA integrity, check NanoDrop UV spectrums and if possible RNA degradation with Bioanalyzer®			

Table that comprises relevant experimental design parameters, to carefully consider before applying this methodology

Table QG13.2 Available software recommendations

Software/resource	Notes	Method /language/ platform	Input	Results output	Results format	URL references
CD-HIT-est	The preferred selection for gene clustering at established cut-off thresholds	Executables	MultiFASTA file	Clusters file, and representative sequences	Raw text	http://weizhongli-lab.org/cd-hit/
		Linux/UNIX			Multi FASTA file	Huang et al. (2010)
Experimental Design in Differential Abundance analysis (EDDA)	Here you can design your RNAseq experiment	Web	Number of samples, replicates	Raw text	Raw text/html	http://edda.gis.a-star.edu.sg/dad/
		R packages				Luo et al. (2014)
		Any OS				
		Linux/UNIX				
FastX Toolkit	A useful software suite that allows to process initial quality control, screening, and filtering of sequencing files.	Executables	FASTQ	FASTA	MultiFASTA	http://hannonlab.cshl.edu/fastx_toolkit
		Linux/UNIX		plot images	png, pdf images (for quality boxplots).	–
Galaxy Server/ Tutorial	Galaxy allows to process and manipulate from raw sequences/data/ mapping files to plots.	Source	FASTA	Plots, alignments, mappings.	Plots (png, pdf), raw text, alignments (SAM, BAM)	https://usegalaxy.org/ https://usegalaxy.org/u/aun1/p/galaxy101 –
		Web server				Giardine et al. (2005)
		Linux/UNIX				
		Any OS	FASTQ and virtually all the known			

Glimmer-Gm	Metagenomic gene prediction	scripts Linux/UNIX	FASTA	MultiFASTA file with gene predictions Gene calling scores file	MultiFASTA Raw text	http://www.cbcb.umd.edu/software/glimmer-mg Kelley et al. (2012)
R	Is a programming language, intended for statistical computations and analyses.	Source, and executables Any OS	Tables, raw text, csv, JSON, BIOM, FASTQ	Raw text plots	Raw text PDF, png	http://cran.r-project.org/ Team (2008)
Bioconductor	Tools for the analysis and comprehension of high-throughput genomic data	R packages Linux/UNIX	FASTA FASTQ and virtually all the known formats	Raw text Html multiFASTA alignments	Txt html multiFASTA alignment files	http://bioconductor.org/install/ Huber et al. (2015)
Metagenomeseq	Determine differential abundant species, genes and features	Bioconductor R package	Abundance table (raw text, csv file)	Tables with p-values and False Discovery Rate (FDR) corrected values	Raw text	http://metastats.cbcb.umd.edu/software.html
Metastats		Web server Linux/UNIX Any OS	R object	Plots	png, pdf	Paulson et al. (2011)
KEGG's Automatic Annotation Server	Automated annotation server using Kyoto Encyclopedia of Genes and Genomes	Web server Any OS	MultiFASTA files	MultiFASTA files with annotated datasets Metabolic pathways maps	MultiFASTA raw text/html png images	http://www.genome.jp/tools/kaas/ Moriya et al. (2007)
Metagenemark	Metagenomic gene prediction	Web server Any OS	FASTA	MultiFASTA file with gene predictions Gene calling scores file	MultiFASTA Raw text	http://exon.gatech.edu/meta_gmhmmp.cgi Zhu et al. (2010)

(continued)

Table QG13.2 (continued)

Software/resource	Notes	Method /language/ platform	Input	Results output	Results format	URL references
Metagenomics analysis server MG-RAST	Web server to annotate high-throughput metagenomic experiments in both taxonomic and functional features. Integrates information from multiple databases into its M5NR DB	Web server	FASTQ, FASTA files	MultiFASTA files	MultiFASTA	http://metagenomics.anl.gov/
		Any OS, works best with Firefox			BIOM, txt, html	Meyer et al. (2008)
				Abundance tables	png, pdf	
				Heatmaps		
				Ordination plots		
NCBI NR DB	The reference DB to perform annotation, it includes all the known proteins deposited on GenBank.	Database	Already formatted DB ready to use with BLAST standalone versions	BLAST DB	Raw text/html	ftp.ncbi.nih.gov/blast/db/
NCBI's Transcriptome Shotgun Assembly (TSA)	TSA hosts assembled sequences of transcriptomes, by any method from traditional cDNA clone/sequencing to NGS datasets	Database	BAM unannotated assembly.	TSA record (Contigs)	GenBank file	http://www.ncbi.nlm.nih.gov/genbank/tsa/
			Sequences of at least 200 b			
			No more than 10 % ambiguous bases		FASTA file	
					ASN.1 file	
NCBI's Short Read Archive SRA	Stores raw sequence data from NGS, is the primary archive of NGS data.	Database	FASTQ	SRA files	Dump to FASTQ is available (FASTQ-dump)	http://www.ncbi.nlm.nih.gov/sra
			BAM	Raw sequencing reads		
			qseq			
			srf			
RNAfold Web Server	RNA secondary structure prediction	Web server	RNA sequence in FASTA format	Minimum free energy structure calculation	raw text, html	http://rna.tbi.univie.ac.at/cgi-bin/RNAfold.cgi
		Standalone version Linux/UNIX			PDF, png	Zuker and Stiegler (1981)

SOAPdenovo	Short Oligonucleotide Analysis Package, Designed as an Illumina's short read genome assembler	Executables	FASTA	Contigs	Raw text	http://soap.genomics.org.cn/soapdenovo.html
			FASTQ	Scaffolds		
			BAM	Mappings		
		64-bit Linux, minimum 5 G RAM		Pregraph		Luo et al. (2012)
Velvet assembler	Short read assembler	Scripts	FASTA	Contigs file	FASTA	https://www.ebi.ac.uk/~zerbino/velvet/
		Linux/UNIX			raw text	Zerbino and Birney (2008)
			FASTQ	Stats file	asm file (open with AMOS)	
				Velvet assembly file		
tRNAscan-SE	tRNA prediction software	Web server	FASTA	tRNA predictions	Raw text	http://lowelab.ucsc.edu/tRNAscan-SE/
		Scripts				
		Linux/UNIX	GenBank	Run statistics		Lowe and Eddy (1997)
			EMBL	Predicted tRNA structures.		
			GC			
			G			
			IG			
			Raw sequence			

Table displaying a selection of the recommended software available for the computational analysis of data yielded by this technique

References

Altschul SF et al (1997) Gapped BLAST and PSI-BLAST: a new generation of protein database search programs. Nucleic Acids Res, 25(17):3389–3402

Alberti A et al (2014) Comparison of library preparation methods reveals their impact on interpretation of metatranscriptomic data. BMC Genomics 15(1):912

Anders S, Huber W (2010) Differential expression analysis for sequence count data. Genome Biol 11(10):R106

Benítez-Páez A et al (2014) Microbiota diversity and gene expression dynamics in human oral biofilms. BMC Genomics 15(1):311

Berman HM (2000) The Protein Data Bank. Nucleic Acids Res 28(1):235–242

Cole JR et al (2009) The Ribosomal Database Project: improved alignments and new tools for rRNA analysis. Nucleic Acids Res 37(November 2008):141–145

Denman RB (1993) Using RNAFOLD to predict the activity of small catalytic RNAs. Biotechniques 15(6):1090–1095

DeSantis TZ et al (2006) Greengenes, a chimera-checked 16S rRNA gene database and workbench compatible with ARB. Appl Environ Microbiol 72(7):5069–5072

Finn RD et al (2008) The Pfam protein families database. Nucleic Acids Res 36(Database issue):D281–D288

Franzosa EA et al (2014) Relating the metatranscriptome and metagenome of the human gut. Proc Natl Acad Sci U S A 111(22):E2329–E2338

Frias-Lopez J et al (2008) Microbial community gene expression in ocean surface waters. Proc Natl Acad Sci U S A 105(10):3805–3810

Giardine B et al (2005) Galaxy: a platform for interactive large-scale genome analysis. Genome Res 15(10):1451–1455

Gilbert JA et al (2008) Detection of large numbers of novel sequences in the metatranscriptomes of complex marine microbial communities. PLoS One 3(8):e3042

Gilbert JA, Hughes M (2011) Gene Expression Profiling: Metatranscriptomics. Methods in Molecular Biology 733:195–205

Glass EM, Meyer F (2012) 13. Analysis of metagenomics data. In: Rodríguez-Ezpeleta N, Hackenberg M, Aransay AM (eds) Bioinformatics for high throughput sequencing. Springer, New York, NY, pp 219–229

Gosalbes MJ et al (2011) Metatranscriptomic approach to analyze the functional human gut microbiota. PLoS One 6(3):e17447

Grabherr MG et al (2011) Full-length transcriptome assembly from RNA-Seq data without a reference genome. Nat Biotechnol 29(7):644–652

Güell M et al (2011) Bacterial transcriptomics: what is beyond the RNA horiz-ome? Nat Rev Microbiol 9(9):658–669

Hardcastle TJ, Kelly KA (2010) baySeq: empirical Bayesian methods for identifying differential expression in sequence count data. BMC Bioinformatics 11(1):422

Hewson I et al (2009) Microbial community gene expression within colonies of the diazotroph, Trichodesmium, from the Southwest Pacific Ocean. ISME J 3(11):1286–1300

Huang Y et al (2010) CD-HIT Suite: a web server for clustering and comparing biological sequences. Bioinformatics (Oxford, England) 26(5):680–682

Huber W et al (2015) Orchestrating high-throughput genomic analysis with Bioconductor. Nat Methods 12(2):115–121

Hunter S et al (2012) InterPro in 2011: new developments in the family and domain prediction database. Nucleic Acids Res 40(Database issue):D306–D312

Kanehisa M, Goto S (2000) KEGG: kyoto encyclopedia of genes and genomes. Nucleic Acids Res 28(1):27–30

Karp PD et al (2002) The EcoCyc database. Nucleic Acids Res 30(1):56–58

Kelley DR et al (2012) Gene prediction with Glimmer for metagenomic sequences augmented by classification and clustering. Nucleic Acids Res 40(1):e9

Knight R et al (2012) Unlocking the potential of metagenomics through replicated experimental design. Nat Biotechnol 30(6):513–520

Kosakovsky Pond S et al (2009) Windshield splatter analysis with the Galaxy metagenomic pipeline. Genome Res 19(11):2144–2153

Langmead B et al (2009) Ultrafast and memory-efficient alignment of short DNA sequences to the human genome. Genome Biol 10(3):R25

Li H, Durbin R (2009) Fast and accurate short read alignment with Burrows-Wheeler transform. Bioinformatics (Oxford, England) 25(14):1754–1760

Li R et al (2009) SOAP2: an improved ultrafast tool for short read alignment. Bioinformatics (Oxford, England) 25(15):1966–1967

Li S et al (2014) Multi-platform assessment of transcriptome profiling using RNA-seq in the ABRF next-generation sequencing study. Nat Biotechnol 32(9):915–925

Li S-K et al (2013) Organism-specific rRNA capture system for application in next-generation sequencing. PLoS One 8(9):e74286

Love MI, Huber W, Anders S (2014) Moderated estimation of fold change and dispersion for RNA-seq data with DESeq2. Genome Biol 15(12):550

Lowe TM, Eddy SR (1997) tRNAscan-SE: a program for improved detection of transfer RNA genes in genomic sequence. Nucleic Acids Res 25(5):955–964

Lozupone C et al (2011) UniFrac: an effective distance metric for microbial community comparison. ISME J 5(2):169–172

Luo H et al (2014) The importance of study design for detecting differentially abundant features in high-throughput experiments. Genome Biol 15(12):527

Luo R et al (2012) SOAPdenovo2: an empirically improved memory-efficient short-read de novo assembler. GigaScience 1(1):18

Markowitz VM et al (2008) IMG/M: a data management and analysis system for metagenomes. Nucleic Acids Res 36(October 2007):534–538

Meyer F et al (2008) The metagenomics RAST server—a public resource for the automatic phylogenetic and functional analysis of metagenomes. BMC Bioinformatics 9:386

Moriya Y et al (2007) KAAS: an automatic genome annotation and pathway reconstruction server. Nucleic Acids Res 35(Web Server issue):W182–W185

Nawrocki EP (2009) Structural RNA homology search and alignment using Covariance Models. Washington University, St. Louis

Nawrocki EP, Kolbe DL, Eddy SR (2009) Infernal 1.0: inference of RNA alignments. Bioinformatics (Oxford, England) 25(10):1335–1337

Overbeek R et al (2014) The SEED and the Rapid Annotation of microbial genomes using Subsystems Technology (RAST). Nucleic Acids Res 42(5):1–9

Parkhomchuk D et al (2009) Transcriptome analysis by strand-specific sequencing of complementary DNA. Nucleic Acids Res 37(18):e123

Paulson J, Pop M, Bravo H (2011) Metastats: an improved statistical method for analysis of metagenomic data. Genome Biol 12(Suppl 1):P17

Powell S et al (2014) eggNOG v4.0: nested orthology inference across 3686 organisms. Nucleic Acids Res 42(Database issue):D231–D239

Pruitt KD, Tatusova T, Maglott DR (2005) NCBI Reference Sequence (RefSeq): a curated non-redundant sequence database of genomes, transcripts and proteins. Nucleic Acids Res 33(Database issue):D501–D504

R Development Core Team (2008). R: A language and environment for statistical computing. R Foundation for Statistical Computing, Vienna, Austria. ISBN 3-900051-07-0, http://www.R-project.orgR

Robinson MD, McCarthy DJ, Smyth GK (2010) edgeR: a Bioconductor package for differential expression analysis of digital gene expression data. Bioinformatics (Oxford, England) 26(1):139–140

Sambrook J, Russell D (2012) Molecular cloning: a laboratory manual, 4th edn. Cold Spring Harbor Laboratory Press, Cold Spring Harbor

Schloss PD (2010) The effects of alignment quality, distance calculation method, sequence filtering, and region on the analysis of 16S rRNA gene-based studies. PLoS Comput Biol 6(7):e1000844

Sorek R, Cossart P (2010) Prokaryotic transcriptomics: a new view on regulation, physiology and pathogenicity. Nat Rev Genet 11(1):9–16

Szklarczyk D et al (2011) The STRING database in 2011: functional interaction networks of proteins, globally integrated and scored. Nucleic Acids Res 39(Database issue):D561–D568

Tarazona S et al (2011) Differential expression in RNA-seq: a matter of depth. Genome Res 21(12):2213–2223

Tatusov RL et al (2000) The COG database: a tool for genome-scale analysis of protein functions and evolution. Nucleic Acids Res 28(1):33–36

The Gene Ontology Consortium (2014) Gene Ontology Consortium: going forward. Nucleic Acids Res 43(D1):D1049–D1056

Trapnell C et al (2013) Differential analysis of gene regulation at transcript resolution with RNA-seq. Nat Biotechnol 31(1):46–53

UniProt Consortium (2008) The universal protein resource (UniProt). Nucleic Acids Res 36(Database issue):D190–D195

Wattam AR et al (2014) PATRIC, the bacterial bioinformatics database and analysis resource. Nucleic Acids Res 42(Database issue):D581–D591

Westermann AJ, Gorski SA, Vogel J (2012) Dual RNA-seq of pathogen and host. Nat Rev Microbiol 10(9):618–630

Wilke A et al (2012) The M5nr: a novel non-redundant database containing protein sequences and annotations from multiple sources and associated tools. BMC Bioinformatics 13:141

Yandell M, Ence D (2012) A beginner's guide to eukaryotic genome annotation. Nat Rev Genet 13(5):329–342

Zerbino DR, Birney E (2008) Velvet: algorithms for de novo short read assembly using de bruijn graphs. Genome Res 18(5):821–829

Zhu W, Lomsadze A, Borodovsky M (2010) Ab initio gene identification in metagenomic sequences. Nucleic Acids Res 38(12):e132

Zuker M, Stiegler P (1981) Optimal computer folding of large RNA sequences using thermodynamics and auxiliary information. Nucleic Acids Res 9(1):133–148

Chapter 14
Eukaryotic Single-Cell mRNA Sequencing

Kenneth J. Livak

14.1 Introduction

Single-cell sequencing of the transcriptome and genome is entering the scientific mainstream after years when cost and technical obstacles made it a mere vision in the eyes of researchers in the field. Driving the change are new systems for the rapid, efficient processing of large numbers of individual cells, improved methods for amplifying genetic material, and the emergence of reliable high-throughput sequencing instruments.

Designating single-cell sequencing its Method of the Year in early 2014, Nature Methods said the choice marked a "turning point in the widespread adoption" of these techniques. The ability to examine DNA and RNA at single-cell resolution is already yielding major advances in a wide range of critical areas including the examination of clonal structures of tumors, the unbiased identification and characterization of distinct cell types, the decoding of transcriptomes of rare cells such as circulating tumor cells, the revelation of broad genetic diversity in neurons and other somatic cells, analysis of genomes of uncultivable microbes, and improvement of preimplantation screening of human embryos.

K.J. Livak, Ph.D. (✉)
Fluidigm Corporation, South San Francisco, CA, USA
e-mail: ken.livak@fluidigm.com

A.M. Aransay, J.L. Lavín Trueba (eds.), *Field Guidelines for Genetic Experimental Designs in High-Throughput Sequencing*,
DOI 10.1007/978-3-319-31350-4_14

The advances are part of a growing recognition of the extent of cellular heterogeneity. The wave of new data from single-cell research is unveiling cell-to-cell differences in genetic makeup, and to an even larger degree in gene expression, that had been invisible to standard bulk-cell analytic methods. This is because bulk methods yield results averaging all cells in a population or tissue sample. Examination of these differences and their causes and outcomes is leading to further revelations about cell function, cell states, cell type, and cell signaling and interaction.

This chapter reviews the most prevalent methods for obtaining mRNA sequence information from eukaryotic cells. The following chapter covers single-cell DNA sequencing. The steps in mRNA sequencing are: cell isolation, cell lysis, synthesis of cDNA using reverse transcriptase, initial amplification, fragmentation, attachment of sequencing adaptors and optional barcodes, library amplification, sequencing, initial processing of data, and analysis of quantitative results. These steps are discussed in more detail below.

This chapter has a limited scope: it focuses on single-cell mRNA sequencing from mammalian cells. Furthermore, much of the mRNA sequencing work has used cells in suspension, with studies on cells from tissue just beginning to emerge. Single-cell libraries can be prepared using adaptors for any sequencing platform; in practice, though, the vast majority of single-cell mRNA Seq publications have used Illumina-based sequencing. For this reason, this chapter focuses on preparing libraries for the Illumina® platforms.

14.2 Cell Isolation Methods

Four main methods are available for isolating cells for single-cell sequencing—micromanipulation, laser-capture microdissection (LCM), fluorescence-activated cell sorting (FACS), and microfluidic systems. Only the last two methods can provide the high throughput needed for the large sample numbers required to reliably reflect a cell population's heterogeneity. Here we discuss the features, strengths, and limitations of the four methods.

14.2.1 Micromanipulation

Micromanipulation is a common choice because of the ease of use and low equipment cost of the mainly manual process. Because it relies on observation with a microscope, it allows visual inspection of each cell. This enables focusing on just the cells of interest, but is also subject to human error. Limited throughput can be addressed by using devices that add a degree of automation and allow handling of up to 100 cells an hour (Choi et al. 2010). Optical tweezers employing laser technology enable measurement as well as manipulation of individual cells (Zhang and Liu 2008). Finally, patch pipettes can be used to draw RNA from individual cells but may fail to capture all of the desired material.

14.2.2 Laser-Capture Microdissection

Laser-capture microdissection (Frumkin et al. 2008), in which cells are cut away from sections of tissue with the assistance of high-resolution microscopy, is the only one of these four methods that retains spatial location information and thus allows more detailed analysis of how the cell relates to its microenvironment. But it requires skilled manual operation, and the process of first slicing the tissue with a laser and then extracting the target material may result in unintentional capture of adjacent material or failure to capture all of the target cell (Espina et al. 2006). Some extraction methods involve heat or adhesive material, which may compromise genetic sample integrity. With these challenges as well as limited throughput, LCM is a lesser-used option.

14.2.3 Fluorescence-Activated Cell Sorting

Fluorescence-activated cell sorting (FACS) is distinguished by the choice it provides between untargeted and targeted sampling. In untargeted capture, cells can be counted and measured by the scattering of laser light projected at them, then sorted as single cells into 96- or 384-well plates. In targeted capture, specific cell types are tagged with fluorescent probes and sorted based on the presence or absence of the specific markers. Targeted capture enables focusing sequencing costs on just the cell type (or types) of interest, as long as the cell type can be identified using cell-surface markers. Though it provides automated and high-throughput sorting (Dalerba et al. 2011), FACS relies on a high number of cells in suspension, and samples are later processed in multiple-well plates, incurring costs associated with both reagent volumes and the manual labor or robotics needed for well plate workflow. The advantage of accommodating high cell numbers is that a large number of single cells (up to thousands) can be collected in a few hours. This is valuable when an important sample is available only during a limited time window. Furthermore, these single-cell samples are stored at −80 °C, permitting processing and sequencing of a small batch of single cells and having the remaining samples as an archive that can be used if the preliminary results indicate more cells should be analyzed. Dependence on high cell number limits one of the key benefits of single-cell sequencing, analysis of rare cells. Finally, cells may be damaged by high flow rate in the instrument.

14.2.4 *Microfluidic Devices*

Microfluidic devices use liquid flow-through, micrometer-scale channels to isolate and capture cells. Unlike other methods, these systems incorporate reagent handling and enable execution of up to thousands of discrete reactions at a time in nanoliter chambers on a single instrument. These features combine very high throughput and high precision in reaction control, and have been shown to have correspondingly positive effects on accuracy and reliability. By increasing the effective concentration of rare samples, nanoliter-scale processing can also improve sensitivity. The precision and automation of microfluidic processing have been extended in recent years to additional workflow steps, now including culturing, lysing, amplification, and downstream analysis (Wang et al. 2012; White et al. 2011; Landry et al. 2013; Kellogg et al. 2014). The PDMS (polydimethylsiloxane) technology used in these publications is also the basis for the commercial C1™ system (Fluidigm, https://www.fluidigm.com/products/c1-system). The efficiencies of these systems can drastically reduce costs related to the reagent quantities, personnel, and time needed to perform experiments.

Whether by detaching cells from substrate material and surrounding cells or starting with cells in culture, cell isolation by definition occurs outside a natural microenvironment and may involve trauma, bringing inevitable and potentially substantial changes both to the transcriptomic material and to the laboratory results it yields. Studying, limiting, and accounting for these effects remains a central challenge in design of experiment and the handling of sample cells.

14.3 Cell Lysis

One advantage of working with the RNA from a single cell is that generally the RNA does not have to be purified. The volume of a single cell is on the picoliter scale. Even at the nanoliter scale used in microfluidics, the dilution factor is great enough that enzymatic reactions can be performed directly in the cell lysate. Each of the methods in Table 14.1 includes a lysis protocol, and there are many more in the literature. Commercially available lysis solutions include CelluLyser™ (TATAA Biocenter), RealTime ready™ Cell Lysis (Roche Diagnostics), and Single Cell-to-CT™ (Life Technologies). Han et al. (2014) describe a procedure for evaluating the efficiency of lysis. Cells are stained with a live cell cytosolic dye such as Calcein AM (Life Technologies). Then, under the microscope, the effects of different lysis agents are observed to see if the cells lyse and the dye is completely dispersed. It is important to check the efficacy of lysis because some types of cells will lyse with just water and others require harsher treatment (Ståhlberg et al. 2011).

A simple lysis solution to start with is 0.5 % NP-40, 50 mM Tris-HCl, pH 8.4, 1 mM EDTA. This provides a pH that is optimal for reverse transcriptase, and the NP-40 does not inhibit reverse transcriptase. Inclusion of a brief incubation (1–2 min) at 65–70 °C is sufficient to lyse many mammalian cells. If harsher lysis is

Table 14.1 Single-cell mRNA sequencing methods

Shorthand	Reference	Commercial kit	Library content	Initial amplification	Attachment of 5′ adaptor
Tang	Tang et al. (2009)		Whole transcript	PCR	TdT addition of As
STRT	Islam et al. (2012, 2014)		5′ end	PCR	Template switch
CEL-Seq	Hashimshony et al. (2012)	MessageAmp™ II aRNA Amplification Kit	3′ end	IVT	Ligation to 3′ end of aRNA
SMART-Seq	Ramsköld et al. (2012)	SMARTer® original and v3	Whole transcript	PCR	Template switch
Quartz-Seq	Sasagawa et al. (2013)		Whole transcript	PCR	TdT addition of As
Smart-Seq2	Picelli et al. (2013, 2014)		Whole transcript	PCR	Template switch
SCRB-Seq	Soumillon et al. (2014)		3′ end	PCR	Template switch

required, this solution can be supplemented with 30 μg/mL proteinase K and incubation at 50 °C for 30 min followed by 70 °C for 1 min. In this case, the proteinase K inhibitor AAPF (Cat. No. 539470, EMD Millipore) is included in the reverse transcriptase reaction at a concentration of 0.33 mM. Bengtsson et al. (2008) describe lysis with an even stronger agent, 0.5 M guanidine thiocyanate, but this requires a tenfold dilution to avoid inhibiting reverse transcriptase.

14.4 Library Preparation

Table 14.1 lists many of the single-cell mRNA Seq methods that have been published in the last few years. Figure QG14.1 in the Annex provides a flowchart of the wet lab workflow for the most widely used options. All methods to date use oligo dT priming to synthesize the first strand of cDNA. This has the benefit of greatly reducing the contribution of ribosomal RNA to the libraries but focuses the analysis on just mRNA. These methods can be distinguished by a number of characteristics.

14.4.1 Whole Transcript Versus End-Tag

In terms of content, the most important distinction is whether the method analyzes the whole transcript or is an end-tagging method. Whole transcript analysis is more comprehensive because it provides information about splice variants, the presence of mutations throughout the transcript, and the occurrence of transcripts that cross translocation or inversion breakpoints. In end-tagging methods, the library consists of fragments derived only from the 5′ end or only from the 3′ end of transcripts. Thus, each transcript contributes only one fragment to the library prior to amplification of the library. This simplifies quantification of RNA expression because the number of reads per gene can be more directly related to a count of the number of transcripts than with whole transcript methods. Because end-tagging methods result in the counting of the number of transcripts, they are sometimes referred to as digital transcript quantification or digital gene expression analysis. Although end-tagging methods do not provide information about the full-length transcript, there are advantages in the accuracy of quantification and workflow that are detailed below. The first decision to make in embarking on a single-cell mRNA Seq project is whether to use a whole transcript or end-tagging method. If the purpose of the study is to quantify RNA expression, then end-tagging methods are preferred. If it is important to detect and analyze splice variants and/or mutations, then a whole transcript method needs to be used.

Regardless of which route is chosen, it is prudent to include spike-in controls. The generally accepted control is the External RNA Controls Consortium (ERCC) mRNA spike-ins (Cat. No. 4456740, Life Technologies). Jiang et al. (2011) provide guidelines on the use of these spike-ins, and most of the methods in Table 14.1 describe how they are incorporated into each particular protocol. It is important to add the spike-ins as part of the cell lysis mix so that the reverse transcriptase and subsequent processing steps are identical for the cell RNA and the control RNA.

14.4.2 PCR Versus In Vitro Transcription

Adding adaptors for PCR. Because the amount of RNA in a single cell is so small (on the order of 10 pg total RNA per cell), single-cell libraries need to be amplified in order to generate enough template for sequencing. The next classification made in Table 14.1 is whether the initial amplification is performed using PCR or in vitro transcription (IVT). PCR has been the predominant method used, but PCR requires that adaptor sequences are appended to both ends of double-stranded (ds) cDNA. Using a primer for first-strand cDNA synthesis that has sequences added 5′ to the oligo dT segment enables adding an adaptor sequence to one end of the cDNA. This end will be called the 3′ end because it corresponds to the 3′ end of the original mRNA. The harder task is attaching an adaptor sequence to the 5′ end of the ds cDNA. The first single-cell mRNA Seq publication (Tang et al. 2009) accomplished this by using terminal transferase to add dAs to the 3′ end of the first-strand cDNA. This enabled using an oligo dT primer to synthesize the second-strand cDNA. Again, by appending sequences 5′ to the oligo dT segment, an adaptor sequence can be placed on the 5′ end of the ds cDNA as well. Although this works to some extent, this method requires multiple steps in order to generate cDNA molecules with adaptor sequences at both ends.

Template switch. This process is simplified by making use of the template switch mechanism in the initial reverse transcriptase reaction (Zhu et al. 2001). In template switch, the reverse transcriptase reaction not only contains an oligo dT primer for the initial priming of cDNA but also a template switch oligo with appropriate adaptor sequences. The template switch oligo is designed to hybridize to the 3′ end of the newly synthesized cDNA strand. When this occurs, reverse transcriptase can use the template switch oligo as a template and extend the first-strand cDNA to include the complement of the adaptor sequence in the template switch oligo. Template switch is very convenient because it enables adding adaptor sequences to both ends of cDNA in a single reverse transcriptase reaction. PCR primers that hybridize to the adaptor sequences on each end of the cDNA are then used to achieve the initial amplification of the cDNA library by PCR. Because of the convenience of adding adaptor sequences to both ends of cDNA in a single reaction, template switch has predominated in single-cell mRNA Seq publications.

In vitro transcription. CEL-Seq differs from the other methods in Table 14.1 because it uses in vitro transcription for the initial amplification of the cDNA library rather than PCR. CEL-Seq still uses an oligo dT primer with adaptor sequences for first-strand cDNA, but in this case the T7 promoter sequence is added to the 5′ end of the oligo dT primer. The use of the T7 promoter means that adaptor sequences only need to be added to the 3′ end of ds cDNA before the initial amplification is performed. After conventional second-strand cDNA synthesis, T7 RNA polymerase is used to make multiple RNA copies (termed aRNA) of the cDNA. These aRNA molecules are complementary to the original mRNA molecules. The 5′ ends of these aRNA molecules have adaptor sequences from the oligo dT primer. Adaptor sequences are added to the 3′ end of aRNA in a subsequent step of library construction.

IVT amplification is a linear process, as opposed to the exponential amplification of PCR. This means that the amplification factor is much smaller (a few hundred copies per original cDNA molecule as opposed to up to a million copies for PCR), but the opportunities for amplification bias are reduced. The exponential nature of PCR has the potential to generate extreme bias, although when properly optimized the actual occurrence of bias is greatly reduced (Devonshire et al. 2011). Still, there are those who feel safer using the linear process of IVT than the exponential process of PCR. For single-cell mRNA Seq, this argument has become moot with the advent of unique molecular identifiers (UMIs, see below).

14.4.3 Completion of Library Construction

After the initial amplification, the steps to complete library construction are the same as for conventional RNA Seq. For methods with PCR as the initial amplification, these steps are fragmentation, attachment of adaptors with optional barcodes, and PCR to append the P5 and P7 tags required for immobilization and amplification in the Illumina flow cells. The most convenient method for performing these tasks is to use the Nextera® XT DNA Sample Preparation Kit and Index Kit (Illumina). It is also possible to use the more conventional route of fragmentation by sonication or enzymatic treatment, end repair, A-tailing, ligation of Illumina sequencing adaptor, and PCR. For the IVT-based method, the library completion steps are optional fragmentation, ligation of an adaptor to the 3′ end of the aRNA, reverse transcriptase reaction to generate DNA, and PCR to append the P5 and P7 tags.

Each of the methods in Table 14.1 has a particular protocol that is followed for library completion. At this point, though, it is possible to substitute alternative protocols. To be clear, for single-cell mRNA Seq, it is critical to follow one of the methods in Table 14.1 through the initial amplification step. After initial amplification, there is leeway in how to complete library construction.

14.4.4 Sample Barcodes

For single-cell mRNA Seq, it is desirable to use sample barcodes so that libraries from multiple single cells can be pooled before adding to an Illumina flow cell. This reduces the overall sequencing cost of a project by amortizing the cost of running an Illumina sequencing lane over many single-cell samples. This is especially true now that it has been demonstrated that as few as 50,000 reads per cell are sufficient to distinguish different cell types (Pollen et al. 2014). For whole transcript methods, sample barcodes are generally added after fragmentation. This means that single-cell libraries have to be processed as individual samples through the steps of reverse transcriptase generation of cDNA, initial amplification, fragmentation, and attachment of adaptors with sample barcodes. For end-tagging methods, sample barcodes can be attached to one end of the cDNA during the reverse transcriptase step. This has a great benefit in terms of workflow because it means that at any point after the reverse transcriptase step, multiple samples can be pooled and any subsequent processing is performed on a single sample. Typically, 96 different barcodes are used, enabling pooling of 96 single-cell libraries. This requires synthesis of 96 different oligos, each with a unique sample barcode. For 5′-end-tagging, this would be 96 different template switch oligos. For 3′-end-tagging, this would be 96 different oligo dT oligos. As these oligos are expensive to synthesize, this is an upfront expense that needs to be considered. If it is anticipated that thousands of single cells will eventually be analyzed, the oligo cost per single cell is minimal.

The capacity for pooling single-cell libraries can be increased by attaching sample barcodes to both ends of the ds cDNA to be sequenced. Consider the example of 3′-end-tagging performed with 96 different oligo dT primers, each with a unique sample barcode. This places a sample barcode at the 3′ end of ds cDNA and enables pooling 96 single-cell libraries at any point after the reverse transcriptase step. After fragmentation, the adaptor attached to the 5′ end of the ds cDNA can also have a barcode. If adaptors with eight different barcodes are used, then eight pools of 96 samples can be pooled after the adaptor attachment step. This enables sequencing $8\times96=768$ single-cell libraries on one lane of an Illumina sequencer. This dual barcoding is what is used in the Nextera XT DNA Sample Preparation Kit for tagmentation, and is termed dual-indexing by Illumina.

14.4.5 Unique Molecular Identifiers (UMIs)

Kivioja et al. (2012) describe the process of counting the absolute number of molecules using unique molecular identifiers (UMIs). It is simplest to think of a UMI as a random sequence label (generally N_4 to N_{10}) attached to a molecule to be sequenced prior to any amplification step. For conventional RNA Seq, quantification is inferred from the total number of reads that map to each particular transcript. With UMIs, quantification is done by counting the number of unique UMIs observed for each particular transcript regardless of how many times each UMI is read. For the whole transcript methods in Table 14.1, fragmentation to sequencing-sized pieces occurs after the initial amplification. This precludes using UMIs with any of these methods. Thus, for the methods in Table 14.1, the use of UMIs is restricted to 5′-end-tagging and 3′-end-tagging methods. For 5′-end-tagging, the UMI is incorporated into the template switch oligo. For 3′-end-tagging, the UMI is incorporated into the oligo dT primer.

The concept of UMIs was introduced by Hug and Schuler (2003), who referred to them as tag sequences. In addition to Kivioja et al. (2012), Shiroguchi et al. (2012), Islam et al. (2014), and Fu et al. (2014) describe how to apply UMIs to RNA Seq. Shiroguchi et al. (2012) carefully designed 20 nt (nucleotide) UMIs to minimize miscategorization due to sequencing errors, avoid secondary structure, eliminate significant overlap or complementarity with each other or with primer and adaptor sequences, and avoid sequence motifs known to be problematic for sequencing chemistries. Islam et al. (2014), on the other hand, used a 5 nt random sequence UMI. The marginal benefit of the careful design is offset by the fact that it uses 20 bases of sequencing capacity to read the UMI. Thus, the use of a random-nucleotide UMI is indicated. For very abundant transcripts, there is some chance that the same UMI sequence will be used more than once. This effect can be corrected for based on probability calculations, but it does introduce some error into the quantification of very abundant transcripts. In terms of the number of random nucleotides to use for the UMI, N_5 should be adequate for most mammalian cells. For larger cells, which would be expected to have a greater number of total transcripts, the use of N_6 or N_7 might be prudent.

14.4.6 Identifiers and End-Tagging

The use of sample barcodes and UMIs provides advantages for end-tagging methods compared to whole transcript methods. For sample barcodes, the advantage is in terms of workflow. Pooling 96 single-cell samples after the reverse transcriptase step means that all subsequent steps are done on a single sample, reducing sample handling complexity, labor, and the cost of reagents. The whole transcript methods add sample barcodes at a much later step, and thus much of the library processing steps must be done on individual single-cell samples. For UMIs, the advantage is in terms of the accuracy of quantification. UMIs are introduced during the reverse transcriptase step before any amplification of the cDNA library. By basing quantification on counting the number of unique UMIs per transcript, the possibility that amplification bias might distort quantification is greatly reduced. Thus, the tradeoff between end-tagging and whole transcript methods is better quantitative information versus information about the structure and sequence of the entire transcript.

14.5 Sequencing

One of the surprises in comparing bulk mRNA Seq to single-cell mRNA Seq is that read depth per cell does not need to be particularly high. There are two reasons for this. The first stems from the nature of eukaryotic transcription. A growing body of evidence accumulated over the last 10 years indicates eukaryotic transcription occurs in bursts or pulses (e.g., Dar et al. 2012). Consequently, single-cell RNA expression data is inherently noisy. The correlation coefficient comparing transcript levels for two seemingly identical cells can be on the order of 0.6. Only by averaging data from multiple cells is it possible to achieve correlations with bulk data above 0.9. The variability in sampling due to shallow read depth is insignificant compared to the variability between single cells. The way to address this noise is to obtain data from many single cells (on the order of hundreds to thousands). Thus, funds spent on sequencing are better used to obtain data from a larger number of single cells than to sequence any particular cell to great depth. The need to process a large number of cells is also why high-throughput methods are so critical for single-cell mRNA Seq analysis.

The second reason is empirical. Two studies (Jaitin et al. 2014; Pollen et al. 2014) have shown that as few as 50,000 reads per cell are sufficient to distinguish distinct cell types. This is true even for closely related neural cell types (Pollen et al 2014). This finding seems to be independent of the mRNA Seq protocol, as Jaitin et al. (2014) used an end-tagging method and Pollen et al. (2014) used a whole transcript method. Because many studies involve more than just distinguishing cell types, sequencing at depths greater than 50,000 reads per cell is usually warranted. A practical guide is to pool 96 single-cell libraries and analyze on a single lane of a HiSeq®. This should generate on the order of one million to two million reads per cell. If 50,000 reads per cell are sufficient for a study, then an end-tagging method with dual barcodes can be used to combine multiple 96-cell pools. For example, 20 pools can be generated that have one of 96 sample barcodes introduced in the reverse transcriptase step. If 20 distinct barcodes are added to the other end of the library fragments, then all 20 pools can be combined to generate data from 1920 (20×96) cells on a single lane of a HiSeq with a read depth of 50,000–100,000 reads per cell.

Table QG14.1 in the Annex summarizes the experimental design options for the most widely used methods in Table 14.1. For whole transcript methods, typically paired-end reads at 50 nt per read are performed. Depending on how sample barcodes are incorporated into the libraries, one or two index reads may also be required. For STRT as published, single-end reads of 50 nt and an index read of 8 nt were used. For CEL-Seq and SCRB-Seq, paired-end reads are used but the read length is not equally divided between the two reads. Both of these methods incorporate the initial sample barcode and the UMI into the oligo dT primer to be in line with read 1 (i.e., not as an index read). Read 1 is used to read the sample barcode and UMI but then stopped because the subsequent bases are the Ts of the oligo dT segment. Thus, the length of read 1 is determined by the length of the sample barcode plus the length of the UMI. If dual barcodes are being used, then there is an index read. Finally, the remainder of the reads is used for read 2. The published account of SCRB-Seq used 17 cycles on read 1 to decode the sample barcode and UMI, an 8-cycle index read to decode the second barcode, and a 34-cycle read 2 to sequence the cDNA. This is possible because the Illumina kits nominally labeled 50 cycles can actually perform slightly more than 50 cycles.

14.6 Initial Processing of Data

Table 14.2 lists tools that can be used for the initial processing of the raw sequence data obtained from the sequencer, and Fig. QG14.2 in the Annex is a flowchart of the steps. This list is not meant to be exhaustive, but rather to present the programs that are most widely used to get from raw sequence data to quantification of transcript levels.

14.6.1 Early Screening of Results

Generally, reads not considered valid by the Illumina software are discarded. Then, FastQC can be used as a simple way to perform a preliminary quality check on the raw sequence data. This may indicate whether there have been problems in library construction. For example, on the FastQC website there is a report for a run contaminated with adaptor dimer. It is sometimes necessary to trim adaptor sequences, and this can be easily done with Cutadapt. PRINSEQ combines quality assessment with the ability to filter or trim reads with user-defined options, including by quality score. Finally, sample barcodes and UMIs need to be extracted for downstream binning of reads and then removed from the reads prior to alignment.

14.6.2 Reference Alignment

After removal of known extraneous sequences, the reads need to be aligned to a reference. The most general alignment is to the genome using programs like BWA and Bowtie. If ERCC spike-ins have been used, the indexed genome reference file should be appended to include an artificial chromosome consisting of a concatemer of the ERCC control sequences. Because of splicing, there will be reads that do not align to the genome but are still derived from mRNA. Tools like TopHat, STAR, and

Table 14.2 mRNA sequencing data processing tools

Step	Tool	Reference	Link
Grooming	FastQC		http://www.bioinformatics.bbsrc.ac.uk/projects/fastqc/
	Cutadapt	Martin (2011)	http://code.google.com/p/cutadapt/
	PRINSEQ	Schmieder and Edwards (2011)	http://prinseq.sourceforge.net/
Alignment	BWA	Li and Durbin (2010)	http://bio-bwa.sourceforge.net/
	Bowtie	Langmead et al. (2009), Langmead and Salzberg (2012)	http://bowtie-bio.sourceforge.net/index.shtml
	TopHat	Trapnell et al. (2009), Kim et al. (2013)	http://ccb.jhu.edu/software/tophat/index.shtml
	STAR	Dobin et al. (2013)	https://github.com/alexdobin/STAR/releases
	HISAT	Kim et al. (2015)	https://github.com/infphilo/hisat
Quantification of levels	RSEM	Li and Dewey (2011)	http://deweylab.biostat.wisc.edu/rsem/
	rpkmforgenes	Ramsköld et al. (2009)	http://sandberg.cmb.ki.se/media/data/rnaseq/instructions-rpkmforgenes.html
	Cufflinks	Trapnell et al. (2012, 2013)	http://cole-trapnell-lab.github.io/cufflinks/

HISAT are used to align these reads to the genome and enable mapping to both known and novel splice sites. It is also possible to use BWA and Bowtie to align directly to transcripts by using an indexed RefSeq reference file. This provides quicker alignment and annotation to known transcripts, but precludes the detection of novel transcripts and splice variants.

14.6.3 Assembly of Transcripts

The end result of alignment to the genome is reads mapped to genomic coordinates. These mappings need to be assembled into transcripts and then a quantitative measure for each transcript can be determined. The assembly can be to known transcript models, such as those available from the UCSC Genome Browser (Meyer et al. 2013). The tool rpkmforgenes performs the task of comparing aligned reads to a reference file of known transcripts and determining an RPKM value for each of the annotated transcripts (see below for a discussion of units). Cufflinks and RSEM provide the additional capability of de novo transcript assembly, enabling the detection and quantification of novel transcripts. Each of these programs has its own methods for handling reads that map to multiple genes or isoforms. The output from Cufflinks is RPKM (reads per kilobase per million reads) or FPKM (fragments per kilobase per million reads), and the output from RSEM is TPM (transcripts per million, see below for a discussion of units).

14.6.4 Quantitative Units

The original quantitative unit for RNA Seq data is RPKM (Mortazavi et al. 2008). This is defined as reads mapped to a feature (transcript or exon) divided by the length of the feature in kilobases and also divided by the total number of reads in million-read units. Thus, RPKM attempts to normalize for both read depth of the library and for length of the transcript (or exon). RPKM is used for single-end reads. With paired-end reads, RPKM becomes FPKM (Trapnell et al. 2010). This is because two reads are used to determine a single mapped fragment. As pointed out by Wagner et al. (2012), though, there is a flaw in the composition of the denominator of RPKM (or FPKM) that leads to inconsistencies in comparing RPKM values between different samples. They propose the unit TPM, where normalization is to the estimated number of transcripts sampled in a sequencing run rather than the total number of reads. Wagner et al. (2012) provide a formula for converting between RPKM and TPM. TPM is becoming the preferred unit for expressing the quantitative results of RNA Seq. The need to decide on units is obviated by the use of UMIs. The number of unique UMIs for each transcript is a digital count of the number of transcript molecules detected per cell.

14.6.5 Reporting of Data

The end result of primary processing is a giant table where each row is a different gene and each column is a single cell. For whole transcript methods, each entry is the TPM value for that particular gene and cell. For end-tagging methods with UMIs, each entry is the number of unique UMIs observed. This table is similar to the data obtained from microarrays and can be analyzed and displayed using the tools developed for microarrays. The only difference of note is that the data are extremely sparse. That is, most entries are zero. Because of the zeroes, conversion to log space is performed by first adding one to the transcript count, e.g., $\log_2(\text{TPM}+1)$ or $\log_2(\text{UMI count}+1)$.

14.7 Analysis of Quantitative Results

With growing recognition of the potential of single-cell RNA Seq for a range of applications will come the need to measure its technical strengths and limits in those lines of research. Here we review means of comparing the accuracy, sensitivity, reproducibility, and other aspects of single-cell versus bulk sequencing, and of different single-cell methods.

The most straightforward measure of sensitivity is number of genes detected per cell. This will, of course, depend on the type of cell being analyzed. For the references in Table 14.1, the range reported was 2000–10,000 genes detected per cell. This metric can be used to assess the quality of any particular single-cell library. If the number of genes per cell is significantly below the median for all cells of that type in the study, this can be used as a criterion for removing that cell from further analysis.

Assessing reproducibility is complicated by the fact that each single cell is a biological replicate, not a technical replicate. Thus, how well the results of two single cells correlate is not a good measure of the reproducibility of the method. This is why the ERCC spike-ins are useful. Because the same amount of spike-ins is added to each single-cell lysate, the spike-ins provide the technical replication required to measure reproducibility. Another good way to assess reproducibility is to pool the results from multiple single cells. The average result from 100 single cells should show a good correlation to bulk RNA sequencing data for the same type of cells.

Wu et al. (2014) is a useful example of how to appraise single-cell mRNA Seq results to compare different methods and different preparation techniques, and to compare single-cell with bulk results. To evaluate accuracy, they focused on 40 transcripts and compared results determined by single-cell qPCR with those determined by single-cell mRNA Seq. The correlations were sufficient to indicate that single-cell mRNA Seq can indeed be used for the quantitative measurement of RNA expression. Interestingly, they found that the nanoliter reaction volumes enabled by microfluidics improve accuracy as well as sensitivity and reproducibility.

14.8 Downstream Quantitative Analysis

It is often useful to obtain an initial look at the data before delving into the more sophisticated analyses described below. For this purpose, multidimensional methods of analysis are preferred because of the large number of genes involved and because of the large cell-to-cell variation observed for any particular gene transcript. The most widely used methods are hierarchical clustering and principal component analysis. A number of freeware and commercial packages can be used for this task. The Singular™ Analysis Toolset (https://www.fluidigm.com/software) is particularly effective because it is designed specifically for the display of single-cell data.

The greatest challenge in analyzing single-cell mRNA Seq results is how to handle the noise, both biological and technical, in the data. This is a dynamic field, so it is impossible to state categorically how best to do this at the present time. Table 14.3 lists recent publications that have tried to address the challenges of noise in single-cell RNA expression data. Table QG14.2 in the Annex lists additional information about readily available software packages from a subset of these publications.

Brennecke et al. (2013) provide a quantitative statistical tool to distinguish true biological variability from technical noise. This method exploits spike-ins to quantify how technical noise varies with expression level and then uses this information to assess the statistical significance of cell-to-cell variation. Basically, this method

Table 14.3 Single-cell mRNA sequencing downstream analysis tools

Title	Reference	Link
Accounting for technical noise in single-cell RNA-seq experiments	Brennecke et al. (2013)	http://www.nature.com/nmeth/journal/v10/n11/extref/nmeth.2645-S2.pdf
Inferring the kinetics of stochastic gene expression from single-cell RNA-sequencing data	Kim and Marioni (2013)	
The dynamics and regulators of cell fate decisions are revealed by pseudotemporal ordering of single cells	Trapnell et al. (2014)	http://cole-trapnell-lab.github.io/monocle-release/
Validation of noise models for single-cell transcriptomics	Grün et al. (2014)	
Bayesian approach to single-cell differential expression analysis	Kharchenko et al. (2014)	http://pklab.med.harvard.edu/scde/index.html
Single-cell RNA-seq reveals dynamic paracrine control of cellular variation	Shalek et al. (2014)	
Cell types in the mouse cortex and hippocampus revealed by single-cell RNA-seq	Zeisel et al. (2015)	https://github.com/linnarsson-lab/BackSPIN

provides a statistical yardstick for deciding whether expression noise exceeds technical noise.

Kim and Marioni (2013) present a statistical framework based on a Poisson-beta model that specifically addresses the transcriptional bursting of eukaryotic RNA expression. Their analysis enables estimating parameters such as burst size and burst frequency from single-cell mRNA Seq data.

Monocle (Trapnell et al. 2014) addresses a different aspect of single-cell noise. This is the asynchrony noise observed when a developmental process is induced and monitored in vitro. Using single-cell mRNA Seq data collected at different time points, Monocle reorders the individual cells based on developmental progress (pseudotime axis) rather than along the experimental time axis. This enables identification of genes that share distinct variation motifs and improves resolution in dissecting regulatory pathways.

Grün et al. (2014) use control "single cells" to assess technical noise. These "single-cell" controls were created by taking 20 pg aliquots from RNA extracted from 1 million cells. Analysis of these controls identified two sources of technical noise: Poisson sampling for low-expression transcripts, and global sample-to-sample variation in sequencing efficiency for high-expression transcripts. These insights were used to develop three noise models to correct for the technical noise in mRNA Seq results. The accuracy of using these models was validated by comparing to single-cell results determined by smFISH (single-molecule fluorescent in situ hybridization).

In single-cell RNA expression experiments, failure to detect the transcript from a particular gene in a cell is an ambiguous result. It could mean the gene is inactive in that cell, the gene is active but the transcript was not detected because of the burst kinetics of eukaryotic transcription, or technical noise interfered with detection. Kharchenko et al. (2014) use Bayesian statistics to develop a probabilistic model to correct for the distortions in single-cell expression measurements due to dropout (non-detected) events. This model can be used to detect differential expression and to identify subpopulations in a manner that is more tolerant of single-cell noise.

The prevalence of non-detection events means that single-cell RNA expression data has a dualistic nature. In cells where a particular transcript is detected, there is a continuous distribution of expression levels. This can be thought of as the analog aspect of single-cell data. Yet, there is often a group of cells where the transcript is not detected and expression can be characterized as on or off. This can be viewed as the digital aspect of single-cell data. Similar to the way that McDavid et al. (2013) analyzed single-cell qPCR data, Shalek et al. (2014) use three parameters to characterize the expression profile of each transcript. For cells where the transcript is detected, the mean (μ) and variance (σ^2) are used to capture the analog aspect of the profile. The third parameter (α) is the fraction of cells where the transcript is detected, and thus captures the digital aspect of the profile. The analog aspect of the data could also be expressed by fitting the detected data to a gamma distribution and using the shape parameter α and inverse scale parameter β to characterize the distribution.

The most prevalent application of single-cell RNA Seq to date has been the unbiased identification of distinct cell types. Standard hierarchical clustering algorithms have difficulty dealing with the sparse nature of single-cell RNA Seq data sets. This is because most genes are not informative in most pairwise comparisons, and only introduce noise into the analysis. Zeisel et al. (2015) developed BackSPIN, a program based on a method called divisive biclustering, and used it to identify 47 molecularly distinct subclasses of cells in the mammalian cerebral cortex.

14.9 Concluding Remarks

It is difficult to write a chapter like this in such a fast-moving field. Improvements and innovations that are being worked on will soon make some of the material in this chapter obsolete. Two particular technical advances are worth noting. The first is use of microfabrication or droplets to increase the number of reaction vessels for generating single-cell RNA libraries (e.g., Fan et al. 2015; Macosko et al. 2015; Klein et al. 2015). This will enable efficient processing, in terms of workflow and expense, of thousands of cells. These methods are not yet generally accessible but appear certain to come to prominence during 2015. The second technical advance is the capability to analyze multiple analytes from the same single cell. A good example of this is DR-Seq (Dey et al. 2015), which is a method for sequencing both the genome and the transcriptome from the same cell.

Annex: Quick Reference Guide

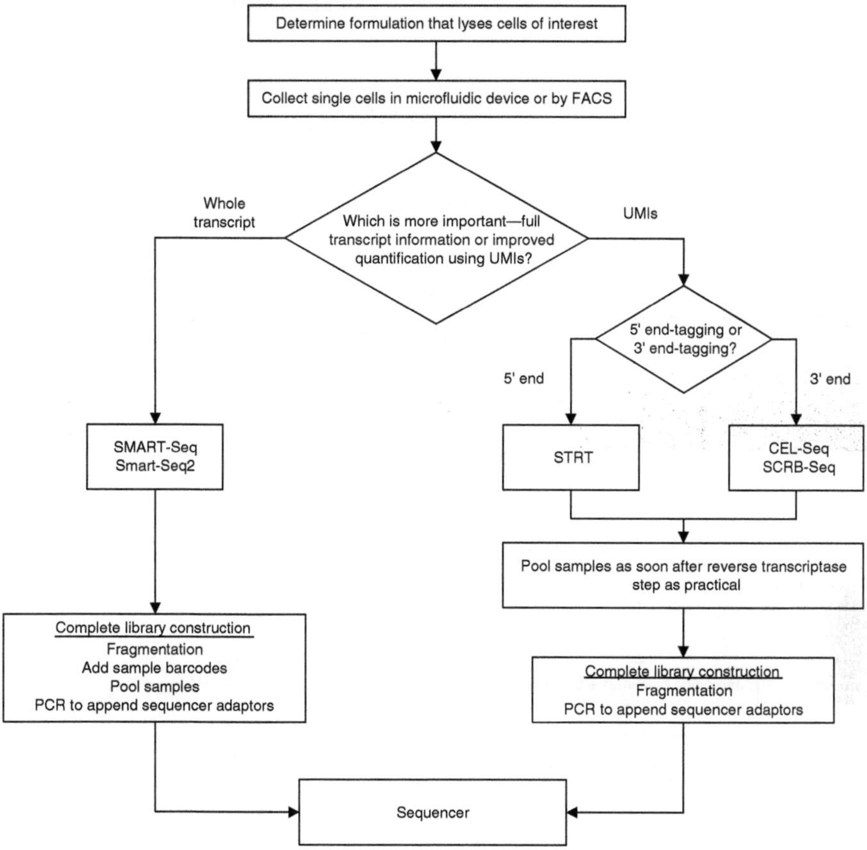

Fig. QG14.1 Representation of the wet-lab procedure workflow

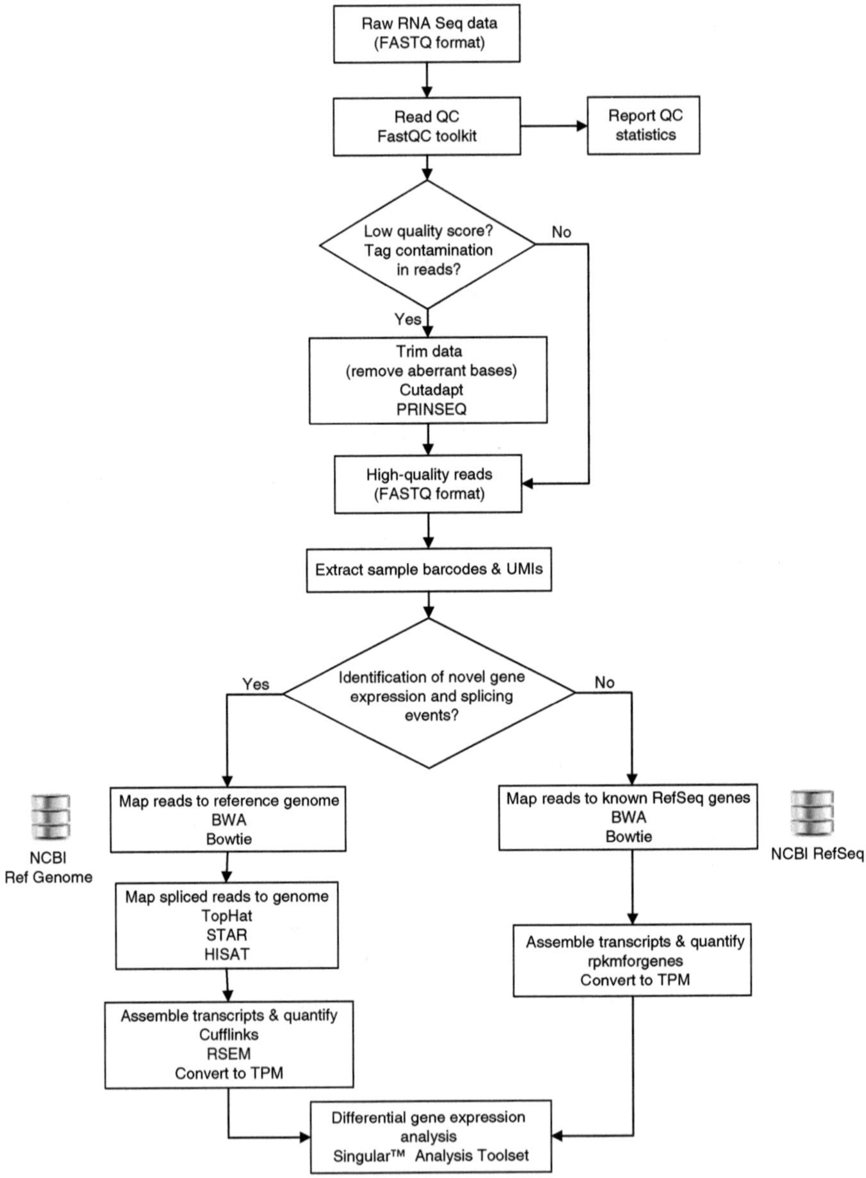

Fig. QG14.2 Main steps of the computational analysis pipeline

Table QG14.1 Experimental design considerations

Method	Controls	Sequencing depth	Sequencing setup	Sequencing reads	References
SMART-Seq and Smart-Seq2	ERCC RNA controls spiked into lysis buffer	1–2 million reads per cell	Pool 96 cells and load onto one lane of HiSeq	50 nt paired-end reads and 2 index reads	Ramsköld et al. (2012)
					Picelli et al. (2013, 2014)
STRT	ERCC RNA controls spiked into lysis buffer	50,000 to 2 million reads per cell	Pool 96–1536 cells and load onto one lane of HiSeq	50 nt single-end read and 1 index read (second index read if pooling is >96)	Islam et al. (2012, 2014)
CEL-Seq and SCRB-Seq	ERCC RNA controls spiked into lysis buffer	50,000 to 2 million reads per cell	Pool 96–1536 cells and load onto one lane of HiSeq	10–17 nt read 1 (for sample barcode and UMI); 34–50 nt read 2 (for transcript identity); and 1 index read if pooling is >96	Hashimshony et al. (2012)
					Soumillon et al. (2014)

Table that comprises relevant experimental design parameters, to carefully consider before applying this methodology

Table QG14.2 Available software recommendations

Software	Title	Reference	Link	Results output	Results format
	Accounting for technical noise in single-cell RNA-seq experiments	Brennecke et al. (2013)	http://www.nature.com/nmeth/journal/v10/n11/extref/nmeth.2645-S2.pdf	• Technical noise fit • Inference of highly variable genes	• Plot of average normalized read count versus cv^2
Monocle	The dynamics and regulators of cell fate decisions are revealed by pseudotemporal ordering of single cells	Trapnell et al. (2014)	http://cole-trapnell-lab.github.io/monocle-release/	• Differentially expressed genes • Pseudotemporal expression patterns	• Differential expression tables and graphics
SCDE	Bayesian approach to single-cell differential expression analysis	Kharchenko et al. (2014)	http://pklab.med.harvard.edu/scde/index.html	• Differentially expressed genes • Genome Browser-compatible graphics	• Differential expression tables and graphics
BackSPIN	Cell types in the mouse cortex and hippocampus revealed by single-cell RNA-seq	Zeisel et al. (2015)	https://github.com/linnarsson-lab/BackSPIN	• Clustering CEF files	• CEF files (tab delimited text)

Table displaying a selection of the recommended software available for the computational analysis of data yielded by this technique

References

Bengtsson M, Hemberg M, Rorsman P et al (2008) Quantification of mRNA in single cells and modelling of RT-qPCR induced noise. BMC Mol Biol 9:63
Brennecke P, Anders S, Kim JK et al (2013) Accounting for technical noise in single-cell RNA-seq experiments. Nat Methods 10:1093–1095
Choi JH, Ogunniyi AO, Du M et al (2010) Development and optimization of a process for automated recovery of single cells identified by microengraving. Biotechnol Prog 26:888–895
Dalerba P, Kalisky T, Sahoo D et al (2011) Single-cell dissection of transcriptional heterogeneity in human colon tumors. Nat Biotechnol 29:1120–1127
Dar RD, Razooky BS, Singh A et al (2012) Transcriptional burst frequency and burst size are equally modulated across the human genome. Proc Natl Acad Sci U S A 109:17454–17459
Devonshire AS, Elaswarapu R, Foy CA (2011) Applicability of RNA standards for evaluating RT-qPCR assays and platforms. BMC Genomics 12:118–127
Dey SS, Kester L, Spanjaard B et al (2015) Integrated genome and transcriptome sequencing of the same cell. Nat Biotechnol 33:285–289
Dobin A, Davis CA, Schlesinger F et al (2013) STAR: ultrafast universal RNA-seq aligner. Bioinformatics 29:15–21
Espina V, Wulfkuhle JD, Calvert VS et al (2006) Laser-capture microdissection. Nat Protoc 1:586–603
Fan HC, Fu GK, Fodor SP et al (2015) Combinatorial labeling of single cells for gene expression cytometry. Science 347:1258367
Frumkin D, Wasserstrom A, Itzkovitz S et al (2008) Amplification of multiple genomic loci from single cells isolated by laser micro-dissection of tissues. BMC Biotechnol 8:17
Fu GK, Xu W, Wilhelmy J et al (2014) Molecular indexing enables quantitative targeted RNA sequencing and reveals poor efficiencies in standard library preparations. Proc Natl Acad Sci U S A 111:1891–1896
Grün D, Kester L, van Oudenaarden A (2014) Validation of noise models for single-cell transcriptomics. Nat Methods 11:637–640
Han L, Zi X, Garmire LX et al (2014) Co-detection and sequencing of genes and transcripts from the same single cells facilitated by a microfluidics platform. Sci Rep 4:6485
Hashimshony T, Wagner F, Sher N et al (2012) CEL-Seq: single-cell RNA-Seq by multiplexed linear amplification. Cell Rep 2:666–673
Hug H, Schuler R (2003) Measurement of the number of molecules of a single mRNA species in a complex mRNA preparation. J Theor Biol 221:615–624
Islam S, Kjällquist U, Moliner A et al (2012) Highly multiplexed and strand-specific single-cell RNA 5′ end sequencing. Nat Protoc 7:813–828
Islam S, Zeisel A, Joost S et al (2014) Quantitative single-cell RNA-seq with unique molecular identifiers. Nat Methods 11:163–166
Jaitin DA, Kenigsberg E, Keren-Shaul H et al (2014) Massively parallel single-cell RNA-Seq for marker-free decomposition of tissues into cell types. Science 343:776–779
Jiang L, Schlesinger F, Davis CA et al (2011) Synthetic spike-in standards for RNA-seq experiments. Genome Res 21:1543–1551
Kellogg RA, Gomez-Sjoberg R, Leyrat AA et al (2014) High-throughput microfluidic single-cell analysis pipeline for studies of signaling dynamics. Nat Protoc 9:1713–1726
Kharchenko PV, Silberstein L, Scadden DT (2014) Bayesian approach to single-cell differential expression analysis. Nat Methods 11:740–742
Kim D, Lansmead B, Salzberg SL (2015) HISAT: a fast spliced aligner with low memory requirements. Nat Methods 12:357–360
Kim D, Pertea G, Trapnell C et al (2013) TopHat2: accurate alignment of transcriptomes in the presence of insertions, deletions and gene fusions. Genome Biol 14:R36
Kim JK, Marioni JC (2013) Inferring the kinetics of stochastic gene expression from single-cell RNA-sequencing data. Genome Biol 14:R7

Kivioja T, Vähärautio A, Karlsson K et al (2012) Counting absolute numbers of molecules using unique molecular identifiers. Nat Methods 9:72–74
Klein AM, Mazutis L, Akartuna I et al (2015) Droplet barcoding for single-cell transcriptomics applied to embryonic stem cells. Cell 161:1187–1201
Landry ZC, Giovanonni SJ, Quake SR et al (2013) Optofluidic cell selection from complex microbial communities for single-genome analysis. Methods Enzymol 531:61–90
Langmead B, Salzberg SL (2012) Fast gapped-read alignment with Bowtie 2. Nat Methods 9:357–359
Langmead B, Trapnell C, Pop M et al (2009) Ultrafast and memory-efficient alignment of short DNA sequences to the human genome. Genome Biol 10:R25
Li B, Dewey CN (2011) RSEM: accurate transcript quantification from RNA-Seq data with or without a reference genome. BMC Bioinformatics 12:323
Li H, Durbin R (2010) Fast and accurate long-read alignment with Burrows–Wheeler transform. Bioinformatics 26:589–595
Macosko EZ, Basu A, Satija R et al (2015) Highly parallel genome-wide expression profiling of individual cells using nanoliter droplets. Cell 161:1202–1214
Martin M (2011) Cutadapt removes adapter sequences from high-throughput sequencing reads. EMBnet J 17:10–12
McDavid A, Finak G, Chattopadyay PK et al (2013) Data exploration, quality control and testing in single-cell qPCR-based gene expression experiments. Bioinformatics 29:461–467
Meyer LR, Zweig AS, Hinrichs AS (2013) The UCSC Genome Browser database: extensions and updates 2013. Nucleic Acids Res 41:D64–D69
Mortazavi A, Williams BA, McCue K et al (2008) Mapping and quantifying mammalian transcriptomes by RNA-Seq. Nat Methods 5:621–628
Picelli S, Björklund ÅK, Faridani OR et al (2013) Smart-seq2 for sensitive full-length transcriptome profiling in single cells. Nat Methods 10:1096–1098
Picelli S, Faridani OR, Björklund ÅK et al (2014) Full-length RNA-seq from single cells using Smart-seq2. Nat Protoc 9:171–181
Pollen AA, Nowakowski TJ, Shuga J et al (2014) Low-coverage single-cell mRNA sequencing reveals cellular heterogeneity and activated signaling pathways in developing cerebral cortex. Nat Biotechnol 32:1053–1058
Ramsköld D, Luo S, Wang YC et al (2012) Full-length mRNA-Seq from single-cell levels of RNA and individual circulating tumor cells. Nat Biotechnol 30:777–782
Ramsköld D, Wang ET, Burge CB et al (2009) An abundance of ubiquitously expressed genes revealed by tissue transcriptome sequence data. PLoS Comput Biol 5:e1000598
Sasagawa Y, Nikaido I, Hayashi T et al (2013) Quartz-Seq: a highly reproducible and sensitive single-cell RNA sequencing method reveals non-genetic gene expression heterogeneity. Genome Biol 14:R31
Schmieder R, Edwards R (2011) Quality control and preprocessing of metagenomic datasets. Bioinformatics 27:863–864
Shalek AK, Satija R, Shuga J et al (2014) Single-cell RNA-seq reveals dynamic paracrine control of cellular variation. Nature 510:363–369
Shiroguchi K, Jia TZ, Sims PA et al (2012) Digital RNA sequencing minimizes sequence-dependent bias and amplification noise with optimized single-molecule barcodes. Proc Natl Acad Sci U S A 109:1347–1352
Soumillon M, Cacchiarelli D, Semrau S et al (2014) Characterization of directed differentiation by high-throughput single-cell RNA-Seq. BioRxiv. doi:10.1101/003236
Ståhlberg A, Kubista M, Åman P (2011) Single-cell gene-expression profiling and its potential diagnostic applications. Expert Rev Mol Diagn 11:735–740
Tang F, Barbacioru C, Wang Y et al (2009) mRNA-Seq whole-transcriptome analysis of a single cell. Nat Methods 6:377–382
Trapnell C, Cacchiarelli D, Grimsby J et al (2014) The dynamics and regulators of cell fate decisions are revealed by pseudotemporal ordering of single cells. Nat Biotechnol 32:381–386

Trapnell C, Hendrickson DG, Sauvageau M et al (2013) Differential analysis of gene regulation at transcript resolution with RNA-seq. Nat Biotechnol 31:46–53

Trapnell C, Pachter L, Salzberg SL (2009) TopHat: discovering splice junctions with RNA-Seq. Bioinformatics 25:1105–1111

Trapnell C, Roberts A, Goff L et al (2012) Differential gene and transcript expression analysis of RNA-seq experiments with TopHat and Cufflinks. Nat Protoc 7:562–578

Trapnell C, Williams BA, Pertea G et al (2010) Transcript assembly and quantification by RNA-Seq reveals unannotated transcripts and isoform switching during cell differentiation. Nat Biotechnol 28:511–515

Wagner GP, Kin K, Lynch VJ (2012) Measurement of mRNA abundance using RNA-seq data: RPKM measure is inconsistent among samples. Theory Biosci 131:281–285

Wang J, Fan HC, Behr B et al (2012) Genome-wide single-cell analysis of recombination activity and de novo mutation rates in human sperm. Cell 150:402–412

White AK, Vaninsberghe M, Petriv OI et al (2011) High-throughput microfluidic single-cell RT-qPCR. Proc Natl Acad Sci U S A 108:13999–14004

Wu AR, Neff NF, Kalisky T et al (2014) Quantitative assessment of single-cell RNA-sequencing methods. Nat Methods 11:41–46

Zeisel A, Muñoz Manchado AB, Codeluppi S et al (2015) Cell types in the mouse cortex and hippocampus revealed by single-cell RNA-seq. Science 347:1138–1142

Zhang H, Liu KK (2008) Optical tweezers for single cells. J R Soc Interface 5:671–690

Zhu YY, Machleder EM, Chenchik A et al (2001) Reverse transcriptase template switching: a SMART approach for full-length cDNA library construction. Biotechniques 30:892–897

Chapter 15
Eukaryotic Single-Cell DNA Sequencing

Keith E. Szulwach and Kenneth J. Livak

15.1 Introduction

Significant advances in the ability to isolate and interrogate the genomes of individual cells have recently led to great strides in our understanding of how somatic variation can affect normal development and disease. To interrogate the genomes of individual cells, two basic technical decisions need to be made. The first is which method to use for whole genome amplification (WGA). The second is how to convert the WGA product into a sequencing library. The second choice depends on the downstream application and, to some extent, the sequencing budget. Finally, once sequencing has been completed, it is also important to consider how technical artifacts arising during amplification impact the identification of mutations. Applications of single-cell DNA sequencing include revealing clonality in tumor samples, detecting lower-frequency mutations with improved sensitivity, tracing cell lineages, analyzing rare cell types, and studying somatic mosaicism in cancer, neuronal diversity, and other tissues. Applications not addressed in this chapter include microbial ecology and prenatal genetic diagnosis.

15.2 Applications

Known for extreme heterogeneity, cancer cell populations are one area where single-cell methods have profoundly affected research. By sequencing hundreds of individual cells to detect copy number variation, Navin et al. (2011) identified clonal evolution as central to tumor growth in two breast cancer patients. Their data tracing

K.E. Szulwach, Ph.D. • K.J. Livak, Ph.D. (✉)
Fluidigm Corporation, South San Francisco, CA, USA
e-mail: keith.szulwach@fluidigm.com; ken.livak@fluidigm.com

A.M. Aransay, J.L. Lavín Trueba (eds.), *Field Guidelines for Genetic Experimental Designs in High-Throughput Sequencing*,
DOI 10.1007/978-3-319-31350-4_15

the lineage of these clonal populations indicated relatively brief periods of intense genomic change followed by more even growth of the cancer mass, in contrast to canonical models of tumorigenesis that posit steady accumulation of mutations. Other studies in which clonal structure has been a primary element include Jan et al. (2012), suggesting that relapse of acute myeloid leukemia (AML) patients may be rooted in a history of multiple hematopoietic stem cell mutations preceding full onset of AML, and Anderson et al. (2011), which linked genetic abnormalities in individual cells to clonal patterns with implications for treatment and risk of relapse. Attempts have been made to infer clonal structure from bulk tumor DNA sequencing, but the analysis of these results requires simplifying assumptions, such as that all mutations are heterozygous. As demonstrated by Paguirigan et al. (2015), single-cell data are needed in order to make unambiguous assignment of genotype combinations. Thus, relying on bulk data to determine clonal structure will generate an incomplete, and often misleading, picture of clonal heterogeneity in tumors. These paths of inquiry could lead to answers on fundamental topics such as how metastasis occurs and whether it begins in particular cell types or even in cell fusion.

Among the key agents in metastasis are circulating tumor cells (CTCs), whose rarity prevents bulk analysis and thus makes them a target for single-cell sequencing. Working with a few CTCs, Ni et al. (2013) reported insertions and deletions in exomes and single-nucleotide variations that were particular to individual cells and thus might inform therapy for specific patients. They also reported similar copy number variation (CNV) patterns among lung cancer adenocarcinoma patients, and different CNV patterns among small-cell lung cancer patients, a finding that could benefit CTC-based diagnostics.

Intriguingly, extensive somatic genetic heterogeneity has also been observed during normal development, where it has been proposed to play important roles in cellular fitness and disease etiologies. Beyond tumorigenesis, the presence and role of somatic mosaicism has been recognized within the central nervous system (Cai et al. 2014; Coufal et al. 2011; Evrony et al. 2015; McConnell et al. 2013; Muotri et al. 2010). Diversity in large-scale copy number alterations and transposable element mobility has been observed during normal neuronal development as well as in the context of human neurological diseases. Moving forward, the influence of genetic diversity on cellular fitness even during normal development and how such processes ultimately impacts the progression of human diseases will provide a high-definition perspective previously unattainable.

15.3 Cell Isolation Methods

Three of the four methods discussed in the chapter "Eukaryotic Single-Cell mRNA Sequencing" have been used to isolate cells for single-cell DNA sequencing—micromanipulation, fluorescence-activated cell sorting (FACS), and microfluidic systems. In addition, DNA sequencing libraries can be prepared from individual nuclei (Navin et al. 2011; Baslan et al. 2012; Wang et al. 2014). This is especially useful in analyzing solid tumors where it is difficult to obtain clean, single-cell suspensions. Furthermore, the use of nuclei enables analysis of flash-frozen samples, which may have been stored for decades (Leung et al. 2015). It has also been used extensively in analyzing neural samples because of the difficulty of obtaining intact cells from these highly networked tissues. Single nuclei can be obtained using FACS or microfluidic capture.

15.4 WGA

Figure QG15.1 in the Annex is a flowchart of the wet lab workflow used for single-cell DNA sequencing. Table QG15.1 in the Annex lists more details about the methods used for whole genome amplification. Figure 4 in Blainey (2013) and the accompanying text (pp. 416–419) provide an excellent summary of these different WGA methods. For mammalian single cells, three methods now are mainly used: MDA (multiple displacement amplification, Dean et al. 2002); PicoPLEX™ (based on degenerate-oligonucleotide primer PCR, DOP-PCR), commercialized by Rubicon Genomics, and MALBAC (multiple annealing and looping-based amplification cycles, Zong et al. 2012). MDA is a single-step isothermal reaction. Following denaturation of genomic DNA, random 3′-protected 6-mers are extended on the genomic template using a polymerase with strong strand-displacing activity, generally phi29 DNA polymerase. PicoPLEX and MALBAC are two-step processes that use thermal cycling. The first step, often termed preamplification, uses primers with degenerate bases at the 3′ end, a polymerase with strand-displacing activity, and a limited number of thermal cycles. The second step is essentially conventional PCR with a single primer corresponding to the 5′ end of the primers used in the preamplification step. PicoPLEX and MALBAC differ in the DNA polymerases used, the structure of the primers, and details of the thermal cycling protocols. Other variations that use degenerate primers and PCR are termed degenerate oligonucleotide-primed PCR (DOP-PCR).

De Bourcy et al. (2014) compared these three methods on single-cell bacterial genomes; the results are instructive for amplification of mammalian genomes as well. The main criteria used to judge the quality of WGA are coverage, uniformity, and error rate. Their results indicate that a key factor influencing WGA quality can be amplification gain. Regardless of the WGA method, WGA quality tends to worsen as amplification gain increases. However, not all characteristics are affected equally. PCR-based methods (PicoPLEX and MALBAC) perform better in terms of uniformity (lack of bias) in comparison to MDA across all levels of amplification gain tested, whereas MDA amplification uniformity deteriorates as amplification gain increases. In terms of errors introduced during WGA, MDA exhibits a tenfold lower rate of single-nucleotide errors than PicoPLEX and MALBAC, presumably due to the higher fidelity of phi29 DNA polymerase compared to the polymerases used in the PCR-based methods. Although these particular results demonstrated a high breadth of genomic coverage in bacterial genomes for all methods tested, it has been found by multiple other groups that MDA tends to amplify a much larger fraction of mammalian genomes than PCR-based methods. Typically, MDA has been found to amplify ≥90 % of single-cell mammalian genomes accessible by conventional whole genome sequencing. MALBAC has been reported to amplify up to 70 % of mammalian genomes, while DOP-PCR amplifies only ~10 %. For these reasons, De Bourcy et al. (2014) concluded that PCR-based methods may be especially well suited for analysis of copy number variation and MDA preferred for analysis of single-nucleotide variation (SNV).

Labs have reported that the use of nanoliter reaction volumes is particularly advantageous for MDA in terms of coverage uniformity and low error rates (Wang et al. 2012; De Bourcy et al. 2014; Gole et al. 2013). Therefore, a combination of microfluidics and MDA may be the best general method that can be used for both CNV and SNV.

15.5 Library Construction

The three choices for library construction are influenced by the method by which the DNA will be interrogated downstream. Starting with whole-genome amplified DNA, whole genome sequencing (WGS), whole exome sequencing (WES), and targeted sequencing can be performed. WGS library construction is the most straightforward. For WGS, the processing steps for each WGA sample are fragmentation, attachment of adaptors (with optional barcodes), and PCR to append the P5 and P7 tags required for immobilization and amplification in the Illumina® flow cells. The most convenient method for accomplishing these tasks is to perform tagmentation using the Illumina Nextera® System (e.g., Nextera Rapid Capture Kit, FC-140-1003). It is also possible to use the more conventional route of fragmentation by sonication or enzymatic treatment, end repair, A-tailing, ligation of Illumina sequencing adaptor, and PCR. Illumina and New England BioLabs have kits for performing these tasks. WGS on a few single cells is the most comprehensive way to assess the quality of the WGA method. Because of the complexity involved, this sequencing is generally performed on a HiSeq® instrument. For more than a few cells, it is much more cost efficient to use WES or targeted sequencing. This makes it practical to sequence the large number of cells that are required to effectively characterize clonal heterogeneity and detect rarer subpopulations. However, one application where WGS is not cost-prohibitive for analysis of a moderate number of cells is detection and quantification of CNV. As pioneered by Navin et al. (2011) and utilized by others (Francis et al. 2014; Baslan et al. 2015), read coverage as shallow as 0.06× per cell can provide useful information on CNV in tumors. In this case, it is important to barcode the single-cell libraries during library construction so that multiple libraries can be pooled together prior to sequencing.

For WES, libraries are prepared from single-cell WGA products just as for WGS. At this point, exonic sequences are selected from the libraries. Incorporating barcodes as part of the adaptor attachment step enables pooling multiple single-cell libraries prior to exome capture. This is more cost-effective for experiments aimed at identifying protein-coding SNVs as sequencing efforts can be focused on the 1–2 % of the genome that is exonic. Hybridization pullout has become the method of choice for exome capture. For single-cell libraries, solution-based rather than array-based hybridization is used for pullout because of the small amounts of DNA involved. To date, alternatives such as AmpliSeq™ from Life Technologies® and HaloPlex™ from Agilent Technologies have not been widely adopted for whole exome enrichment. Chilamakuri et al. (2014) compared the four most popular kits for exome capture by hybridization pullout: Agilent® SureSelect™ Human All Exon, NimbleGen™ SeqCap™ EZ Exome Library, Illumina TruSeq™ Exome Enrichment, and Illumina Nextera Exome Enrichment. All technologies performed reproducibly and well. Slight, but consistent, variations in coverage, GC bias, and SNV detection were observed, which might make one of the methods more suitable for a particular application than the others. Following exome capture, a limited PCR amplification is performed to restore double-stranded DNA fragments suitable for loading on a MiSeq™ or HiSeq instrument.

Leung et al. (2015) incorporated a very stringent quality control measure for deciding whether a WGA library should be processed further for WES. They designed a PCR panel consisting of 22 primer pairs, one for each autosome. A single-cell WGA library must show PCR products for all 22 assays in order to qualify the library for exome capture.

Libraries for targeted sequencing can be prepared in a similar manner to what is used for WES. The only difference is that fewer oligonucleotide baits are used in the capture step. In addition, there are many variations of multiplex PCR that have been used for target enrichment. Finally, there are methods based on selective circularization of probes (Dahl et al. 2007; Hiatt et al. 2013; HaloPlex is a commercial version) that combine elements of hybridization capture and PCR. Mamanova et al. (2010) and Mertes et al. (2011) provide technical details on how these three general methods work. Altmüller et al. (2014) list the commercially available options across the three categories and provide guidelines on how to choose among the options. In general, PCR methods have better specificity than hybridization. This is an advantage in discriminating homologous targets and in minimizing the contribution of repetitive elements to the sequencing libraries. High specificity can sometimes be a disadvantage because polymorphisms in primer binding sites can lead to reduced yield for particular segments. Coverage is characterized by the amount of the genome interrogated and the number of target regions. Typically, PCR methods have been used for coverage of up to about 500 kb and hybridization methods for above 500 kb. Read depth for targeted single-cell libraries does not need to be as great as for bulk libraries. This is because bulk sequencing is generally trying to detect variants at low frequency, whereas variants in single cells are at least hemizygous, meaning they should be detected in about 50 % of the reads. For PCR-based methods, sequencing to an average depth of 100× should be more than adequate to compensate for locus-specific variation in amplification uniformity.

Targeted sequencing is the method of choice for analyzing hundreds or thousands of single cells because focusing sequencing on a relatively small proportion of the genome can dramatically reduce sequencing costs. Thus, cost per sample is a critical parameter to consider when comparing different methods, depending on the breadth of genetic information desired. Targeted DNA sequencing from single-cell whole genome amplified DNA can also provide an excellent means to cost-effectively validate mutations detected by WGS and WES. It is also important to have a large number of barcodes available so that many single-cell samples can be pooled together in parallel sequencing experiments. Standard panels are the most economical option for targeted sequencing because the cost of oligonucleotide synthesis is amortized over many customers. Standard panels also have the advantage that they are thoroughly validated. Of course, standard panels may not contain all the targets that are important for a particular study. Custom panels have a large upfront cost. This drawback becomes negligible, though, when a large number of samples are analyzed.

15.6 Data Analysis

Identification of mutations in single-cell data can be accomplished in much the same manner as conventional approaches using bulk genomic DNA, with a few considerations to account for the types of technical artifacts introduced during whole genome amplification (WGA). Following the identification of mutations across a population of cells, relationships between mutation profiles can be used to dissect clonality and phylogeny of the population. Here we summarize the key steps unique to identifying single-cell mutations and examples of how single-cell mutation profiles can be used to reconstruct the clonal phylogeny of cell populations. For data processing steps that overlap with conventional genetic analyses from bulk genomic DNA, we refer the reader to more comprehensive descriptions.

Table 15.1 lists popular programs that are used for alignment to the genome followed by the identification of variants. Ruffalo et al. (2011) and Yu et al. (2012) assess the performance of the alignment tools in Table 15.1 plus others using simulated data. The different types of mutations that can be detected are single-nucleotide variants (SNVs), insertion/deletions (INDELs), copy number variants (CNVs), and structural variants (SVs). Single-cell analysis focuses predominantly on SNVs and CNVs. Figure QG15.2 in the Appendix is a flowchart of the steps used to process single-cell data to detect SNVs and CNVs. Van der Auwera et al. (2013) is a detailed guide on the best practices to use in establishing this type of computational pipeline.

Table 15.1 DNA sequencing data processing tools

Step	Tool	Reference	Link
Alignment	BWA	Li and Durbin (2009a)	http://bio-bwa.sourceforge.net/
	Bowtie	Langmead et al. (2009), Langmead and Salzberg (2012)	http://bowtie-bio.sourceforge.net/index.shtml
	SOAPaligner/soap2	Li et al. (2009c)	http://soap.genomics.org.cn/#down2
	Novoalign		http://www.novocraft.com/support/download/
SNV and INDEL	GATK	McKenna et al. (2010), DePristo et al. (2011)	https://www.broadinstitute.org/gatk/download/
	VarScan	Koboldt et al. (2009), Koboldt et al. (2012)	http://varscan.sourceforge.net/
	SAMtools	Li et al. (2009b)	http://samtools.sourceforge.net/
	SNVer	Wei et al. (2011)	http://snver.sourceforge.net/
	CRISP	Bansal (2010)	https://sites.google.com/site/vibansal/software/crisp
CNV	SegSeq	Chiang et al. (2009)	http://www.broadinstitute.org/software/cprg/?q=node/39
	CNVnator	Abyzov et al. (2011)	http://sv.gersteinlab.org/cnvnator/
	Gingko	Garvin et al. (2015)	http://qb.cshl.edu/ginkgo
SV	BreakDancer	Chen et al. (2009)	http://breakdancer.sourceforge.net/
	BreakPointer	Drier et al. (2013)	https://www.broadinstitute.org/cancer/cga/breakpointer
	CLEVER	Marschall et al. (2012)	https://code.google.com/p/clever-sv/
	GASVPro	Sindi et al. (2012)	http://compbio.cs.brown.edu/projects/gasv/
	SVMerge	Wong et al. (2010)	http://svmerge.sourceforge.net/

15.6.1 SNV

With single-cell DNA sequencing, it is important to account for the technical artifacts introduced during whole genome amplification. Such artifacts can be classified into four main categories; genomic coverage, SNV detection efficiency, allelic dropout rate (ADR), and SNV false positive rate (FPR). Genomic coverage

Table 15.2 Metrics defining the technical performance of single-cell whole genome amplification

Metric	Calculation	Description
Genomic coverage	$\frac{\text{Bases with} \geq 1\times \text{coverage}}{\text{Total number of bases}}$	Fraction of the genome covered at ≥1×
Allelic dropout rate (ADR)	$\frac{1}{n}\sum_{n}^{i=1}\frac{\text{Hom}_{\text{SingleCell}}}{\text{Het}_{\text{Bulk}}}$	Mean fraction of sites called homozygous in single cells that were heterozygous in bulk genomic DNA
SNV false positive rate (FPR)	$\frac{1}{n}\sum_{n}^{i=1}\frac{\text{Het}_{\text{SingleCell}}}{\text{Hom}_{\text{Bulk}}}$	Mean fraction of sites called heterozygous in single-cells that were called homozygous in bulk genomic DNA

influences SNV detection efficiency, as mutations within regions of the genome that were either not amplified or not sequenced will be missed. Single-nucleotide errors introduced by the polymerase during WGA will lead to false positive mutation calls, while bias in the amplification of one allele over another at heterozygous loci will lead to allelic dropout. Table 15.2 describes how each of the metrics is quantified.

Once these values have been determined empirically for a given experiment, they can be taken into consideration when identifying mutations in single-cell whole genome amplified DNA. As an example of this, Hou et al (2013) applied the SNV false discovery rate to a binomial test to determine the probability of a given SNV being false given the number of cells that variant was observed in amongst the total number of cells tested. Because polymerase-induced errors occurring during WGA are random (as evidenced by their distribution across the genome), the chance of an error occurring at the same position in the genome in two independent single cells is low, allowing for probabilities to be assigned to the set of variants detected across the population of cells tested. As the field of single-cell DNA sequencing moves forward, we expect further development of these metrics in parallel with their incorporation into analysis packages specifically geared toward handling single-cell genetic data. Fluidigm has developed one such solution, the Singular™ Analysis Toolset, based on open-source R software code.

The possibility of false positives means it is important to validate mutations identified by single-cell DNA sequencing. Wang et al. (2014) accomplished this by applying duplex sequencing (Schmitt et al. 2012) to the putative mutations detected in their single-nuclei libraries. A duplex library was prepared from a bulk sample and target segments were selected covering the set of putative mutations. By attaching different random molecular tags to each end of each DNA fragment, the duplex library enables maintenance of strand identity. The targeted duplex library was then sequenced to very high depth. Reading to an average depth of approximately 100,000× generated approximately 5000× single-molecule coverage. True mutations are indicated by concordance on both strands. The results also measure precise allele frequencies in the bulk sample.

15.6.2 CNV

For CNV analysis, a comprehensive study using single-cell paired-end DNA sequencing to obtain accurate, high-resolution copy number profiles was reported by Voet et al. (2013). They identified the different types of artifacts generated by MDA- and PCR-based WGA and developed analysis tools to robustly distinguish these artifacts from true copy number variants. This enabled them to detect copy number changes occurring in a single cell cycle. Furthermore, they reported more accurate copy number profiles using PicoPLEX for WGA than MDA.

Identification of CNVs is generally performed using methods similar to those used for sequencing DNA isolated from large populations of cells. Typically, the genome is divided into bins of a defined size dictated by the resolution at which a CNV is to be called. Aligned reads are then assigned to these bins to determine the read density of each. As not all portions of the genome are equally mappable, it is often useful to employ a variable-size binning approach (Navin et al. 2011) to ensure that all bins have an equal probability of reads being assigned to them. In this approach, bin size is adjusted to account for regions of the genome to which reads cannot be uniquely assigned so that all bins have an equal portion of mappable sequence. The GC content of bins can also impact mappability as a result of under-representation of reads from GC-rich and AT-rich regions of the genome that are prone to bias during WGA, PCR, and sequencing. Generally, such bias can be corrected for by LOESS regression. Using normalized read densities, copy numbers can be calculated across the genome, usually assuming that the median read density corresponds to a copy number of two for largely diploid genomes. With copy numbers determined, the genome can be segmented by identifying bins that exhibit similar copy numbers. In single-cell analysis, an important consideration is that copy number changes should occur at discrete intervals due to the fact that they will have integer copy number states. Finally, a number of statistical approaches can be used to segment copy number profiles from single-cell data, including those implemented in the approaches described in Table 15.1. In single-cell analysis, these have included circular binary segmentation (CBS) (Olshen et al. 2004; Venkatraman and Olshen 2007), Kolmogorov–Smirnov segmentation (KS) (Navin et al. 2011), and piecewise constant fitting (PCF) (Voet et al. 2013).

15.6.3 Reporting of Data

The end result of primary processing of sequence reads obtained from a single-cell DNA sequencing experiment is dependent on the mode of mutation calling performed. For SNV and INDEL analysis, variant callers typically report genotypes in a variant call format (VCF) text file. A VCF file is in text format and contains a header at the top that tells the user meta-information about the parameters and filters used during variant calling as well as a series of definitions for the abbreviations used for the reporting of information on the genotypes. Following the header will be the genotype information. This information includes the genomic position of the variant, the reference allele, the alternate allele, the quality score of the variant call, a flag indicating whether or not the variant passed the applied filters defined during variant calling, and a series of data associated with the variant call (homozygous or heterozygous). The data associated with the variant call generally includes metrics such as sequencing depths, reads supporting each allele detected, and genotype quality. For CNV and SV analysis, variants are usually reported in a straightforward manner. For CNVs this will typically include the genomic coordinates for the beginning and end of a portion of a contiguous portion of the genome over which copy number was determined to be the same as well a statistical value that indicates the probability of the identified segment being false. For SVs, the output will include similar values except the genomic coordinates of both break points will usually be specified.

15.6.4 Tertiary Analysis

The ultimate goal of single-cell DNA sequencing is to identify the clones present in the population of cells analyzed and then infer the phylogeny that relates these clones to one another. This type of analysis, though, is still in its infancy and so there are no easily accessible software packages to accomplish this task. Accurately distinguishing the clonal structure and phylogeny across a population of cells can be impacted by technical artifacts introduced during WGA. In general, definition of clonal structures is performed by relating single-cell mutation profiles to one another, while accounting for the probability of mutations being missed due to lack of genomic coverage or being incorrectly called as a result of ADR and FPR. Although, at present, software packages are not available for such analyses, an example of a statistical approach to define clonal structure from single-cell data was reported by Gawad et al. (2014). They used multiple methods to examine the clonal structures of cancer cell populations isolated from six acute lymphoblastic leukemia patients. In the first, single-cell mutation profiles were used in a probabilistic modeling-based approach. Mutation calls were considered binary and then applied to a multivariate Bernoulli model that considers genomic coverage and ADR to quantify the probability of an observed single-cell mutation profile for each cell. The finite mixture of the multivariate Bernoulli distributions was then used to represent distinct clones. Mutation profiles across cells were also subjected to hierarchical clustering based on Jaccard distances and compared to the probabilistic method, yielding similar results. Finally, mutation profiles were used in a multiple correspondence analysis (MCA) as independent assessment of the underlying clonal structures from each patient. Similar to a principal component analysis (PCA) often applied to RNA expression profiles, MCA can be used to characterize categorical data, such as binary mutation calls. Single-cell mutation profiles are then represented as individual points in a two-dimensional Euclidean space, with similar profiles clustering together.

Annex: Quick Reference Guide

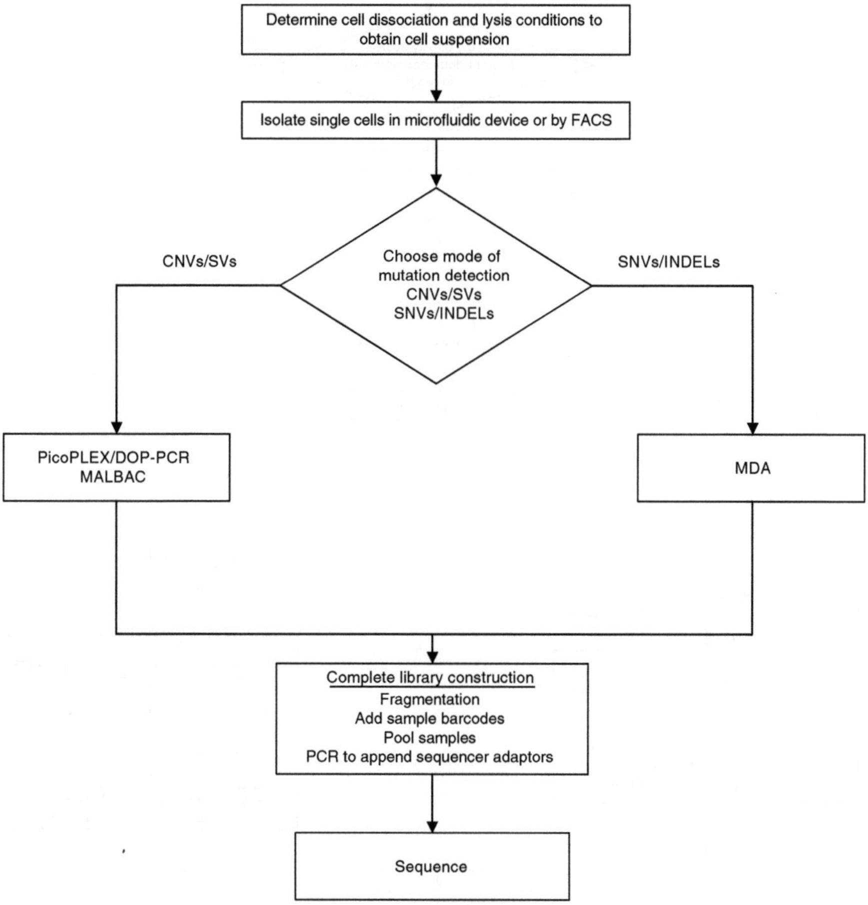

Fig. QG15.1 Representation of the wet-lab procedure workflow

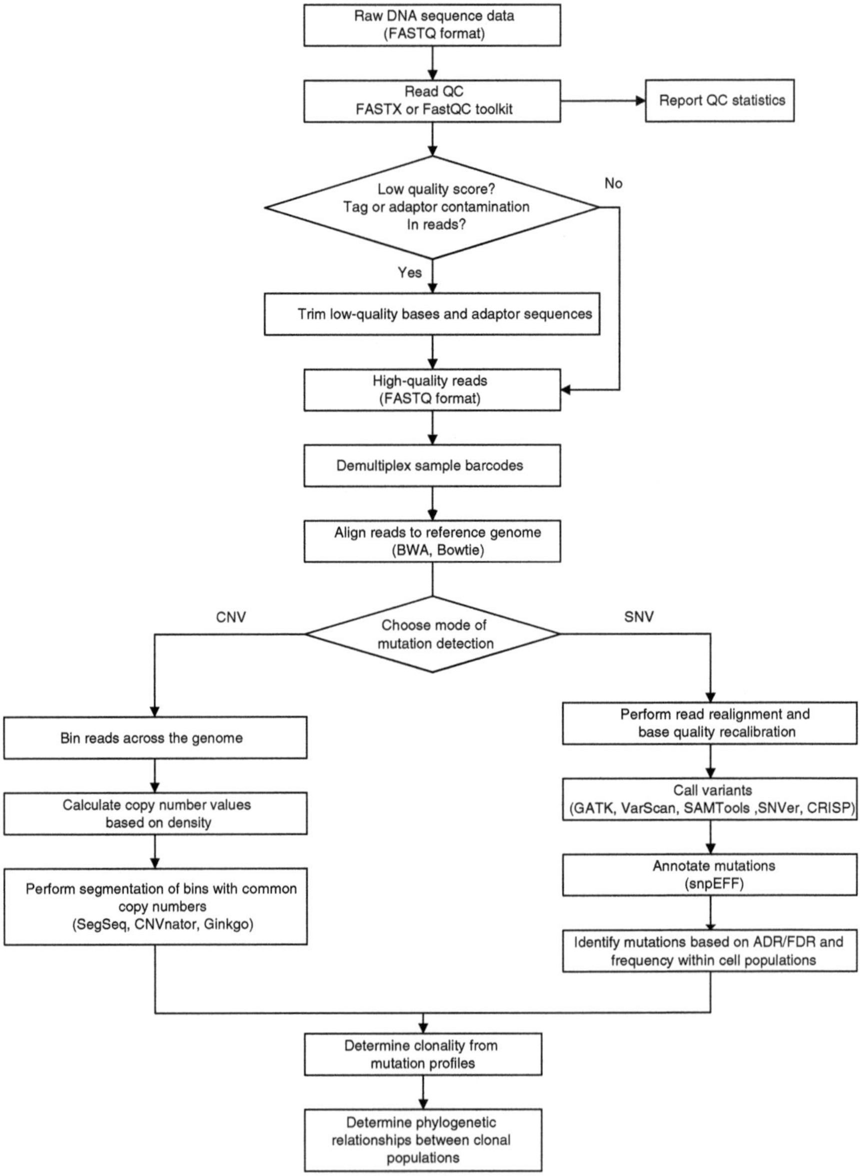

Fig. QG15.2 Main steps of the computational analysis pipeline

Table QG15.1 Experimental design considerations

Method	Expected breadth of genomic coverage	Best-suited mode of mutation detection	Sequencing throughput required	Sequencing setup	Sequencing reads	References
Multiple Displacement Amplification (MDA)	≥90 %	SNPs/INDELs	≥50×, ~150 Gb for WGS, ~2–5 Gb for WES	≥50 bp paired-end reads,	≥50 bp paired-end reads	Wang et al. (2014)
PicoPLEX (DOP-PCR)	~10 %	CNVs/SVs	≥2 million reads	Pool up to 96 cells and load onto one lane of HiSeq	≥50 bp reads, paired-end for SVs	Baslan et al. (2015)
Multiple Annealing and Looping-Based Amplification Cycles (MALBAC)	~70 %	CNVs/SVs	≥2 million reads	Pool up to 96 cells and load onto one lane of HiSeq	≥50 bp reads, paired-end for SVs	Zong et al. (2012); Ni et al. (2013)

Table that comprises relevant experimental design parameters, to carefully consider before applying this methodology

Table QG15.2 Available software recommendations

Software	Title	Reference	Link	Results output	Results format
GATK	A framework for variation discovery and genotyping using next-generation DNA sequencing data	DePristo et al. (2011)	https://www.broadinstitute.org/gatk/download/	• List of variants (SNVs, INDELs) relative to reference genome	• VCF, summarizing genomic coordinates, data quality, and other annotations
VarScan	VarScan 2: Somatic mutation and copy number alteration discovery in cancer by exome sequencing	Koboldt et al. (2012)	http://varscan.sourceforge.net/	• List of variants (SNVs, INDELs, CNVs) relative to reference genome or control sample	• VCF, summarizing genomic coordinates, data quality, and other annotations or tab delimited file listing genomic coordinates and other annotation
SAMtools	The sequence alignment/map (SAM) format and SAMtools	Li et al. (2009b)	http://samtools.sourceforge.net/	• List of variants (SNVs, INDELs) relative to reference genome	• VCF, summarizing genomic coordinates, data quality, and other annotations
SNVer	SNVer: a statistical tool for variant calling in analysis of pooled or individual next-generation sequencing data	Wei et al. (2011)	http://snver.sourceforge.net/	• List of variants (SNVs, INDELs) relative to reference genome	• VCF, summarizing genomic coordinates, data quality, and other annotations
CRISP	A statistical method for the detection of variants from next-generation resequencing of DNA pools	Bansal (2010)	https://sites.google.com/site/vibansal/software/crisp	• List of variants (SNVs, INDELs) relative to reference genome	• VCF, summarizing genomic coordinates, data quality, and other annotations
SegSeq	High-resolution mapping of copy-number alterations with massively parallel sequencing	Chiang et al. (2009)	http://www.broadinstitute.org/software/cprg/?q=node/39	• Summary of CNV breakpoints marking chromosomal segments with differing CNVs in case-control samples	• Tab delimited file listing genomic coordinates for each segment with corresponding copy ratio and *p*-value

CNVnator	CNVnator: an approach to discover, genotype, and characterize typical and atypical CNVs from family and population genome sequencing	Abyzov et al. (2011)	http://sv.gersteinlab.org/cnvnator/	• Identification of CNV segments from read densities (RD)	• Tab delimited file listing CNV type, genomic coordinates, CNV size, normalized RD, and statistical values
Ginkgo	Interactive analysis and quality assessment of single-cell copy-number variations	Garvin, et al. (2015)	http://qb.cshl.edu/ginkgo	• Copy number profiles of individual cells and phylogenetic trees of related cells	• Graphical results of CNV profiles, phylogenetic trees, and heatmaps of CNVs across cells populations
Singular Analysis Toolset	Analysis and visualization of single-cell variants	NA	https://www.fluidigm.com/software	• List of variants relative to reference genome or a population of single-cells. Clustering and heatmap visualization of defined sets of variants	• Tab delimited files summarizing single-cell variants, VCFs, dendrograms, and heatmap visualizations of variant profiles

Table displaying a selection of the recommended software available for the computational analysis of data yielded by this technique

References

Abyzov A, Urban AE, Snyder M et al (2011) CNVnator: an approach to discover, genotype, and characterize typical and atypical CNVs from family and population genome sequencing. Genome Res 21:974–984

Altmüller J, Budde BS, Nürnberg P (2014) Enrichment of target sequences for next-generation sequencing applications in research and diagnostics. Biol Chem 395:231–237

Anderson K, Lutz C, van Delft FW et al (2011) Genetic variegation of clonal architecture and propagating cells in leukaemia. Nature 469:356–361

Bansal V (2010) A statistical method for the detection of variants from next-generation resequencing of DNA pools. Bioinformatics 26:i318–i324

Baslan T, Kendall J, Rodgers L et al (2012) Genome-wide copy number analysis of single cells. Nat Protoc 7:1024–1041

Baslan T, Kendall J, Ward B et al (2015) Optimizing sparse sequencing of single cells for highly multiplex copy number profiling. Genome Res 25:714–724

Blainey PC (2013) The future is now: single-cell genomics of bacteria and archaea. FEMS Microbiol Rev 37:407–427

Cai X, Evrony GD, Lehmann HS et al (2014) Single-cell, genome-wide sequencing identifies clonal somatic copy-number variation in the human brain. Cell Rep 8:1280–1289

Chen K, Wallis JW, McLellan MD et al (2009) BreakDancer: an algorithm for high-resolution mapping of genomic structural variation. Nat Methods 6:677–681

Chiang DY, Getz G, Jaffe DB et al (2009) High-resolution mapping of copy-number alterations with massively parallel sequencing. Nat Methods 6:99–103

Chilamakuri CS, Lorenz S, Madoui MA et al (2014) Performance comparison of four exome capture systems for deep sequencing. BMC Genomics 15:449

Coufal NG, Garcia-Perez JL, Peng GE et al (2011) Ataxia telangiectasia mutated (ATM) modulates long interspersed element-1 (L1) retrotransposition in human neural stem cells. Proc Natl Acad Sci U S A 108:20382–20387

De Bourcy CF, De Vlaminck I, Kanbar JN et al (2014) A quantitative comparison of single-cell whole genome amplification methods. PLoS One 9:e105585

Dahl F, Stenberg J, Fredriksson S et al (2007) Multigene amplification and massively parallel sequencing for cancer mutation discovery. Proc Natl Acad Sci U S A 104:9387–9392

Dean FB, Hosono S, Fang L et al (2002) Comprehensive human genome amplification using multiple displacement amplification. Proc Natl Acad Sci U S A 99:5261–5266

DePristo MA, Banks E, Poplin R et al (2011) A framework for variation discovery and genotyping using next-generation DNA sequencing data. Nat Genet 43:491–498

Drier Y, Lawrence MS, Carter SL (2013) Somatic rearrangements across cancer reveal classes of samples with distinct patterns of DNA breakage and rearrangement-induced hypermutability. Genome Res 23:228–235

Evrony GD, Lee E, Mehta BK et al (2015) Cell lineage analysis in human brain using endogenous retroelements. Neuron 85:49–59

Francis JM, Zhang CZ, Maire CL et al (2014) EGFR variant heterogeneity in glioblastoma resolved through single-nucleus sequencing. Cancer Discov 4:956–971

Garvin T, Aboukhalil R, Kendall J et al (2015) Interactive analysis and quality assessment of single-cell copy-number variations. bioRxiv. doi:10.1101/011346

Gawad C, Koh W, Quake SR (2014) Dissecting the clonal origins of childhood acute lymphoblastic leukemia by single-cell genomics. Proc Natl Acad Sci U S A 111:17947–17952

Gole J, Gore A, Richards A et al (2013) Massively parallel polymerase cloning and genome sequencing of single cells using nanoliter microwells. Nat Biotechnol 31:1126–1132

Hiatt JB, Pritchard CC, Salipante SJ et al (2013) Single molecule molecular inversion probes for targeted, high-accuracy detection of low-frequency variation. Genome Res 23:843–854

Hou Y, Fan W, Yan L et al (2013) Genome analyses of single human oocytes. Cell 155:1492–1506

Jan M, Snyder TM, Corces-Zimmerman MR et al (2012) Clonal evolution of preleukemic hematopoietic stem cells precedes human acute myeloid leukemia. Sci Transl Med 4:149ra118
Koboldt DC, Chen K, Wylie T et al (2009) VarScan: variant detection in massively parallel sequencing of individual and pooled samples. Bioinformatics 25:2283–2285
Koboldt DC, Zhang Q, Larson DE et al (2012) VarScan 2: somatic mutation and copy number alteration discovery in cancer by exome sequencing. Genome Res 22:568–576
Langmead B, Salzberg SL (2012) Fast gapped-read alignment with Bowtie 2. Nat Methods 9:357–359
Langmead B, Trapnell C, Pop M et al (2009) Ultrafast and memory-efficient alignment of short DNA sequences to the human genome. Genome Biol 10:R25
Leung ML, Wang Y, Waters J et al (2015) SNES: single nucleus exome sequencing. Genome Biol 16:55
Li H, Durbin R (2009) Fast and accurate short read alignment with Burrows-Wheeler transform. Bioinformatics 25:1754–1760
Li H, Handsaker B, Wysoker A et al (2009a) The sequence alignment/map (SAM) format and SAMtools. Bioinformatics 25:2078–2079
Li R, Yu C, Li Y et al (2009b) SOAP2: an improved ultrafast tool for short read alignment. Bioinformatics 25:1966–1967
Mamanova L, Coffey AJ, Scott CE et al (2010) Target-enrichment strategies for next-generation sequencing. Nat Methods 7:111–118
Marschall T, Costa IG, Canzar S et al (2012) CLEVER: clique-enumerating variant finder. Bioinformatics 28:2875–2882
McConnell MJ, Lindberg MR, Brennand KJ et al (2013) Mosaic copy number variation in human neurons. Science 342:632–637
McKenna A, Hanna M, Banks E et al (2010) The Genome Analysis Toolkit: a MapReduce framework for analyzing next-generation DNA sequencing data. Genome Res 20:1297–1303
Mertes F, Elsharawy A, Sauer S et al (2011) Targeted enrichment of genomic DNA regions for next-generation sequencing. Brief Funct Genomics 10:374–386
Muotri AR, Marchetto MC, Coufal NG et al (2010) L1 retrotransposition in neurons is modulated by MeCP2. Nature 468:443–446
Navin N, Kendall J, Troge J et al (2011) Tumour evolution inferred by single-cell sequencing. Nature 472:90–94
Ni X, Zhuo M, Su Z et al (2013) Reproducible copy number variation patterns among single circulating tumor cells of lung cancer patients. Proc Natl Acad Sci U S A 110:21083–21088
Olshen AB, Venkatraman ES, Lucito R et al (2004) Circular binary segmentation for the analysis of array-based DNA copy number data. Biostatistics 5:557–572
Paguirigan AL, Smith J, Meshinchi S et al (2015) Single-cell genotyping demonstrates complex clonal diversity in acute myeloid leukemia. Sci Transl Med 7:281re2
Ruffalo M, LaFramboise T, Koyutürk M (2011) Comparative analysis of algorithms for next-generation sequencing read alignment. Bioinformatics 27:2790–2796
Schmitt MW, Kennedy SR, Salk JJ et al (2012) Detection of ultra-rare mutations by next-generation sequencing. Proc Natl Acad Sci U S A 109:14508–14513
Sindi SS, Onal S, Peng LC et al (2012) An integrative probabilistic model for identification of structural variation in sequencing data. Genome Biol 13:R22
Van der Auwera GA, Carneiro MO, Hartl C et al (2013) From FastQ data to high-confidence variant calls: the Genome Analysis Toolkit best practices pipeline. Curr Protoc Bioinformatics 43:11.10.1–11.10.33
Venkatraman ES, Olshen AB (2007) A faster circular binary segmentation algorithm for the analysis of array CGH data. Bioinformatics 23:657–663
Voet T, Kumar P, Van Loo P et al (2013) Single-cell paired-end genome sequencing reveals structural variation per cell cycle. Nucleic Acids Res 41:6119–6138
Wang J, Fan HC, Behr B et al (2012) Genome-wide single-cell analysis of recombination activity and de novo mutation rates in human sperm. Cell 150:402–412

Wang Y, Waters J, Leung ML et al (2014) Clonal evolution in breast cancer revealed by single nucleus genome sequencing. Nature 512:155–160

Wei Z, Wang W, Hu P et al (2011) SNVer: a statistical tool for variant calling in analysis of pooled or individual next-generation sequencing data. Nucleic Acids Res 39:e132

Wong K, Keane TM, Stalker J et al (2010) Enhanced structural variant and breakpoint detection using SVMerge by integration of multiple detection methods and local assembly. Genome Biol 11:R128

Yu X, Guda K, Willis J et al (2012) How do alignment programs perform on sequencing data with varying qualities and from repetitive regions? BioData Min 5:6

Zong C, Lu S, Chapman AR et al (2012) Genome-wide detection of single-nucleotide and copy-number variations of a single human cell. Science 338:1622–1626

Chapter 16
Submitting Data to a Public Repository, the Final Step of a Successful HTS Experiment

Christopher O'Sullivan and Jonathan Trow

16.1 Submitting Data to SRA

The Sequence Read Archive (SRA) forms the base of the NCBI archive stack. It holds raw data supporting downstream analysis and value-added data types like genome assemblies and annotation. Submissions should contain raw data suitable for reanalysis. Metadata specific to SRA is contained in the SRA Experiment and Run. The SRA Experiment contains details describing sequence library preparation, molecular and bioinformatics workflows, and sequencing instruments. The SRA Run contains sequence data from a specific library prep for a single biological sample. Multiple sequencer runs from a library should be split into distinct SRA Runs or submitted as distinct Read Groups in bam format in order to retain batch information.

We capture submitted details describing your research effort, funding, and publication in NCBI's BioProject resource, and describe the biological samples used to prepare sequencing libraries in NCBI's BioSample resource.

Note: Each SRA Experiment points to a single BioProject and BioSample.

16.1.1 Submission of Protected Human Data

Data which may contain human sequence and which does not have proper consent for public display in an unrestricted database should be submitted to the database of Genotypes and Phenotypes (dbGaP). Submitters should contact the dbGaP helpdesk (dbgap-help@ncbi.nlm.nih.gov) for assistance with beginning the submission process.

C. O'Sullivan • J. Trow (✉)
National Center for Biotechnology Information, U.S. National Library of Medicine, 8600 Rockville Pike, Bethesda, MD 20894, USA
e-mail: osulliva@ncbi.nlm.nih.gov; trowja@ncbi.nlm.nih.gov

A.M. Aransay, J.L. Lavín Trueba (eds.), *Field Guidelines for Genetic Experimental Designs in High-Throughput Sequencing*,
DOI 10.1007/978-3-319-31350-4_16

16.1.2 Bulk Center Submission

Bulk SRA metadata may be submitted via XML by registered submitting Centers. Submitting centers will need to set up a dedicated upload account using Aspera (recommended) or FTP. Centers planning to programmatically submit to SRA should contact the SRA helpdesk (sra@ncbi.nlm.nih.gov) for resources and assistance formatting submission XML and setting up an upload account.

16.1.3 Formatting of Submitted Files

SRA accepts a variety of file formats; following best formatting practices will help prevent delays or errors during data loading:

- It is best to submit fastq files with the original header formatting. Modification or replacement of the systematic identifiers generated by the instrument may lead to errors or delays in Submission processing.
- Bam files are the preferred submission format. Please ensure that submitted bam files have robust header information, including Program (@PG) and Read group (@RG) fields. In addition, alignments to high quality (chromosome level) genomic reference assemblies are strongly recommended. Pre-submission validation of submitted bams can be completed using ValidateSamFile (The Broad Institute 2015), which should ensure successful loading to SRA with no modification.

16.1.4 Gather Information Prior to Starting Submission

16.1.4.1 BioProject: Why Did You Perform Your Analysis?

(a) Project title and abstract
(b) Aims and Objectives
(c) Organisms Sequenced
(d) Funding Sources, Publications, etc.

16.1.4.2 BioSample: What Did You Sequence?

(a) Descriptive sample information
(b) Tabular format is ideal
(c) Examples: Organism(s), age(s), gender(s), location data, cell line(s), etc.

16.1.4.3 SRA Experiment: How Did You Sequence Your Samples?

(a) Sequencing methods you used
(b) The kits you used
(c) The model number(s) of the instrument(s) you used

16.1.4.4 SRA Run: What Is Your Data File Format?

(a) Files must be in an acceptable format: BAM, FASTQ, etc.
(b) MD5 Checksum for each file
(c) Minimum of one unique dataset per sample

16.1.5 SRA Submission Workflow

16.1.5.1 Select or Create BioProject and BioSample(s)

BioProject is a description of your research effort.

BioSample records describe the biologically unique specimens used in your research effort.

If you have already created a BioSample(s) and BioProject(s) as a part of WGS, Genome or Transcriptome Shotgun Assembly (TSA) submission, use those in your SRA submission.

16.1.5.2 SRA-Specific Metadata

- SRA Experiment
  - Describes the sequencing library derived from a single biological specimen
  - Explains "How" you performed the sequencing
  - Multiple Experiments can point to a single Sample, but not vice versa
- *SRA Run*
  - All files specified in a Run are merged into a single dataset
  - We extract sequence, quality, and alignment information from your submitted files and convert them to SRA archive file format

16.1.5.3 Create a New Submission

1. Go to the SRA Batch Submission portal: https://submit.ncbi.nlm.nih.gov/subs/sra/.
2. Click the "New Submission" button.

3. Go to the "Submitter" tab and fill in all requested fields.
4. Click "continue" once you have completed all the "Submitter" Tab fields.
5. Go to the "General Info" tab.
6. Select a BioProject by typing in part of the BioProject name (either the PRJNA#### accession or the BioProject title).

   (a) If you do not have a BioProject registered, go to https://submit.ncbi.nlm.nih.gov/subs/bioproject/ and create a BioProject, then continue your submission from this point.

7. Set the Release Date.
8. Register Samples:

   The "General Info" tab will ask if you are registering new samples:

   (a) If you answer "Yes" the form will display 2 tabs for registering the new samples.
   (b) If you answer "No" you will skip steps 10 and 11 and will go directly to the SRA metadata tab. Use the Metadata tab if you wish to add additional data to existing samples.

9. Click "continue" when you are satisfied with your selections.
10. Go to the "BioSample Type" tab.

    We designed the "BioSample Type" tab to help you select the appropriate BioSample type and the correct spreadsheet that goes with your selected BioSample Type. After you select your BioSample type, you will be directed to the "BioSample Attributes" tab.
11. Once you arrive at the "BioSample Attributes" tab you will upload a tab-separated file describing your samples:

    (a) A unique set of attributes describes a sample, and your identifier for the sample is the sample name. Since we do not use sample name, title, or description to validate unique sample records, your samples should be unique even if these three fields are ignored.
    (b) Fill out the BioSample spreadsheet and include as much metadata as you can. If you have filled out all required columns and your samples are still not unique enough, you can add your own columns containing unique metadata for each sample. When you enter the dates, make sure to enter them in "DD-MM-YYYY" (e.g., 30-Oct-2010) format, the standard "YYYY-MM-DD" format (e.g., 2010-10-30), or the "YYYY-MM" format (e.g., 2010-10).
    (c) You can add additional attributes by creating another column with a new header.

12. Go to the "SRA Metadata" tab.

    The "SRA Metadata" tab accepts the tab-separated file containing your SRA metadata table. If you answered 'no' at step 8 above, change the "sample_name" column heading to "biosample_accession" and enter your SAMN# accessions in this column.

Table 16.1 List of common experiment library strategy, source and selection descriptors

Strategy	Sequencing strategy used in the experiment
WGS	Random sequencing of the whole genome
WXS	Target enrichment for expressed subset of genome
RNA-Seq	sequencing of whole transcriptome
AMPLICON	Amplification of a target loci using PCR
ChIP-Seq	Direct sequencing of chromatin immunoprecipitates
Bisulfite-Seq	Sequencing following treatment of DNA with bisulfite to convert cytosine residues to uracil depending on methylation status
Source	**Type of genetic source material sequenced**
GENOMIC	Genomic DNA (includes PCR products from genomic DNA)
TRANSCRIPTOMIC	Transcription products or non-genomic DNA (EST, cDNA, RT-PCR, screened libraries)
METAGENOMIC	Mixed material from metagenomic or environmental samples
METATRANSCRIPTOMIC	RNA extracted from environmental samples
SYNTHETIC	Synthetic DNA
VIRAL RNA	Viral RNA
Selection	**Method of selection or enrichment used in the experiment**
RANDOM	Random selection by shearing or other method
polyA	enrichment for messenger RNA (mRNA)
ChIP	Chromatin immunoprecipitation
MNase	Micrococcal nuclease (MNase) digestion
Hybrid Selection	target enrichment via complementary hybridization
Restriction Digest	DNA fractionation using restriction enzymes

(a) Library Strategy, Source and Selection (see Table 16.1) are required fields and use controlled vocabularies. The template contains the currently valid values for these fields.
(b) To add a custom attribute in a key-value pair configuration, you can add additional columns by entering a column header in a blank column and then entering the data for that column.
(c) Multiple sequence files from a library are specified by creating new columns with the name "filename#", where # is the file count such as "filename1", "filename2".

13. To upload files using your Web browser, go to the "Files" tab, then click the "Browse" button to select the file(s) you want to upload. We encourage the use of the Aspera plugin for faster data transfer (Aspera, Inc. 2015).

(a) Make sure the plugin is running after you install it by launching the Aspera Connect application you installed with the plugin.
(b) When you use Aspera for the first time, a pop-up should appear below the URL bar that will ask for permissions. If you do not see the pop-up, refresh the page, select a single file and then look for the pop-up.

14. Uploaded files will be validated against the filenames specified in the spreadsheet prior to the completion of your submission.
    (a) Using a Web browser for uploads may not be appropriate for large submissions.
    (b) Aspera command line or ftp client can also be used for uploads.
    (c) Write to sra@ncbi.nlm.nih.gov for instructions, passwords, and secure shell (ssh) keys.
15. Go to the "Overview" tab.
    Use the "Overview" tab to review your submission. If your submission looks correct, click the "Submit" button.
16. Go to the SRA submitter interface to view your submission status, make edits or corrections.

Example 1: Data submission for a study with three distinct experimental conditions, each with three replicates.

At *step 8* register three BioSamples, one for each condition. Make sure to include a distinguishing attribute(s), such as "Treatment" which would be unique for all three in this example case. Then at *step 12*, fill in one row of the spreadsheet for each replicate, making sure that each library_ID is unique and descriptive (see Table 16.2). Finally, continue with the submission process as indicated.

Table 16.2 Example SRA metadata table

bioproject_access	sample_name	library_ID	filename	filename2
PRJNA1000000	Control mouse, untreated	Control rep. 1	Ctrl1_R1.fq	Ctrl1_R2.fq
PRJNA1000000	Control mouse, untreated	Control rep. 2	Ctrl2_R1.fq	Ctrl2_R2.fq
PRJNA1000000	Control mouse, untreated	Control rep. 3	Ctrl3_R1.fq	Ctrl3_R2.fq
PRJNA1000000	Treated mouse, 50mg Compound A	A Treated 50mg. rep. 1	ComA1_R1.fq	ComA1_R2.fq
PRJNA1000000	Treated mouse, 50mg Compound A	A Treated 50mg. rep. 2	ComA2_R1.fq	ComA2_R2.fq
PRJNA1000000	Treated mouse, 50mg Compound A	A Treated 50mg. rep. 3	ComA3_R1.fq	ComA3_R2.fq
PRJNA1000000	Treated mouse, 50mg Compound B	B Treated 50mg. rep. 1	ComB1_R1.fq	ComB1_R2.fq
PRJNA1000000	Treated mouse, 50mg Compound B	B Treated 50mg. rep. 2	ComB2_R1.fq	ComB2_R2.fq
PRJNA1000000	Treated mouse, 50mg Compound B	B Treated 50mg. rep. 3	ComB3_R1.fq	ComB3_R2.fq

References

Aspera, Inc. (2015) Aspera Connect. http://downloads.asperasoft.com/connect2/. Accessed 01 July 2015

The Broad Institute (2015) Picard-ValidateSamFile. http://broadinstitute.github.io/picard/command-line-overview.html#ValidateSamFile. Accessed 01 July 2015

References

Index

A.M. Aransay, J.L. Lavín Trueba (eds.), *Field Guidelines for Genetic Experimental Designs in High-Throughput Sequencing*,
DOI 10.1007/978-3-319-31350-4

Z